Aims and Scope

The goal of the JCMSE is to publish new research results on computational methods employed in all sciences and engineering. Common experience proves that computational methods originally developed in sciences can be of crucial importance to other research areas, including engineering. This potentially beneficial practice of cross-fertilization will be continuously and systematically endorsed by the JCMSE, which shall also simultaneously stimulate similar initiatives within computational methods for knowledge transfer from engineering to sciences.

Currently, there is a communication gap between sciences and engineering. The JCMSE is aimed at bridging this gap. The main rationale for creating the JCMSE was a common observation that there is a genuinely universal link which naturally binds together sciences and engineering. This link is a joint and large set of computational methods that are often invented and reinvented in the two artificially separated fields with an insufficient overlap. The JCMSE was born out from the common experience acquired over the years indicating that progress in the sciences could have been significantly enhanced by a judicious intertwining between the mathematical and the engineering literatures. Inspecting e.g. the abundant and versatile engineering literature one may readily realize that the computational methods developed and successfully used there could be of a direct and great importance to the sciences. The JCMSE shall be the home of such interactions with expected benefit to both the sciences and engineering.

The scope of the JCMSE is primarily within the realm of computational methods, but the aim is also to encompass thorough reporting on the development of theory and models that lead to numerical implementations. The JCMSE will encourage efforts on this indispensable liaison between mathematical modelling and the corresponding computational methods. Moreover, the JCMSE shall provide a continuous support for publishing numerical codes, algorithms and software of either a general purpose or a particular tailoring to certain special tasks with the potential of broader ramifications and applications. Computational methods published by the JCMSE are expected to undergo detailed testing with other existing competitive strategies, as well as with experimental data, whenever possible.

The JCMSE is a journal on theoretical methods, but computational results from real measurements will also be published. A given modern measurement in the sciences may yield a large data base of observables that are customarily digitised and stored for post-experimental studies. The use of various computational methods to analyse these data may well be so powerful that the ensuing results could turn out to be almost equivalent to re-running the original experiment many times, with exploration of new possibilities that might not be accessible to a real measurement. The JCMSE will publish original results along these lines in experimental research, as well.

Ironically, when computers were small, programming was not a problem. However, programming eventually became a serious problem with the availability of large computing facilities. Bridging this latter gap with the advent of new programming languages as well as novel computing modalities, e.g. quantum computers and the like, will also be welcomed by the JCMSE. Overall, the JCMSE is open to all sciences and engineering to facilitate fruitful dialogue among interdisciplinary research fields reporting on new results obtained by means of computational methods.

The journal of Computational Methods in Sciences and Engineering is recognized by the European Society of Computational Methods in Sciences and Engineering (ESCMSE). More information about the ESCMSE can be found on the website *www.uop.gr/escmse/.*

CONTENTS

Special Issue: Computational aspects of electric polarizability calculations: Atoms, Molecules and clusters, Part I

Journal of Computational Methods in Sciences and Engineering 4 (2004) 235
IOS Press

Guest-editorial

Computational aspects of electric polarizability calculations: Atoms, molecules and clusters

George Maroulis
Department of Chemistry, University of Patras, Patras, Greece
E-mail: maroulis@upatras.gr

The theory of electric (hyper)polarizability constitutes the cornerstone of the rational approach to a wide spectrum of phenomena. Such phenomena range from nonlinear optics and phenomena induced by intermolecular interactions to electron scattering. It is also worth emphasizing that polarizability is now routinely associated with important characteristics of electronic structure as hardness/softness, acidity/basicity or properties as the ionization potential. The prediction of reliable values for electric polarizability by rigorous quantum chemical methods has made significant contributions and added new vigor to intensively active fields as the search for new nonlinear optical materials, collision- and interaction-induced spectroscopy, molecular simulation and modeling of fundamental processes and the determination of molecular structures and properties of weakly bonded van der Waals molecules. In addition, electric polarizability is now of some importance to modern pharmacology and is extensively used as a molecular descriptor in QSAR studies. Some of the above fields possess considerable potential for advanced technological applications.

This special issue brings together a representative selection of papers focusing on computational aspects of theoretical determinations of electric polarizability. Our aim is to show the color and content of current investigations by a large number of research groups. We have classified the papers into six categories, divided over two journal issues. The details are noted below. This classification is, as we readily admit, by no means unique. Nevertheless, it brings forth the amazing wealth of significant results made available in recent years.

Professor George Maroulis
Guest-editor, November 2004

Journal of Computational Methods in Sciences and Engineering 4 (2004) 237–250
IOS Press

Polarizability functions of diatomic homonuclear molecules: Semiempirical approach

M.A. Buldakov[a] and V.N. Cherepanov[b,*]
[a]*Laboratory of Ecological Instrument Making, Institute of Monitoring of Climatic and Ecological Systems SB RAS, Tomsk 634055, Russia*
[b]*Department of Physics, Tomsk State University, Tomsk 634050, Russia*

Abstract. A semiempirical method is discussed to construct the polarizability functions for a diatomic homonuclear molecule as a piecewise-continuous function that exhibits physically correct behavior at small and large internuclear distances and agrees with the polarizability function near the nuclear equilibrium position of the molecule. The method is applied to calculate the polarizability functions of N_2 and O_2 molecules in the range of internuclear distances (0, ∞).

Keywords: Polarizability, nitrogen, oxygen

Mathematics Subject Classification: Approximate methods

1. Introduction

The polarizability function conception of a diatomic molecule for any electron state (electron polarizability of the molecule) arises from the adiabatic approximation being used. In this approximation the molecule electron polarizability becomes a function of the molecule internuclear distance R and appears to be more full and important characteristic of the molecule than the polarizability in the vicinity of equilibrium position R_e usually given in scientific articles. The molecule electron polarizability is the second rank tensor that has only two independent components for diatomic molecules and accordingly there are two polarizability functions $\alpha_{zz}(R)$ and $\alpha_{xx}(R) = \alpha_{yy}(R)$ where axis Z is directed along the molecule axis.

Presently, quantum-mechanical and semiempirical methods are used to calculate the polarizability functions of diatomic molecules. The quantum-mechanical calculations of the polarizability functions are usually carried out for relatively small parts of internuclear distances near R_e (for instance [1–22]), and only for the molecule H_2 such calculations have been carried out for full R range [23,24] using special *ab initio* methods being applicable to two-electron molecules only. Moreover, there are many works on *ab initio* calculations of the polarizability functions for two interacting atoms of noble gases

*Corresponding author. E-mail: vnch@phys.tsu.ru.

(i.e. [25–27] and quoted there references). However, these interacting atoms do not form the chemical bonded molecules, and in this work are not considered.

Development of semiempirical approaches is caused by wish to obtain a knowledge about polarizability functions of molecules in all internuclear range using simple analytical expressions. Semiempirical calculations are usually based on the empirical or theoretical information about polarizability functions for separate ranges of internuclear distances. So, for the range of small R the polarizability functions have been found in the work [28] using the united atom model. In a vicinity of equilibrium internuclear position the polarizability functions of diatomic molecules are generally given by Taylor series where expansion parameters (derivatives of the polarizability tensor) may be experimentally found from Raman scattering. To construct the polarizability function at large R the classical electrostatic model by Silberstein (DID model) is used where interacting atoms are considered to be point dipole induced by outer electric field and localized on molecule's nuclei [29,30]. Such approach makes it possible to calculate the molecule polarizability function on a base of known polarizability values for the atoms forming the molecule. Later DID model have been improved by taking into account multipole interactions of atoms [34] and polarizability changing of these when they come closer [31–33]. Presently, there are only a few works [34–37] where the polarizability functions of diatomic molecules are given for the full range of internuclear distances. First, the polarizability functions for all range of internuclear distances were suggested in the work [35] for N_2 molecule. Unfortunately, these functions do not show physically correct behavior neither at small nor at large internuclear distances. The polarizability functions of the work [36] are based on empirically found behavior regularities of these near their maxima and known asymptotic dependencies at large R [29,30,33]. However, the polarizability functions at $R \to 0$ in the frame of this method do not agree with ones theoretically predicted in [28], and versatility of the correlations found needs to be further motivated. The polarizability functions in [37] were obtained using their asymptotes at small [28] and large R [29,30], and the polarizability derivatives empirical found for the N_2 and O_2 molecules at their equilibrium positions. Later, in [34] these functions were improved at the middle range of R by taking into account dispersion [31,32] and multipole interactions of the atoms. Nevertheless, the polarizability functions for the molecule H_2 obtained at the middle internuclear distances in [34] appear to be decreased as to their accurate *ab initio* values [24], that is the result of exchange interaction being ignored.

In this work a semiempirical method is given how the polarizability functions of a homonuclear diatomic molecule may be constructed. In this method the exchange interaction is taken into account only for $\alpha_{zz}(R)$ calculation because this interaction is most important for this component. The functions obtained by this method have been presented as piecewise continuous ones which give physically true polarizability behavior at small and large R and also near the nuclei equilibrium position. The method is applied to calculate the polarizability functions of N_2 and O_2 molecules in full R range for which *ab initio* calculations are lacking.

2. The molecule polarizability at small internuclear distances

The diatomic molecule polarizability at small internuclear distances may be described on the basis of united atom idea [28]. For small internuclear distances the wave function $\Psi_l(r, R)$ of the electron state l is calculated by the methods of perturbation theory where the wave functions of the united atom $\varphi_m(r)$ are used as unperturbed wave functions. In the first order of the perturbation theory

$$\Psi_l(r,R) = \varphi_l(r) + \sum_{m\neq l} \frac{V_{ml}(R)}{E_m^0 - E_l^0}\varphi_m(r), \tag{1}$$

where r is the coordinate set of all electrons of the system and E_k^0 is the k – state energy of the united atom. The index l of the united atom wave function $\varphi_l(r)$ shows the state to which the molecule electron state l comes when the molecule nuclei unite. The matrix element of the "united atom partition operator" is computed in [38] and is a power series of the form

$$V_{ml}(R) = V_{ml}^{(2)} R^2 + V_{ml}^{(3)} R^3 + V_{ml}^{(4)} R^4 + \dots, \tag{2}$$

where the coefficients $V_{ml}^{(j)}$ are fully specified by the electron density distribution of the united atom at m and l states. The energy $E_l(R)$ of the diatomic molecule for small R may be also represented as the series [38,29,40]

$$E_l(R) = E_l^0 + \frac{Z_1 \cdot Z_2}{R} + E_l^{(2)} R^2 + E_l^{(3)} R^3 + E_l^{(4)} R^4 + \dots . \tag{3}$$

Here Z_1 and Z_2 are the charges of the nuclei constituting the united atom and $E_l^{(j)}$ are the constants for the given molecule state.

To calculate the diatomic molecule static polarizability tensor at the electron state n for small internuclear distances we employ the following well known form for tensor components of the electron polarizability

$$\alpha_{ii}^{(n)}(R) = 2 \sum_{m \neq n} \frac{|\langle n(R)|d_i|m(R)\rangle|^2}{E_m(R) - E_n(R)}, \tag{4}$$

where the matrix elements of the dipole moment

$$\langle n(R)|d_i|m(R)\rangle \equiv \langle \Psi_n(r,R)|d_i(r)|\Psi_m(r,R)\rangle, \tag{5}$$

taking into account Eqs (1) and (2), may be represented as the power series in terms R except the linear term. As a result, Eq. (4) may be written as

$$\alpha_{ii}^{(n)}(R) = \alpha_{ii}^{(n)} + A_{ii}^{(n)} R^2 + B_{ii}^{(n)} R^3 + C_{ii}^{(n)} R^4 + \dots , \tag{6}$$

where $\alpha_{ii}^{(n)}$ are the polarizability tensor components of the united atom at the state to which the molecule comes when its nuclei come closer together, and $A_{ii}^{(n)}$, $B_{ii}^{(n)}$, and $C_{ii}^{(n)}$ are the constants for the given molecule electron state.

3. The molecule polarizability at large internuclear distances

The molecule polarizability at large internuclear distances may be described on the basis of an idea about interacting atoms forming the molecule. There are two types of atom interactions: multipole (long-range) and exchange ones for which an analytical description of the molecule polarizability functions is possible. Each of these is predominant for its own range of internuclear distances: the multipole one – at $R \to \infty$ and exchange one – at more small R.

3.1. Multipole interaction of atoms

To take into account the multipole interactions between two identical atoms we apply a model of two interacting isotropic dielectric spheres [41,42]. In this model the polarizability of two dielectric spheres of radius r_0 located on axis Z at distance R each of other and placed into a permanent electric field may be represented in the form of [42]

$$\alpha_{zz}(R) = 2\alpha_0 \frac{\varepsilon+2}{\varepsilon-1}(2\sinh\eta_0)^2 \sum_{p=0}^{\infty}\left[\frac{2}{\varepsilon+1}\right]^p \sum_{n=1}^{\infty} n\chi_n^{(p)}(R), \tag{7}$$

$$\alpha_{xx}(R) = 2\alpha_0 \frac{\varepsilon+2}{\varepsilon-1}(2\sinh\eta_0)^3 \sum_{p=0}^{\infty}\left[\frac{2}{\varepsilon+1}\right]^p \sum_{n=1}^{\infty} \frac{n(n+1)}{2}\omega_n^{(p)}(R), \tag{8}$$

where ε is the sphere permittivity, α_0 is the sphere polarizability, $\cosh\eta_0 = R/(2r_0)$, and the coefficients $\chi_n^{(p)}(R)$ and $\omega_n^{(p)}(R)$ are given by cumbersome recurrent relations (27)–(35) in the work [42]. In the frames of this model the expressions Eqs (7) and (8) are accurate in the range of $R \in (2r_0, \infty)$, however, one can calculate the polarizabilities using these only in a numerical form because of infinite and weakly convergent series.

In the work [34], to represent analytically the polarizability functions $\alpha_{zz}(R)$ and $\alpha_{xx}(R)$ the power series expansions Eqs (7) and (8) with respect to the small parameter $\delta = r_0/R$ have been made and after summation over p and n the following simple asymptotic expressions have been obtained

$$\begin{aligned}\alpha_{zz}(R) = {} & 2\alpha_0 + 4\alpha_0^2\frac{1}{R^3} + 8\alpha_0^3\frac{1}{R^6} + 18\alpha_0^3\left(1+\frac{1}{2\varepsilon+3}\right)\frac{r_0^2}{R^8} + 16\alpha_0^4\frac{1}{R^9} \\ & +32\alpha_0^3\left(1+\frac{2}{3\varepsilon+4}\right)\frac{r_0^4}{R^{10}} + \ldots,\end{aligned} \tag{9}$$

$$\begin{aligned}\alpha_{xx}(R) = {} & 2\alpha_0 - 2\alpha_0^2\frac{1}{R^3} + 2\alpha_0^3\frac{1}{R^6} + 6\alpha_0^3\left(1+\frac{1}{2\varepsilon+3}\right)\frac{r_0^2}{R^8} - 2\alpha_0^4\frac{1}{R^9} \\ & +12\alpha_0^3\left(1+\frac{2}{3\varepsilon+4}\right)\frac{r_0^4}{R^{10}} + \ldots .\end{aligned} \tag{10}$$

Doing so, we used the following relation:

$$\alpha_0 = \frac{\varepsilon-1}{\varepsilon+2}r_0^3 \tag{11}$$

The expressions Eqs (9) and (10) include, except well known terms of DID model, additional terms resulting from multipole interaction. Note, that at $\varepsilon \to \infty$ the expressions obtained correspond to the "metallic-sphere" model [43].

The model above considered for interacting dielectric spheres does not take into account the anisotropy of atom polarizabilities being at the states with the orbital quantum number $L \geqslant 1$ and also the change of atom polarizabilities due to their drawing together. The polarizability anisotropy may be partially taken into account if in the terms of the expressions Eqs (9) and (10), corresponding to DID model, α_0

appears to be replaced by corresponding atoms' polarizability tensor components α^0_{ii} [44] and in the terms including r_0 the value α_0 appears to be assumed being equal to the average atom polarizability (hereinafter r_0 is the van der Waals radius of the interacting atoms). The polarizability change of atoms approaching to one another may be taken into account in the frames of the method suggested in the works [31,32]. Taking into consideration above notes, Eqs (9) and (10) may be written as in [34]

$$\alpha_{zz}(R) = 2\alpha^0_{zz} + 4(\alpha^0_{zz})^2\frac{1}{R^3} + \left[8(\alpha^0_{zz})^3 + \frac{7\gamma C_6}{9\alpha_0}\right]\frac{1}{R^6} + 18\alpha_0^3\left(1 + \frac{1}{2\varepsilon + 3}\right)\frac{r_0^2}{R^8}$$
$$+16(\alpha^0_{zz})^4\frac{1}{R^9} + 32\alpha_0^3\left(1 + \frac{2}{3\varepsilon + 4}\right)\frac{r_0^4}{R^{10}}, \tag{12}$$

$$\alpha_{xx}(R) = 2\alpha^0_{xx} - 2(\alpha^0_{xx})^2\frac{1}{R^3} + \left[2(\alpha^0_{xx})^3 + \frac{4\gamma C_6}{9\alpha_0}\right]\frac{1}{R^6} + 6\alpha_0^3\left(1 + \frac{1}{2\varepsilon + 3}\right)\frac{r_0^2}{R^8}$$
$$-2(\alpha^0_{xx})^4\frac{1}{R^9} + 12\alpha_0^3\left(1 + \frac{2}{3\varepsilon + 4}\right)\frac{r_0^4}{R^{10}}. \tag{13}$$

Here the terms containing van der Waals factor C_6 and the average second hyperpolarizability of atom γ show the polarizability change of atoms approaching to one another due to the dispersion interactions between them.

3.2. *Exchange interaction of atoms*

To take into account the exchange of two atoms' electrons one may use asymptotic methods [45] that are applicable to the range of internuclear distances characterized by weak overlapping of interacting atoms' valence electrons. At such internuclear distances two-electron (one electron from atom "a" and one from atom "b") molecular wave function of the state n may be written in the form

$$\Psi_n(r_1, r_2, R) = c_n^{(1)}\psi_n^{(1)}(r_1, r_2, R) + c_n^{(2)}\psi_n^{(2)}(r_1, r_2, R), \tag{14}$$

where

$$\psi_n^{(1)}(r_1, r_2, R) = [\varphi^{(a)}(r_1, R)\varphi^{(b)}(r_2, R)\chi_I(r_1, r_2, R)]_n,$$
$$\psi_n^{(2)}(r_1, r_2, R) = [\varphi^{(a)}(r_2, R)\varphi^{(b)}(r_1, R)\chi_{II}(r_1, r_2, R)]_n. \tag{15}$$

Here $\varphi^{(a)}(r_1, R), \varphi^{(b)}(r_1, R)$ and $\varphi^{(a)}(r_2, R), \varphi^{(b)}(r_2, R)$ are asymptotic wave function of the first and the second electrons located near corresponding atom cores. The expressions Eqs (14) and (15) are written in the molecular coordinate system where interacting atoms are arranged on axis Z and the origin of coordinates appears to be an equidistant point from both atoms. In this coordinate system r_1 and r_2 are coordinates of the first and second electrons. Explicit form of the functions $\chi_I(r_1, r_2, R)$ and $\chi_{II}(r_1, r_2, R)$ taking into account an interaction of electrons with each other and with extraneous nuclei is given in [45]. The exchange interaction, when the molecule polarizability being calculated, shows itself according to Eq. (4) mainly through its contribution into the matrix element of the electron transition dipole moment. In the approximation applied the exchange interaction is most substantial in the matrix element of the component z of the electron transition dipole moment and, as a result, in the polarizability tensor component $\alpha_{zz}(R)$. An exchange interaction contribution into z-component of the

Table 1
Atomic and molecular polarizabilities

Molecule term	$H_2^1\Sigma_g^+$	$N_2^1\Sigma_g^+$	$O_2^3\Sigma_g^-$	
$(\alpha_{xx})_e$,Å^3	0.679[a)]	1.53 [46]	1.25 [46]	
$(\alpha'_{xx})_e$,Å^3	0.669[a)]	1.12 [46]	0.70 [46]	
$(\alpha''_{xx})_e$,Å^3	0.121[a)]	0.33 [46]	0.80[b)]	
$(\alpha_{zz})_e$,Å^3	0.947[a)]	2.24 [46]	2.33 [46]	
$(\alpha'_{zz})_e$,Å^3	1.372[a)]	3.35 [46]	3.89 [46]	
$(\alpha''_{zz})_e$,Å^3	1.221[a)]	2.93 [46]	8.60[b)]	
United atom	He	Si	S	
Term	^{1}S	^{1}D ($M_L = 0$)	^{3}P ($M_L = 0$)	
$\alpha_{xx}^{(0)}$, Å^3	0.205 [47]	5.62 [48]	2.68 [48]	
$\alpha_{zz}^{(0)}$, Å^3	0.205 [47]	7.50 [48]	3.35 [48]	
Isolated atom	H	N	O	
Term	^{1}S	^{4}S	^{3}P ($M_L = 0$)	^{3}P ($M_L = \pm 1$)
α_{xx}^0, Å^3	0.6668 [47]	1.101[c)]	0.755[c)]	0.825[c)]
α_{zz}^0, Å^3	0.6668 [47]	1.101[c)]	0.895[c)]	0.755[c)]
α_0,Å^3	0.6668	1.101	0.802	

[a)] Calculated using data from [24]; [b)] values refined in [46]; [c)] taken from [37].

dipole moment calculated by means of Eqs (14) and (15) for homonuclear molecule may be represented as

$$\langle n(R)|d_z|m(R)\rangle_{ch} \sim \langle \psi_n^{(1)}(r_1, r_2, R)|d_z|\psi_m^{(2)}(r_1, r_2, R)\rangle = \tilde{A}_{nm}(\beta_n, \beta_m, R)R^{\delta} \times \exp[-(3\beta_n + \beta_m)R/2], \tag{16}$$

where $\tilde{A}_{nm}(\beta_n, \beta_m, R)$ is the function weakly dependent on R in the region of small electron shells overlapping, $(-\beta_n^2/2)$ and $(-\beta_m^2/2)$ are atom's ionization potentials at ground and excited states accordingly, and

$$\delta = \frac{11}{4\beta_n} + \frac{1}{\beta_m} - \frac{1}{2(\beta_n + \beta_m)} + 1. \tag{17}$$

Then substituting Eq. (16) into Eq. (4) and replacing β_m by some effective value $\overline{\beta}$ the expression for the exchange interaction contribution into polarizability of the molecule takes the form

$$[\alpha_{zz}^{(n)}(R)]_{ch} = B(\beta_n, \overline{\beta}, R)R^{2\delta} \exp[-(3\beta_n + \overline{\beta})R]. \tag{18}$$

Here $B(\beta_n, \overline{\beta}, R)$ is also the function weakly dependent on R.

The exchange interaction taken into account gives some contribution to the $\alpha_{xx}(R)$ function too. However, this contribution is significantly less than into $\alpha_{zz}(R)$ and, as a result, is not considered in this work.

Note that in the range of internuclear distances considered the multipole interactions give a contribution into the molecular polarizability function too.

4. Method of the polarizability function construction

Suggested semiempirical method to construct the diatomic molecules polarizability functions at the ground electron state (hereinafter, the index n for the molecular parameters is omited) comprises three conditions appear to be satisfied.

Table 2
Parameters of the polarizability functions of the molecules H_2, N_2 and O_2

Molecule term	$H_2^1\Sigma_g^+$	$N_2^1\Sigma_g^+$	$O_2^3\Sigma_g^-$	
$a_x^{(2)}$,Å	1.6337	−23.0176	−7.0505	
$a_x^{(3)}$	−1.3949	28.7222	8.0313	
$a_x^{(4)}$,Å^{-1}	0.4788	−9.8802	−2.4883	
$a_z^{(2)}$, Å	1.7229	−33.3169	−9.2520	
$a_z^{(3)}$	−0.7294	42.2630	10.7974	
$a_z^{(4)}$, Å^{-1}	0.3065	−14.4742	−3.0763	
			^{3}P ($M_L = 0$)	^{3}P ($M_L = \pm 1$)
$b_x^{(0)}$,Å^3	2.4136	2.8077	6.6210	5.3588
$b_x^{1)}$, Å^2	−8.3935	−6.6211	−17.0528	−13.1341
$b_x^{(2)}$, Å^1	13.4832	9.1354	19.5771	14.9209
$b_x^{(3)}$	−9.0154	−5.0068	−10.2911	−7.6741
$b_x^{(4)}$, Å^{-1}	2.7498	1.2293	2.5462	1.8575
$b_x^{(5)}$, Å^{-2}	−0.3171	−0.1129	−0.2410	−0.1724
$b_z^{(0)}$, Å^3	2.0607	−0.1026	16.1332	20.1111
$b_z^{(1)}$, Å^2	−5.6482	3.4183	−40.6216	−53.0261
$b_z^{(2)}$, Å^1	6.5615	−5.3786	39.1681	53.9935
$b_z^{(3)}$	−0.4758	6.2371	−14.9730	−23.3714
$b_z^{(4)}$, Å^{-1}	−1.3288	−2.5503	2.2070	4.4403
$b_z^{(5)}$, Å^{-2}	0.3325	0.3291	−0.7489	−0.2999
r_0,Å	1.1 [49]	1.5 [49]	1.4 [49]	
ε	4.01[a)]	2.45[a)]	2.24[a)]	
γC_6, Å^{12}	4.18[b)]	10.51[b)]	5.03[b)]	
β_0, a.u.	1.000 [52]	1.033 [52]	1.000 [52]	
$\overline{\beta}$, a.u.	0.41	0.31	0.31	
A, a.u.	2.00 [52]	1.49 [52]	1.32 [52]	
B, Å$^{3-2\delta}$	108.81	167.44	80.94	
R_1, Å	0.9	1.1	1.2	
R_2, Å	2.2	3.0	2.8	

[a)]Calculated by Eq. (11); [b)]the values of γ and C_6 for H were taken from [32], γ for N and O were taken from [50], C_6 for N and O were taken from [51].

1. The molecule polarizability functions $\alpha_{zz}(R)$ and $\alpha_{xx}(R) = \alpha_{yy}(R)$ in a range of small R are given by polynomial like Eq. (6)

$$\alpha_{ii}(R) = \alpha_{ii}^{(0)} + a_i^{(2)}R^2 + a_i^{(3)}R^3 + a_i^{(4)}R^4 + \ldots, \tag{19}$$

where $\alpha_{ii}^{(0)}$ are the polarizability components of the united atom, and the constants $a_i^{(k)}$ $(k \geqslant 2)$ are found using the known values of the polarizability functions and their derivatives at R_e. These coefficients for each component of the polarizability tensor are solution of the system of linear equations. The system of linear equations is determined by equality condition at R_e for derivatives of the polarizability function in the vicinity of equilibrium internuclear position of a molecule

$$\alpha_{ii}(R) = (\alpha_{ii})_e + (\alpha'_{ii})_e\xi + \frac{1}{2}(\alpha''_{ii})_e\xi^2 + \ldots, \tag{20}$$

and the polynomial Eq. (19). In Eq. (20) the polarizability tensor components $(\alpha_{ii})_e$ and their first derivatives $(\alpha'_{ii})_e$, and second derivatives $(\alpha''_{ii})_e$ with respect to $\xi = (R - R_e)/R_e$ are determined at

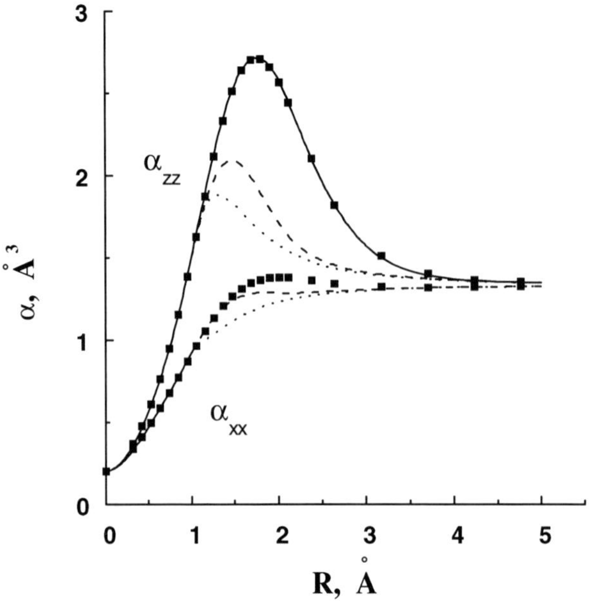

Fig. 1. Polarizability functions of H_2 molecule: solid curve – calculated taking into account the exchange interactions (this work); dashed curves – and calculated without taking into account the exchange interactions (this work); dotted curves – calculated in [37]; squares – *ab initio* calculation of [24].

R_e. Note, that the number of terms in Eq. (19) is determined by the number of known polarizability derivatives of the molecule in Eq. (20). As a result, the polynomials Eq. (19) together with the coefficients $a_i^{(k)}$ obtained describe the polarizability functions of a molecule in a range of small R including the vicinity of the equilibrium internuclear position.

2. The molecule polarizability function $\alpha_{zz}(R)$ in a range of large R is given as a sum of contributions from multipole Eq. (12) and exchange Eq. (18) interactions:

$$\begin{aligned}\alpha_{zz}(R) = 2\alpha_{zz}^0 + 4(\alpha_{zz}^0)^2\frac{1}{R^3} + \left[8(\alpha_{zz}^0)^3 + \frac{6\gamma C_6}{9\alpha_0}\right]\frac{1}{R^6} + 18\alpha_0^3\left(1+\frac{1}{2\varepsilon+3}\right)\frac{r_0^2}{R^8} \\ +16(\alpha_{zz}^0)^4\frac{1}{R^9} + 32\alpha_0^3\left(1+\frac{2}{3\varepsilon+4}\right)\frac{r_0^4}{R^{10}} + BR^{2\delta}\exp[-(3\beta_0+\overline{\beta})R],\end{aligned} \tag{21}$$

and it is assumed that $B(\beta_0, \overline{\beta}, R)$ in Eq. (18) does not depend on R and is a parameter B. The unknown parameter B is defined by fitting the function $\alpha_{zz}(R)$ to *ab initio* data for the molecule H_2 [24] when R corresponds to small electron shells overlapping. Then, this parameter is used for other homonuclear molecules taking into account a scaling factor which depends on a size of atoms electron shells. The scaling factor is found using the values of the parameter A of the electron asymptotic radial wave function in the atom [24] and taking into consideration that the angular wave function of p-electron with zero projection $(m_l = 0)$ on Z axis is $\sqrt{3}$ times as large as the angular function of s-electron. The polarizability function $\alpha_{xx}(R)$ for large R is determined mainly by multipole interactions because the exchange interaction is small in this case. So, in this work the polarizability function $\alpha_{xx}(R)$ is determined by Eq. (13).

3. The molecule polarizability functions in the middle internuclear distances are found joining the polarizability functions of a molecule for small and large R. These joining functions are given by a

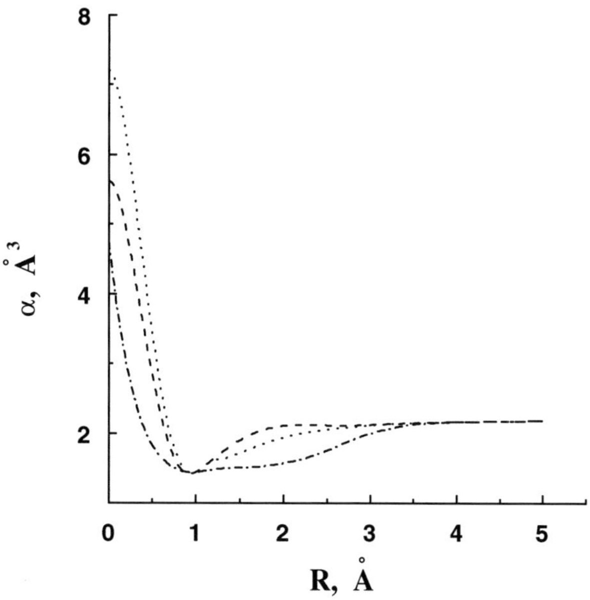

Fig. 2. Polarizability functions $\alpha_{xx}(R)$ of N_2 molecule: dashed curves – calculated without taking into account the exchange interactions (this work); dotted curves – calculated in [37]; dot-and-dashed curve – calculated in [36].

polynomial of degree five in R

$$\alpha_{ii}(R) = \sum_{j=0}^{5} b_i^{(j)} R^j, \tag{22}$$

where the coefficients $b_i^{(j)}$ are found from the join conditions with an accuracy up to second derivatives inclusive. The choice of the join points is determined by the model the polarizability functions to be constructed: the join point R_1 for the polarizability functions Eq. (19) with the polynomial Eq. (22) is to be in the vicinity of equilibrium internuclear position of a molecule; the join point R_2 for the polynomial Eq. (22) with the polarizability functions $\alpha_{zz}(R)$ Eq. 21 and $\alpha_{xx}(R)$ Eq. (13) is to be in a internuclear range corresponding to weak overlapping of the electron shells for the interacting atoms. As a result, in this work the join points were chosen by the way: $R_1 \approx R_e$, $R_2 \approx 2r_0$. The choice of the join points allows some arbitrariness that does not change essentially the form of the polarizability functions.

5. Polarizability functions of the molecules H_2, N_2 and O_2

In this work the polarizability functions are calculated for the homonuclear molecules H_2, N_2 and O_2. There were used reliable *ab initio* values of the molecule H_2 [24] to test the model suggested and to find unknown parameter B in Eq. (21). The values of atomic and molecular constants and also ones of parameters $a_i^{(k)}, b_i^{(j)}, \varepsilon, B, \overline{\beta}, R_1$, and R_2 used to calculate the polarizability functions of the molecules H_2, N_2 and O_2 are given in Tables 1 and 2. The coefficient $B(\equiv B_H)$ of the exchange interaction of the atoms H was found by the method of least squares using *ab initio* data for polarizability function $\alpha_{zz}(R)$ [24] in the range of $2.4 \leqslant R < 4.0$ Å. For the atoms N and O the coefficients B were calculated

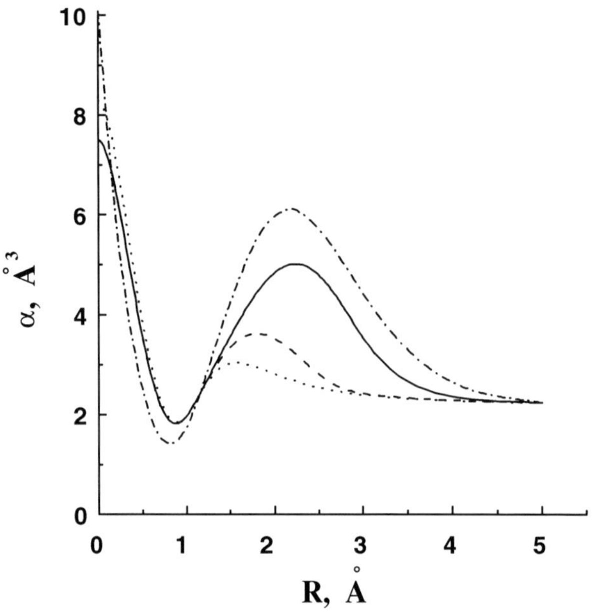

Fig. 3. Polarizability functions $\alpha_{zz}(R)$ of N_2 molecule: solid curve – calculated taking into account the exchange interactions (this work); dashed curve – calculated without taking into account the exchange interactions (this work); dotted curve – calculated in [37]; dot-and-dashed curve – calculated in [36].

from the formula $B_{(N,O)} = 9B_{(H)}A^6_{(N,O)}/A^6_{(H)}$ where $A_{(N,O)}$ is the parameter of the asymptotic radial wave function of the valence p-electron for the atoms N and O respectively, and $A_{(H)}$ is the same parameter of the valence s-electron for the atom H. The values of the parameter $\overline{\beta}$ were estimated using the radiative transitions probabilities in the atoms H, N and O.

Figure 1 shows the polarizability functions of the molecule H_2 calculated which differ due to completeness of the interatomic interactions being taken into account in Eq. (13) and Eq. (21). Analysis of these regularities reveals that the polarizability functions are much better, as compared to [37], in the region of middle R values and are close more to the accurate *ab initio* values [24], if dispersion and multipole interactions between H atoms are taken into account. However, these interactions being taken into account give good agreement with the data from [24] only for the polarizability function $\alpha_{xx}(R)$, while the function $\alpha_{zz}(R)$ is too low in the region of internuclear distances $1.5 \div 3$ Å because the exchange interactions between H atoms have not been taken into account. These interactions contribution into the polarizability function $\alpha_{zz}(R)$ being taken into consideration in Eq. (21) made it possible to come up to good agreement with *ab initio* calculations [24] in full range of internuclear distances. The fact that only one fitting parameter B in $\alpha_{zz}(R)$ (Eq. (21)) needs to be used demonstrates physical correctness of that how an exchange interactions contribution into the polarizability function $\alpha_{zz}(R)$ is taken into account.

The method suggested to calculate the polarizability functions was applied to the molecules N_2 and O_2, herewith for the oxygen molecule the both ways of its decompositions were considered (Figs 2–5). It is good noticeable that, as in a case of the molecule H_2, the dispersion, multipole, and exchange atom interactions taken into account lead the polarizability functions of the molecules N_2 and O_2 to increase in the range of internuclear distances $1 \div 3$ Å. Some differences between the polarizability functions obtained and the functions from [37] in the region of small R is due to other polarizability values for the united atoms Si and S used in this work.

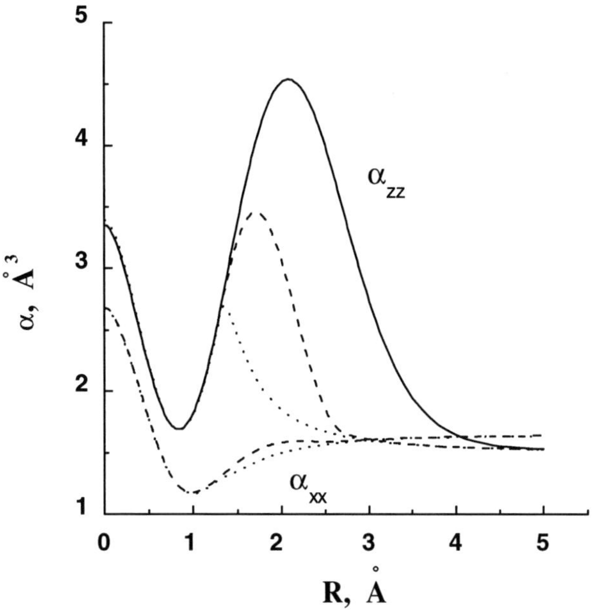

Fig. 4. Polarizability functions of O_2 molecule (atoms O at the states with $|M_L| = 1$): solid curve – calculated taking into account the exchange interactions (this work); dashed curve – calculated without taking into account the exchange interactions (this work); dotted curve – calculated in [37].

It is of interest to compare the N_2 polarizability functions obtained with those calculated by an alternative method in which an attempt was undertaken to take into account the exchange interactions using the correlations found in [36]. Apart from differences in the polarizability functions at small R due to different chosen values of $\alpha_{xx}^{(0)}$ and $\alpha_{zz}^{(0)}$ of Si atom, there should be noted the low values of $\alpha_{xx}(R)$ for $R \approx 1 \div 3$ Å (Fig. 2). This is caused, in our opinion, by the fact that correlations found for the molecules, mainly containing the first-group atoms of the Periodic Table, being without enough motivation applied to the molecule N_2. The polarizability functions from [34] are omitted in Figs 2–3 because they are only of historic interest now.

Note, that in the region of middle R values the polarizability functions $\alpha_{xx}(R)$ obtained in this work represent their low boundary and $\alpha_{zz}(R)$ should be considered as approximate ones because the parameters B and $\overline{\beta}$ for the molecules N_2 and O_2 appear to be only estimated ones. Nevertheless, we suppose the polarizability functions obtained in this work are more accurate than ones in the works [34–37].

6. Conclusion

The semiempirical method considered here allows to obtain in analytical form the diatomic homonuclear molecules polarizability functions in the whole range of internuclear distances. It is shown, that the dispersion, multipole and exchange interatomic interactions taken into account improve noticeably the polarizability functions for middle R values. Unfortunately, the model of two identical dielectric spheres applied for taking into account the multipole interactions restricts the method only by diatomic homonuclear molecules. However, despite the method suggested intends to be used for the molecules

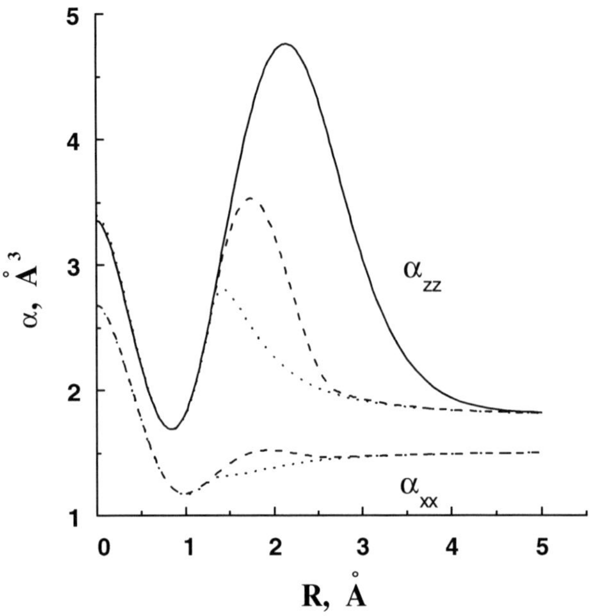

Fig. 5. Polarizability functions of O_2 molecule (atoms O at the states with $|M_L| = 0$): solid curve – calculated taking into account the exchange interactions (this work); dashed curve – calculated without taking into account the exchange interactions (this work); dotted curve – calculated in [37].

being dissociated into atoms at S states, it may also give reasonable results for the atoms which have not a sphere-symmetrical charge distribution (for example, for the molecule O_2).

We realize that the method proposed has typical for semiempirical approaches imperfections. Nevertheless, the method makes it possible to generate the correct idea of physical mechanisms for forming of the diatomic molecule polarizability functions. Quality of the polarizability functions obtained in frames of the method needs to be supported by exact quantum-chemical calculations. So, we hope the work will attract an attention of specialists in the quantum chemistry field and stimulate them for *ab initio* calculations of the molecules polarizability functions.

References

[1] M.A. Morrison and P.J. Hay, Ab initio static polarizabilities of N_2 and linear symmetric CO_2 in the Hartree-Fock approximation: variation with internuclear separation, *J. Phys.* **B10** (1977), 647–652.

[2] M.A. Morrison and P.J. Hay, Molecular properties of N_2 and CO_2 as functions of nuclear geometry: polarizabilities, quadrupole moments, and dipole moments, *J. Chem. Phys.* **70** (1979), 4034–4043.

[3] E.N. Svendsen, J. Oddershede, Ab initio calculation of Raman intensity of the N_2 molecule, *J. Chem. Phys.* **71** (1979), 3000–3005.

[4] J. Oddershede and E.N. Svendsen, Dynamic polarizabilities and Raman intensities of CO, N_2, HCl and Cl_2, *Chem. Phys.* **64** (1982), 359–369.

[5] R.D. Amos, The polarizability derivatives of the nitrogen molecule, *J. Phys.* **B12** (1979), 315–317.

[6] R.D. Amos, SCF and SI calculations of the one-electron properties, polarizabilities and polarizability derivatives of the nitrogen molecule, *Mol. Phys.* **39** (1980), 1–14.

[7] T. Inoue, S. Matsushima and S. Iwata, The CI calculation of the frequency-dependent polarizability and some optical properties for N_2 and O_2, *Mol. Phys.* **56** (1985), 1097–1115.

[8] S.R. Langhoff, S.W. Bauschlicher Ir. and D.P. Chong, Theoretical study of the effects of vibrational-rotational interactions on the Raman spectrum of N_2, *J. Chem. Phys.* **78** (1983), 5287–5292.

[9] H. Hettema, S.P.S. Wormer, P. Jorgensen, H.J.Aa. Jensen and T. Heldaker, Frequency-dependent polarizabilities of O_2 and van der Waals coefficients of dimers containing O_2, *J. Chem. Phys.* **100** (1994), 1297–1302.
[10] P.A. Hyams, J. Cerratt, D.L. Cooper and M. Raimondi, The calculation of molecular response properties using perturbed spin-coupled wave functions. II. Polarizability and magnetic susceptibility of H_2 and LiH as function of internuclear distance, *J. Chem. Phys.* **V. 100** (1994), 4417–4431.
[11] F.A. Gianturco and U.T. Lamanna, Internuclear dependence of static polarizability in hydrogen halides, *J. Phys.* **B12** (1979), 2789–2795.
[12] B.O. Roos and A.J. Sadley, Complete active space (CAS) SCF study of the dipole polarizability function for the $X\,^1\Sigma^+$ state of LiH, *J. Chem. Phys.* **76** (1982), 5444–5451.
[13] A. Hinchliffe, The polarizability of LiBr, *Chem. Phys. Lett.* **70** (1980), 610–612.
[14] D.A. Dixon, R.A. Eades and D.G. Truchlar, Ab initio SCF polarizabilities and electron-molecule adiabatic polarization potentials. 2. Li_2, *J. Phys.* **B12** (1979), 2741–2753.
[15] W. Muller and W. Meye, Static dipole polarizabilities of Li_2, Na_2 and K_2, *J. Chem. Phys.* **85** (1986), 953–957.
[16] R.D. Amos, A configuration-interaction study of the polarizability derivatives of carbon monoxide, *Chem. Phys. Lett.* **70** (1980), 613–617.
[17] G. Maroulis, Accurate dipole polarizability for $Cl_2(X^1\Sigma_g^+)$, *Mol. Phys.* **77** (1992), 1085–1094.
[18] G. Maroulis, Electric polarizability and hyperpolarizability of Cabon Monoxide, *J. Phys. Chem.* **100** (1996), 13466–13473.
[19] G. Maroulis and C. Makris, On the electric properties of Br_2, *Mol. Phys.* **91** (1997), 333–341.
[20] G. Maroulis, A systematic study of basis set, electron correlation, and geometry effects on the electric multipole moments, polarizability, and hyperpolarizability of HCl, *J. Chem. Phys.* **108** (1998), 5432–5448.
[21] G. Maroulis, C. Makris, D. Xenides and P. Karamanis, Electric dipole and quadrupole moment and dipole polarizability of CS, SiO and SiS, *Mol. Phys.* **98** (2000), 481–491.
[22] G. Maroulis, Accurate electric multipole moment, static polarizability and hyperpolarizability derivatives for N_2, *J. Chem. Phys.* **118** (2003), 2673–2687.
[23] W. Kolos and L. Wolniewicz, Polarizability of the hydrogen molecule, *J. Chem. Phys.* **46** (1967), 1426–1432.
[24] J. Ruchlewski, An accurate calculation of the polarizability of the hydrogen molecule and its dependence on rotation, vibration and isotopic substitution, *Mol. Phys.* **41** (1980), 833–842.
[25] B. Fernandez, C. Hatting and H. Koch, A. Rizzo, Ab initio calculation of the frequency-dependent interaction induced hyperpolarizability of Ar_2, *J. Chem. Phys.* **110** (1999), 2872–2882.
[26] G. Maroulis, Computational aspects of interaction hyperpolarizability calculations. A study on $H_2 \ldots H_2$, Ne...HF, Ne... FH, He... He, Ne... Ne, Ar... Ar, and Kr... Kr, *J. Phys. Chem.* **104** (2000), 4772–4779.
[27] G. Maroulis, A. Haskopoulos, Interaction induced (hyper)polarizability in Ne... Ar, *Chem Phys. Lett.* **358** (2002), 64–70.
[28] M.A. Buldakov, N.F. Vasiliev and I.I. Matrosov, Polarizability of diatomic molecules under small internuclear distances, *Opt. Spektrosk. (Russia)* **75** (1993), 597–603.
[29] L. Silberstein, Molecular refractivity and atomic interaction, *1, Phil. Mag.* **33** (1917), 92–128.
[30] L. Silberstein, Molecular refractivity and atomic interaction, *2, Phil. Mag.* **33** (1917), 521–533.
[31] A.D. Buckingham, The polarizability of a pair of interacting atoms, *Trans. Faraday Soc.* **52** (1956), 1035–1041.
[32] A.D. Buckingham and K.L. Clarke, Long-range effects of molecular interactions on the polarizability of atoms, *Chem. Phys. Lett.* **57** (1978), 321–325.
[33] K.L.C. Hunt, B.A. Zilles and J.L. Bohr, Effects of van der Waals interactions on the polarizability of atoms, oscillators, and dipolar rotors at long range, *J. Chem. Phys.* **75** (1981), 3079–3086.
[34] M.A. Buldakov, Polarizability of diatomic homonuclear molecules: internuclear dependence, *Atmos. Oceanic Opt.* **15** (2002), 752–756.
[35] A. Temkin, Internuclear dependence of the polarizability of N_2, *Phys. Rev. A.* **17** (1978), 1232–1235.
[36] A.J.C. Varandas and S.P.J. Rodrigues, Internuclear dependence of static dipole polarizability in diatomic molecules, *Chem. Phys. Lett.* **245** (1995), 66–74.
[37] M.A. Buldakov and I.I. Matrosov, Polarizability of diatomic molecules as a function of internuclear distance, *Optics and Spectroscopy* **78** (1995), 20–24.
[38] W.A. Bingel, United atom treatment of the behavior of potential energy curves of diatomic molecules for small R, *J. Chem. Phys.* **30** (1959), 1250–1253.
[39] W.A. Bingel, Proof of two conjectures of Milliken concerning the electronic energy of diatomic molecules for small R, *J. Chem. Phys.* **49** (1968), 1931–1934.
[40] W. Byers-Brown and J.D. Power, Perturbation theory of short-range atomic interactions, *Proc. Roy. Soc. Lond.* **A317** (1970), 545–574.
[41] A. Goyette and A. Navon, Two dielectric spheres in an electric field, *Phys. Rev. B* **13** (1976), 4320–4327.
[42] J.L. Godet and B. Dumon, Dielectric-sphere-couple model for noble-gas pair polarizability, *Phys. Rev. A.* **46** (1992), 5680–5686.

[43] H.B. Levine and D.A. McQuarrie, Dielectric constant of simple gases, *J. Chem. Phys.* **49** (1968), 4181–4187.
[44] J. Applequist and J.R. Carl, Fung Kwok-Keung, An atom dipole interaction model for molecular polarizability. Application to polyatomic molecules and determination of atom polarizabilities, *J. Am. Chem. Soc.* **94** (1972), 2952–2960.
[45] B.M. Smirnov, *Asymptotic methods in the theory of atomic collisions*, Atomizdat, Moscow, 1973.
[46] M.A. Buldakov, I.I. Ippolitov, B.V. Korolev, I.I. Matrosov, A.E. Cheglokov, V.N. Cherepanov, Y.S. Makushkin and O.N. Ulenikov, Vibration rotation Raman spectroscopy of gas media, *Spectrochimica Acta. Part A.* **52** (1996), 995–1007.
[47] A.A. Radtsig and B.M. Smirnov, *Reference book on atomic and molecular physics*, Atomizdat, Moscow, 1980.
[48] K. Andersson and A.J. Sadley, Electric dipole polarizabilities of atomic valence states, *Phys. Rev. A.* **46** (1992), 2356–2362.
[49] L. Pauling and P. Pauling, *Chemistry*, W.H. Freeman and Company, San Francisco, 1975.
[50] A.K. Das and A.J. Thakkar, Static responce properties of second-period atoms: coupled cluster calculations, *J. Phys. B: At. Mol. Opt. Phys.* **31** (1998), 2215–2223.
[51] G.D. Zeiss and W.J. Meath, Dispersion energy constants $C_6(A, B)$, dipole oscillator strength sums and refractivities for Li, N, O, H_2, NH_3, H_2O, NO and N_2O, *Mol. Phys* **33** (1977), 1155–1176.
[52] E.E. Nikitin and B.M. Smirnov, *Atomic and molecular processes*, Nauka, Moscow, 1988.

Journal of Computational Methods in Sciences and Engineering 4 (2004) 251–265
IOS Press

Dipole polarizability and second hyperpolarizability of difluoroacetylene: Basis set dependence and electron correlation effects

Miroslav Medveď[a,**], Benoît Champagne[a,*], Jozef Noga[b,c] and Eric A. Perpète[a]
[a]*Laboratoire de Chimie Théorique Appliquée, Facultés Universitaires Notre-Dame de la Paix, rue de Bruxelles 61, B-5000 Namur, Belgium*
[b]*Institute of Inorganic Chemistry, Slovak Academy of Science, Dúbravská cesta 9, SK-84536 Bratislava, Slovak Republic*
[c]*Department of Physical and Theoretical Chemistry, Faculty of Science, Comenius University, SK-84215, Bratislava, Slovak Republic*

Abstract. *Ab initio* calculations of the static dipole polarizability and the static and dynamic second hyperpolarizabilities, including the electron correlation effects via the standard Coupled Cluster method with single and double excitations (CCSD) and non-iterative triple excitations (CCSD(T)) as well as the vibrational motion (nuclear relaxation) effects by means of the Finite Field method formulated by Kirtman and co-workers, have been carried out for difluoroacetylene (C_2F_2). The basis set dependence of the geometry parameters and molecular properties using various types of basis sets (ANO-L, POL, HyPOL, aug-cc-pVTZ) is analysed. A particular attention is devoted to the evaluation of the electron correlation effects on the vibrational contributions to the molecular properties.

Keywords: Polarizability, second hyperpolarizability, basis set dependence, electron correlation, coupled cluster method, nuclear relaxation, infinite frequency approximation, finite field method

PACS: 31.15.Dv, 31.25.Qm, 33.15.Kr, 42.65.An

1. Introduction

During the last decades many theoretical and experimental efforts have been directed to the exploration and the design of organic *non-linear optical* (NLO) materials [1,2]. Particular interest is nowadays devoted to the investigation of NLO properties of conjugated organic molecules due to their large NLO susceptibility and their relatively simple synthesis as well as possible chemical modification. The conjugation of such systems makes the theoretical investigations sensitive to the inclusion of electron

*Corresponding author. Tel.: +32 81 724554; Fax: +32 81 724567; E-mail: benoit.champagne@fundp.ac.be.
**Permanent address: Department of Chemistry, Matej Bel University, Tajovského 40, SK-97400 Banská Bystrica, Slovak Republic.

correlation. Although the electron correlation is nowadays generally considered as essential for evaluation of the electronic component of NLO properties, the level of a correlation method necessary to accurately estimate the vibrational component, whose importance particularly for higher-order hyperpolarizabilities is now fully established [2–4], remains open. Up to now there are only a few studies of vibrational hyperpolarizabilities of polyatomic molecules going beyond the MP2 (second-order Møller-Plesset perturbation theory) level (see, e.g. [5]).

In the present study we selected a small model system containing a multiple bond for which the thorough analysis of basis set dependence, electron correlation and vibrational motion effects could be performed. Though there is obviously no conjugated π-chain between carbon atoms in difluoroacetylene, the conjugation between the π-bond and the lone pairs of the fluorine atoms can provide a helpful insight to the relation between the conjugation and the NLO properties. As a matter of fact, difluoroacetylene represents a monomer to conjugated polyynes with polar chain ends that were found to have promising NLO properties [6]. Enhancement of the vibrational NLO responses by replacement of hydrogen atoms by fluorine atoms in methane was reported by Quinet and Champagne [7].

As the system of our interest contains a triple bond, two highly polar single bonds as well as six lone pairs, it can be expected that the correlation and vibrational effects will be significant and should be treated at appropriate levels. The CC method is considered to be probably the most common correlation method of quantum chemistry aimed at accurate predictions of molecular properties of small many electron systems [8–11]. Since the original formulation by Čížek [12,13], many versions of this method have emerged [8,14]. Yet the most practical and usually sufficiently accurate approach, within the family of approximate single reference CC methods, appears to be CCSD(T) [15–18] with single and double excitation CCSD amplitudes solved iteratively [18], while triples are computed perturbatively using converged CCSD amplitudes. There are several alternatives how to assess the reliability of the CCSD(T) method in a particular case. Most often used 'internal' diagnostics methods that can reveal possible quasi-degeneracies are the T_1 diagnostics [11,19–21] and detection of largest t_1 and t_2 CCSD excitation amplitudes [22–24]. Although it was confirmed before by the T_1 diagnostics [25] that the ground state of C_2F_2 in the vicinity of equilibrium can well be described by single reference methods, due to necessity to obtain field-dependent minimum geometries we have checked highest CCSD amplitudes in our wave function.

Although not focused on electric properties, there have been several theoretical and experimental studies done for difluoroacetylene. In spite of its explosive decomposition character the first successful synthesis of C_2F_2 was described by Bürger and Sommer in [26] what enabled more detailed physicochemical investigations of this molecule. Spectroscopic parameters including harmonic as well as anharmonic frequencies, rotation and vibration-rotation constants have been obtained by combination of matrix and high resolution infrared measurements with *ab initio* methods at the Hartree-Fock SCF level and the correlated MP2 level with TZ2P+f basis set by Bürger et al. [27] and improved later by Breidung et al. [25] with TZ2Pf basis set at the CCSD and CCSD(T) levels. We are not aware of any direct experimental data on bond distances in C_2F_2. However, based on the MP2 results corrected in order to be consistent with the experimental rotational constant Bürger predicted the values of r_{C-C} and r_{C-F} equal to 1.1865 and 1.2832 Å, respectively. As concluded in [25], the CCSD(T) results for geometrical parameters lead to slight overestimation of the experimental rotational constant. This overestimation was attributed to the insufficiency of the used basis sets rather than to an inappropriate treatment of the electron correlation. Various DFT (density functional theory) methods have been used for evaluation of the molecular structure parameters, electron affinity, harmonic vibrational frequencies, and both fluorine atom and fluoride anion dissociation energies of C_2F_2 [28]. Isomerisation of difluorovinilidene into difluoroacetylene has been investigated by Jursic [29]. The best results were obtained by hybrid DFT methods [30–34], while the "pure" DFT methods (such as BLYP and BP86) even predict non-linear equilibrium geometries for the ground state. As for other molecular properties, Ruud et al. investigated isotropic as well as anisotropic magnetizability of difluoroacetylene at the Hartree-Fock level with the use of London atomic orbitals [35].

In this paper we report on the electronic and vibrational contributions to the longitudinal dipole polarizability and second hyperpolarizability of difluoroacetylene. Convergence of the molecular properties and molecular structure parameters with respect to the size and quality of the basis sets is analysed at the Hartree-Fock, CCSD and CCSD(T) levels. Vibrational contributions are evaluated by means of the finite field method formulated by Kirtman and co-workers [36]. This approach provides the *nuclear relaxation* term to the static and, within the *infinite frequency approximation*, also dynamic hyperpolarizabilities for the EOKE and dc-ESHG processes. Since difluoroacetylene is a symmetric molecule, the Eckart conditions [37,38] are implicitly taken into account. Finally, the electron correlation effects on the vibrational contribution are evaluated.

2. Theoretical and computational aspects

If a molecule is exposed to external static and/or dynamic electric fields ($\vec{F}$), the field-dependence of its dipole moment can be described by a Taylor series

$$\begin{aligned}\mu_\alpha(\omega_\sigma) = \mu_\alpha^0 + \sum_\beta \alpha_{\alpha\beta}(-\omega_\sigma;\omega_1)F_\beta(\omega_1) + \frac{1}{2}K^{(2)}\sum_{\beta\gamma}\beta_{\alpha\beta\gamma}(-\omega_\sigma;\omega_1,\omega_2)F_\beta(\omega_1)F_\gamma(\omega_2) + \dots \\ +\frac{1}{6}K^{(3)}\sum_{\beta\gamma\delta}\gamma_{\alpha\beta\gamma\delta}(-\omega_\sigma;\omega_1,\omega_2,\omega_3)F_\beta(\omega_1)F_\gamma(\omega_2)F_\delta(\omega_3) + \dots,\end{aligned} \tag{1}$$

where the $K^{(n)}$ factors are such that the nonlinear responses converge towards the same static limit. The numbers in parentheses $(-\omega_\sigma;\omega_1,\omega_2,\omega_3)$, with $\omega_\sigma = \sum_i \omega_i$, are the frequencies of the oscillating electric fields (in the order $F_\beta, F_\gamma, F_\delta, \dots$) and the summations run over the field indices β, γ, and δ associated with the Cartesian coordinates. μ_α^0 is the α-component of the permanent dipole moment. The static, dc-Kerr (EOKE), electric field-induced second harmonic generation (ESHG), third harmonic generation (THG), and degenerate four-wave mixing (DFWM) responses are then given by $\gamma(0;0,0,0)$, $\gamma(-\omega;\omega,0,0)$,$\gamma(-2\omega;\omega,\omega,0)$,$\gamma(-3\omega;\omega,\omega,\omega)$, and $\gamma(-\omega;\omega,-\omega,\omega)$, respectively.

Based on the clamped-nucleus approximation [39] to the exact *summation-over-states* (SOS) method, Bishop and Kirtman (BK) [40–43] have developed a general double-perturbation approach to evaluate the vibrational (hyper)polarizabilities at non-resonant frequencies. In the BK formalism the total static as well as dynamic (hyper)polarizabilities can be decomposed as

$$P = P^e + P^{ZP} + P^v, \tag{2}$$

where P^e is the electronic contribution at the equilibrium geometry, P^{ZP} is the zero-point vibrational averaging contribution, and P^v is the pure vibrational contribution.

For the static limit and within the infinite-frequency approximation [44], an alternative approach for evaluation of the pure vibrational (hyper)polarizability based on the changes of P^e and P^{ZP} generated by a static external electric field, taking the distortion of the equilibrium geometry into account, has been proposed [36,44–47]. In this approach, one can obtain so-called *nuclear relaxation* (NR) and *curvature*

(ZP-C) terms, whose analytical expressions can be compared with BK P^v formulae revealing that NR contribution contains the lowest-order BK terms of each square bracket type [44], while the remaining higher order-terms constitute the ZP-C contribution [45]. In this study, we restricted ourselves to evaluate the electronic contributions in the static limit and the leading vibrational contributions including nuclear relaxation terms only. Although the NR term can be evaluated through analytical formulas taking advantage of field-induced coordinates [47], in this study we follow the numerical finite field (FF) method formulated by Kirtman and co-workers [36]. This method requires determination of μ, α, β (and γ) in the presence of a uniform static electric field (F) with and without allowing the nuclei to relax to the equilibrium positions when the field is present.

If we denote the equilibrium geometry in the presence of a static electric field (so called *pump field*) by $\mathbf{R}_F$, and in the absence of the field by $\mathbf{R}_0$, we can define

$$(\Delta P)_{\mathbf{R}_0} = P(F, \mathbf{R}_0) - P(0, \mathbf{R}_0) \tag{3}$$

and

$$(\Delta P)_{\mathbf{R}_F} = P(F, \mathbf{R}_F) - P(0, \mathbf{R}_0) \tag{4}$$

where $P = \mu, \alpha, \beta, \ldots$. Thanks to the inversion center in C_2F_2, a Taylor expansion of the quantities given by Eqs (3) and (4) in the small field leads to the following expressions

$$(\Delta\mu_\alpha)_{\mathbf{R}_0} = \alpha^0_{\alpha\beta}F_\beta + \frac{1}{6}\gamma^0_{\alpha\beta\gamma\delta}F_\beta F_\gamma F_\delta, \tag{5}$$

$$(\Delta\alpha_{\alpha\beta})_{\mathbf{R}_0} = \frac{1}{2}\gamma^0_{\alpha\beta\gamma\delta}F_\gamma F_\delta, \tag{6}$$

$$(\Delta\beta_{\alpha\beta\gamma})_{\mathbf{R}_0} = \gamma^0_{\alpha\beta\gamma\delta}F_\delta, \tag{7}$$

where P^0 are static electronic properties, i.e. $\alpha^0_{\alpha\beta} = \alpha^e_{\alpha\beta}$ (0;0) and $\gamma^0_{\alpha\beta\gamma\delta} = \gamma^e_{\alpha\beta\gamma\delta}$ (0;0,0,0). Similarly, we can write

$$(\Delta\mu_\alpha)_{\mathbf{R}_F} = a_1 F_\beta + \frac{1}{6}g_1 F_\beta F_\gamma F_\delta, \tag{8}$$

$$(\Delta\alpha_{\alpha\beta})_{\mathbf{R}_F} = \frac{1}{2}g_2 F_\gamma F_\delta, \tag{9}$$

$$(\Delta\beta_{\alpha\beta\gamma})_{\mathbf{R}_F} = g_3 F_\delta. \tag{10}$$

The coefficients a_1, g_1, g_2 and g_3, which can be found by FF fitting techniques, contain the vibrational contribution due to relaxation [48,49] of the molecular geometry caused by the pump field. Once the coefficients are known, the nuclear relaxation (or displacement) contributions P^{nr} can be extracted from the coefficients a and g in the following manner

$$\alpha^{nr}_{\alpha\beta}(0;0) = a_1 - \alpha^e_{\alpha\beta}(0;0). \tag{11}$$

$$\gamma^{nr}_{\alpha\beta\gamma\delta}(0;0,0,0) = g_1 - \gamma^e_{\alpha\beta\gamma\delta}(0;0,0,0), \tag{12}$$

$$\gamma^{nr}_{\alpha\beta\gamma\delta}(-\omega;\omega,0,0)_{\omega\to\infty} = g_2 - \gamma^{e}_{\alpha\beta\gamma\delta}(0;0,0,0), \tag{13}$$

$$\gamma^{nr}_{\alpha\beta\gamma\delta}(-2\omega;\omega,\omega,0)_{\omega\to\infty} = g_3 - \gamma^{e}_{\alpha\beta\gamma\delta}(0;0,0,0). \tag{14}$$

Within the infinite-frequency approximation, $\gamma^{nr}_{\alpha\beta\gamma\delta}(-3\omega;\omega,\omega,\omega)_{\omega\to\infty}$ corresponding to the THG process is equal to zero, and for the diagonal component of the hyperpolarizability tensor (as well as for the mean value), the DFWM hyperpolarizability can be through the first order expressed as follows:

$$\gamma^{nr}_{zzzz}(-\omega;\omega,-\omega,\omega)_{\omega\to\infty} = 4\left[\gamma^{nr}_{zzzz}(-\omega;\omega,0,0)_{\omega\to\infty} - \frac{4}{3}\gamma^{nr}_{zzzz}(-2\omega;\omega,\omega,0)_{\omega\to\infty} \right. \\ \left. -\frac{1}{6}\gamma^{nr}_{zzzz}(0;0,0,0)\right]. \tag{15}$$

The electronic and vibrational (nuclear relaxation) contributions to (hyper)polarizabilities have been evaluated via a finite difference scheme using lower-order properties (energy and/or dipole moment) determined at different field amplitudes. Accurate low-order derivatives require that the field amplitude be small enough to avoid contamination from higher-order hyperpolarizabilities. However, the smaller the field amplitude, the lower the accuracy of the required derivative because the number of significant digits in the field-dependent energy (or other low order property) difference decreases. A good compromise has to be found if a Romberg-type procedure [50,51] is to be successfully applied to remove the higher-order contaminations. Successive improvement of e.g. the static electronic γ estimates starting from $\gamma^{0,k}$ given as (F denotes the smallest field amplitude)

$$\begin{aligned}\gamma^{0,k} &= -\left[\frac{\partial^4 E(2^k F)}{\partial(2^k F)^4}\right]_{F=0} \\ &= \lim_{F\to 0}\frac{-E(-2^{k+1}F) + 4E(-2^k F) - 6E(0) + 4E(2^k F) - E(2^{k+1}F)}{(2^k F)^4},\end{aligned} \tag{16}$$

can be evaluated using the general iterative expression

$$\gamma^{p,k} = \frac{4^p\gamma^{p-1,k} - \gamma^{p-1,k+1}}{4^p - 1}. \tag{17}$$

Here $p > 1$ is the order of the Romberg iteration. Analogous procedure can be used for the evaluation of the $\alpha^0_{\alpha\beta}, (\Delta\mu_\alpha)_{R_F}, (\Delta\alpha_{\alpha\beta})_{R_F}, (\Delta\beta_{\alpha\beta\gamma})_{R_F}$ as well as the quantities a_1, g_1, g_2, and g_3 appearing in Eqs (8)–(10). After testing the numerical stability of the results we have used the *pump* electric fields of 0, 0.0008, 0.0016, 0.0032, 0.0064, and 0.0128 a.u. and the *probe* electric fields of 0, $\pm$ 0.0016, $\pm$ 0.0032, $\pm$ 0.0064, $\pm$ 0.0128, $\pm$ 0.0256 a.u., respectively.

Selection of a suitable basis set appears to be essential for accurate predictions of the electric molecular properties. As evidenced by recent discussions concerning extrapolation to the complete basis set (CBS) limit [52,53] this is not an easy aspect to resolve. It has been shown for small systems [54,55] that for accurate evaluation of (hyper)polarizabilities the use of diffused and polarized basis functions is required because the outer part of the electronic cloud is primarily responsible for the linear and even more for non-linear responses. Dykstra and coworkers [56] have pointed out that incorporating of diffuse and polarization basis functions is more important than doubling or tripling the valence functions. The task is even more challenging in case of linear systems [2,55], where convergence difficulties due to the linear dependency problems disable using of basis sets that otherwise provide reliable results. Due to this reason we were forced to exclude e.g. the doubly augmented series of the correlation consistent polarized valence basis sets (d-aug-cc-pVXZ) [57]. Moreover, we have also encountered convergence problems with the HyPOL basis [58] (see below).

In our study, we have first checked the convergence patterns for the geometry parameters and the electronic (hyper)polarizabilities of C_2F_2 at the Hartree-Fock level for a series of Gaussian type basis sets 6-311G with polarization and diffuse functions running up to 6-311++G(3df, 3pd) [59,60]. Those bases are usually used for evaluating NLO properties of conjugated organic molecules [2,5]. Since it appeared that (mainly) the diffuse part of the latter bases set series is insufficiently described, we have decided for a generally contracted series of Widmark's ANO-L (large atomic natural orbitals) basis sets [61,62]. We have used contraction ranges from [4s3p2d] over [6s5p4d3f] up to [7s6p3d2f]. Though these sets were not specifically optimized for higher-order electric molecular properties, we could expect that they contain sufficiently diffuse functions needed for nonlinear responses. Our assumption is based on our former experience with calculations of dipole polarizabilities of CN, NO and O_2 [24], for which the ANO[6543] basis set provided the results at the CCSD and CCSD(T) levels comparable with those obtained with the d-aug-cc-pVQZ and aug-cc-pV5Z bases. Namely, the avaraged and maximum differences between ANO[6543] and aug-cc-pV5Z values at the CCSD(T) level were found to be 0.03 and 0.05 a.u., respectively. In addition, the ANO-L basis sets are quite stable with respect to linear dependencies. The primary disadvantage of these bases is, however, that a very large number of primitive GTOs is needed for converging towards the basis set limit [63]. Therefore, we have also employed the medium-sized singly augmented triple-zeta polarized valence correlation consistent basis set (aug-cc-pVTZ) [57], which is also commonly used in correlated calculations of electric properties. A smaller set of primitive GTOs is usually necessary in this type of basis sets in order to achieve results of comparable quality as with the larger ANO basis sets. Since calculations with higher cardinal numbers (X) would be too demanding, we have not checked the convergence patterns in this series, which would enable using of various extrapolation formulae to predict the complete basis set (CBS) limit. Finally, we have also used relatively small and flexible Sadlej's polarized basis sets (POL) [64,65] and extended (second-order) polarized basis sets (HyPOL) [58] that were designated for calculation of dipole polarizabilities and hyperpolarizabilities, respectively.

All the geometry optimizations as well as molecular property calculations reported in this study were obtained with the ACES II [66], GAUSSIAN 98 [67] and GAMESS [68] quantum chemistry programs.

3. Results and discussion

3.1. Geometry parameters and the rotational constant

The geometry parameters for difluoroacetylene optimized at the RHF, CCSD and CCSD(T) levels using various basis sets are reported in Table 1. Since there are no experimental data available, the bond distances are compared with previous theoretical studies, and the rotational constant evaluated at the CCSD(T) level is compared with the experimental value, as well.

When compared with the CCSD(T) results, the RHF values generally underestimate the bond distances by about 0.03–0.04 Å for C–C, and 0.02–0.03 Å for C–F. Though CCSD values lead to significant improvement, they are still below the CCSD(T) predictions by about 0.007–0.008 Å for C–C, and 0.004–0.005 Å for C–F, in agreement with the previous study by Breidung et al. [25]. Disabling the correlation of 1s orbitals gives rise to an increase of the bond distances by 0.004–0.006 Å for the ANO-L type

Table 1
Basis set dependence of bond lengths (in Ångström units) and the rotational constant (in cm^{-1}) in equilibrium geometries of C_2F_2. The values written in italics correspond to calculations with uncorrelated 1s orbitals

Basis	$r_{eq}(C \equiv C)$			$r_{eq}(C-F)$			B_e^a
	RHF	CCSD	CCSD(T)	RHF	CCSD	CCSD(T)	
ANO432	1.164	1.188	1.194	1.268	1.289	1.293	0.116838
ANO4321	1.164	1.184	1.191	1.265	1.282	1.287	0.117750
ANO5432	1.160	1.179	1.186	1.263	1.276	1.281	0.118816
ANO6532	1.160	1.179	1.186	1.263	1.279	1.283	0.118578
		1.183	*1.190*		*1.284*	*1.288*	*0.117703*
ANO6543	1.160	1.178	1.185	1.263	1.277	1.281	0.118888
		1.182	*1.190*		*1.283*	*1.287*	*0.117821*
ANO7632	1.160	1.179	1.186	1.263	1.279	1.284	0.118459
		1.183	*1.190*		*1.283*	*1.288*	*0.117703*
POL532	1.169	1.202	1.209	1.267	1.293	1.296	0.115455
		1.203	*1.210*		*1.294*	*1.297*	*0.115272*
HyPOL5332	1.166	1.196	1.204	1.266	1.289	1.293	0.116143
		1.197	*1.204*		*1.291*	*1.295*	*0.115913*
aug-cc-pVTZ (5432)	1.161	1.182	1.189	1.264	1.280	1.284	0.118246
Previous theoretical results							
MP2/TZP2+f[b]			1.188			1.286	0.118080
CCSD/DZP[c]			1.196			1.296	0.116351
CCSD(T)/TZ2Pf[d]			1.191			1.289	0.117475
B3LYP/6-311+G(2d,2p)[e]			1.182			1.286	0.118507
B3P86/6-311+G(2d,2p)[e]			1.182			1.280	0.119223
B3PW91/6-311+G(2d,2p)[e]			1.183			1.281	0.119031
Experiment[b]							0.118516

[a]The rotational constant was evaluated at the CCSD(T) level and respective previous theoretical results considering the molecule of $^{12}C_2{}^{19}F_2$ as a rigid rotor.
[b]Ref. [26].
[c]Ref. [69].
[d]Ref. [25].
[e]Ref. [29].

basis sets, whereas negligible effect of freezing the core can be noticed in calculations with the POL and HyPOL bases.

The bond distances obtained with ANO-L series converge to 1.186 ± 0.001 Å for C–C, and 1.283 ± 0.001 Å for C–F. With [6s5p3d2f] and/or [7s6p3d2f] bases an excellent agreement of rotational constants with the experimental value is accomplished. Our converged results are also in agreement with the predicted values by Burger et al. [27]. Reasonable values can be obtained already with the [5s4p3d2f] contraction. Merely slightly overestimating the length of triple bond, the geometry parameters obtained with the aug-cc-pVTZ basis set fit well to the pattern obtained with the ANO-L basis sets. On the other hand, the POL and HyPOL basis sets fail to predict accurate structural parameters for C_2F_2. Consequently, with the latter bases, too large equilibrium distances lead to an overestimation of the dipole polarizabilities (vide infra). Moreover, one should use these basis sets with care when evaluating the vibrational (hyper)polarizabilities, for which a proper description of the geometry dependence on the external electric field is of outmost importance.

3.2. Electronic contributions to linear polarizability and second hyperpolarizability

Electronic contributions to the static linear polarizability and second hyperpolarizability evaluated for the optimized geometries are presented in Tables 2 and 3. Both quantities were found very sensitive to

Table 2
Basis set dependence of static longitudinal electronic dipole polarizability of C_2F_2 The RHF, CCSD and CCSD(T) values given in a.u. correspond to the respective optimum geometries. In evaluating the electron correlation (EC) and triple excitation (TEX) contributions the CCSD(T) values were considered as reference. The values written in italics correspond to calculations with uncorrelated 1s orbitals

Basis	$\alpha^e_{zz}(0;0)$						
	RHF	CCSD	CCSD(T)	EC (in a.u.)	EC (in %)	TEX (in a.u.)	TEX (in %)
ANO432	30.12	32.87	33.76	3.64	10.78	0.89	2.64
ANO4321	30.10	32.60	33.48	3.38	10.10	0.88	2.63
ANO5432	30.35	32.65	33.54	3.19	9.51	0.89	2.65
ANO6532	30.36	32.73	33.64	3.28	9.75	0.91	2.71
		33.01	*33.94*	*3.58*	*10.55*	*0.93*	*2.74*
ANO6543	30.36	32.64	33.55	3.19	9.52	0.91	2.71
		32.96	*33.89*	*3.53*	*10.42*	*0.93*	*2.74*
ANO7632	30.36	32.74	33.65	3.29	9.78	0.91	2.70
		33.00	*33.93*	*3.57*	*10.52*	*0.93*	*2.74*
POL532	30.88	34.20	35.08	4.20	11.97	0.88	2.51
		34.27	*35.15*	*4.27*	*12.15*	*0.88*	*2.50*
HyPOL5332	30.67	33.96	34.84	4.17	11.97	0.88	2.53
		34.04	*34.93*	*4.26*	*12.20*	*0.89*	*2.55*
aug-cc-pVTZ (5432)	30.41	32.85	33.75	3.34	9.90	0.90	2.67
Recommended value			33.55 $\pm$ 0.10[a]				

[a]As recommended value we take the CCSD(T) result obtained with the ANO6543 basis set. Its uncertainty is estimated from the difference between the ANO6543 and ANO7632 results.

the level of the electron correlation treatment, as well as to the quality of the basis set.

Our preliminary results at the RHF level with the 6-311G type basis set series indicated that at least two d-polarization functions have to be included in the basis set, and that the augmenting by an extra polarization function of d-type is more important than introducing an f-type function. Though neither the linear polarizability or the hyperpolarizability convergence limit was achieved by going up to the 6-311++G(3df,3pd) set, the necessity of a proper description of the diffuse part of the wave function by appropriate diffuse s and p functions became evident. Described trends were confirmed by the use of the ANO-L basis sets. Convergence limit of the polarizability in this hierarchy was achieved already with the [5432] contraction, giving almost the same value as provided by the ANO6543 set. Decreasing of the degree of contraction in the s and p subspaces (ANO6532 and ANO7632) implies increasing of the CCSD and CCSD(T) polarizabilities by about 0.1 a.u., leaving the RHF practically unchanged. Results obtained with the aug-cc-pVTZ basis set slightly overshoot the converged ANO-L value by $\sim$ 0.05 a.u. at the RHF level, and $\sim$ 0.2 a.u. at the CC levels. This corresponds to the fact that the C–C bond length is slightly overestimated with the aug-cc-pVTZ basis set, leading to a larger "volume" of the molecule and consequently a larger linear polarizability. The effect is significantly more pronounced for the CC results with the POL and HyPOL basis sets that overshoot the ANO6543 values by $\sim$ 1.3–1.6 a.u.

As expected, for the second hyperpolarizability, the accuracy is lower. Here we can hardly write about the converged basis set limit values. From the convergence pattern with the ANO-L basis sets one can see that the subspace of s and p functions is not saturated unless going to ANO7632, still differing from ANO6543 by about 100 a.u. at the CCSD(T) level.

In order to roughly estimate the expected limit value, we can proceed as follows. First, we can estimate the sp correction limit from an extrapolation formula,

$$EXC = (\gamma(ANO7632) - \gamma(ANO6532))^2/(\gamma(ANO6532) - \gamma(ANO5432)), \tag{18}$$

Table 3

Basis set dependence of the static longitudinal electronic second hyperpolarizability of C_2F_2 The RHF, CCSD and CCSD(T) values given in a.u. correspond to the respective optimum geometries. In evaluation of the electron correlation (EC) and triple excitation (TEX) contributions the CCSD(T) values were considered as a reference. The values written in italics correspond to the calculations with uncorrelated 1s orbitals

Basis	$\gamma^e_{zzzz}(0;0,0,0)$						
	RHF	CCSD	CCSD(T)	EC (in a.u.)	EC (in %)	TEX (in a.u.)	TEX (in %)
ANO432	401	873	975	575	58.94	102	10.45
ANO4321	396	832	931	536	57.53	100	10.71
ANO5432	533	1007	1124	591	52.58	117	10.40
ANO6532	732	1277	1415	683	48.30	138	9.75
		1295	*1433*	*701*	*48.94*	*138*	*9.64*
ANO6543	740	1293	1431	691	48.26	139	9.68
		1310	*1449*	*709*	*48.91*	*140*	*9.62*
ANO7632	795	1365	1513	718	47.48	148	9.77
		1383	*1530*	*735*	*48.07*	*147*	*9.62*
POL532	647	1306	1418	770	54.35	112	7.90
		1319	*1420*	*773*	*54.43*	*102*	*7.16*
HyPOL5332	914	1727	1843	929	50.42	116	6.32
		1769	*1892*	*978*	*51.69*	*123*	*6.48*
aug-cc-pVTZ (5432)	777	1351	*1491*	715	47.93	140	9.40
Recommended value			1562 ±55[a]				

[a]The recommended value was calculated at the CCSD(T) level as the sum of the ANO7632 result, the difference between the ANO6543 and ANO6532 values (∼ 16 a.u.), and the extrapolation correction (∼ 33 a.u.) estimated from Eq. (18). The uncertainty is given by the sum of these corrections and the numerical uncertainty on the ANO7632 result (± 5 a.u.).

where $\gamma(X)$ is the electronic longitudinal hyperpolarizability obtained with the basis set X. Then we can add the contribution due to increasing the *df* sets as given by the difference between the ANO6532 and the ANO6543 results. These corrections were ∼ 33 a.u (EXC) and ∼ 16 a.u. for the latter. Hence, our final estimate figured out as the sum of the ANO7632 value and the two described corrections is ∼ 1562 ± 55 a.u.. In this context, the value of 1491 a.u. obtained with the aug-cc-pVTZ can be considered fairly good. Though the Sadlej's POL basis set also provides acceptable results, based on the previous discussion on the geometry parameters and the linear polarizability, the agreement is most probably fortuitous. The values obtained with the HyPOL basis set are too large and they suffer from an uncertainty of about 100 a.u.. The latter results from less accurate energies. Indeed, due to diffuse functions in the basis set, the tolerance for the basis set linear dependence has been increased by decreasing the threshold for the eigenvalues of the overlap matrix to 10^{-7}, while the threshold for the Hartree-Fock density matrix had to be increased to 10^{-8}.

Table 2 provides the information on the importance of the electron correlation (EC) effects in evaluation of the electronic linear polarizability. EC contribution to the total value (with CCSD(T) taken as the reference) ranges between 9–12% depending on the used basis set. For the most reliable basis sets the interval is narrowed to 9.5–9.9%, which corresponds to 3.2–3.3 a.u.. At the CCSD level we are able to recover 2.3–2.4 a.u., i.e. about 70% of the total electron correlation contribution. The contribution arising from inclusion of the triple excitation configurations into the wave function expansion appears to be fairly stable and, with its value of 0.89–0.91 a.u., could be considered non-negligible. The core correlation contribution (about −0.3 a.u.) from the 1s-like orbitals seems to be still significant.

An analogous analysis of the EC effects can be done for the electronic second hyperpolarizability based on the data presented in Table 3. The EC effects for this property are definitely much more essential than for the polarizability. The total EC contribution ranges between 47–49% for the higher members of the

Table 4
Static longitudinal electronic, vibrational (nuclear relaxation), and total linear dipole polarizability of C_2F_2

Basis	$\alpha^{e}_{zz}(0;0)$			$\alpha^{nr}_{zz}(0;0)$			$\alpha^{tot}_{zz}(0;0)$		
	RHF	CCSD	CCSD(T)	RHF	CCSD	CCSD(T)	RHF	CCSD	CCSD(T)
POL532	30.88	34.20	35.08	4.60	4.44	4.54	35.48	38.64	39.61
ANO4321	30.10	32.60	33.48	4.76	4.49	4.56	34.86	37.08	38.04
ANO5432	30.35	32.65	33.54	4.63	4.40	4.47	34.98	37.05	38.01
aug-cc-pVTZ (5432)	30.41	32.85	33.75	4.62	4.43	4.50	35.03	37.28	38.25
Best estimate			33.55			4.48			38.03
			± 0.10[a]			± 0.02[b]			± 0.12

[a]The value was obtained at the CCSD(T) level with the ANO6543 basis set (see Table 2).
[b]The value corresponds to the average of the ANO5432 and aug-cc-pVTZ CCSD(T) results. Its uncertainty is the absolute deviation of the two values from the average.

Table 5
Static longitudinal electronic, vibrational (nuclear relaxation), and total second dipole hyperpolarizability of C_2F_2

Basis	$\gamma^{e}_{zzzz}(0;0,0,0)$			$\gamma^{nr}_{zzzz}(0;0,0,0)$			$\gamma^{tot}_{zzzz}(0;0,0,0)$		
	RHF	CCSD	CCSD(T)	RHF	CCSD	CCSD(T)	RHF	CCSD	CCSD(T)
POL532	647	1306	1418	2236	2540	2733	2883	3845	4150
ANO4321	396	832	931	2114	2373	2638	2510	3204	3569
ANO5432	533	1007	1124	2168	2320	2508	2701	3326	3632
aug-cc-pVTZ (5432)	777	1351	1491	2153	2372	2568	2930	3723	4059
Best estimate			1562			2538			4100
			± 55[a]			± 50[b]			± 105

[a]See the comment in Table 3.
[b]The value corresponds to the average of the ANO5432 and aug-cc-pVTZ CCSD(T) results. Its uncertainty includes the absolute deviation of the two values from the average as well as the numerical inaccuracy.

basis sets in the ANO-L hierarchy and the aug-cc-pVTZ basis. With the CCSD method we recovered approximately 80% of the total EC contribution. In accord with the polarizability calculations, the triple excitation contribution is relatively stable throughout the variety of the used basis sets and amounts for about 140–150 a.u. Unlike for the polarizability, the core correlation effects (15–20 a.u. for the ANO-L series) are insignificant, suggesting that the outer part of the electronic cloud of our system almost fully determines the higher-order hyperpolarizability.

3.3. *Vibrational contributions to linear polarizability and second hyperpolarizability*

The nuclear relaxation (NR) contributions to dipole (hyper)polarizabilities for the static limit and the dynamic processes ($\omega \rightarrow \infty$) are presented in Tables 4–6. Since the evaluation of the NR contributions is much more computationally demanding than the mere electronic ones, we restricted ourselves to the smaller basis sets. Namely, the ANO5432 basis set was selected, as it provided very good estimations of both geometry parameters and electronic linear polarizability. Second, the aug-cc-pVTZ was chosen especially due to its reasonable values for the electronic second hyperpolarizability. Finally, two smaller sets including the ANO4321 and the Sadlej's POL basis sets were applied to see the overall sensitivity of the NR contributions with respect to the basis sets.

Table 4 shows that the NR contributions to the linear polarizability are much less sensitive both to the selection of the basis set and to the electron correlation effects than their electronic counterparts. The NR contributions represent about 11–12 % of the total static value, what means that their inclusion is as important as the inclusion of the EC effects. The EC effect on the NR contribution (denoted hereafter as

Table 6
Dynamic longitudinal vibrational (nuclear relaxation) contributions to the second dipole hyperpolarizability corresponding to the EOKE, ESHG and DFWM non-linear optical responses of C_2F_2

Basis	EOKE: $\gamma^{nr}_{zzzz}(-\omega,\omega,0,0)$			ESHG: $\gamma^{nr}_{zzzz}(-2\omega;\omega,\omega,0)$			DFWM: $\gamma^{nr}_{zzzz}(-\omega;\omega,-\omega,\omega)$		
	RHF	CCSD	CCSD(T)	RHF	CCSD	CCSD(T)	RHF	CCSD	CCSD(T)
POL532	506	734	795	27	67	62	390	885	1024
ANO4321	559	685	755	13	63	70	757	824	888
ANO5432	543	655	731	9	68	66	678	708	900
aug-cc-pVTZ (5432)	532	676	743	~ 0	60	68	693	805	897
Best estimate			737 $\pm$ 10[a]			67 $\pm$ 6[a]			899 $\pm$ 25[a]

[a]The value corresponds to the average of the ANO5432 and aug-cc-pVTZ CCSD(T) results. Its uncertainty includes the absolute deviation of the two values from the average as well as the numerical inaccuracy.

NR/EC) appears to be rather insignificant. Our best estimate for the total static longitudinal polarizability is 38.03 $\pm$ 0.12 a.u.. The latter value has been obtained as the sum of the CCSD(T)/ANO6543 electronic contribution and the NR contribution taken as the average of the ANO5432 and aug-cc-pVTZ results.

The data on the static NR contributions presented in Table 5 again demonstrate the crucial role of the electron correlation in evaluating the higher order properties. Though, unlike for the electronic hyperpolarizabilty, the results are less sensitive to the selection of a basis set. The NR contributions represent about 70% of the total static value, i.e. they prevail over the electronic contributions. The magnitude of the NR/EC term calculated as the average of the ANO5432 and aug-cc-pVTZ values is 380 a.u., what represents $\sim$ 15% of the NR contribution. About half of this value corresponds to the contribution from the triple excitations. Our estimated value of the total static longitudinal hyperpolarizability is 4100 $\pm$ 105 a.u..

Table 6 collects the results for the dynamic NR contributions to the longitudinal hyperpolarizability corresponding to the EOKE, ESHG, and DFWM non-linear responses. With an exception of the rather small ESHG values, the magnitude of the NR contributions is comparable with the electronic ones. Stability of the results with respect to the choice of the basis set is, however, much higher. The NR/EC term estimated from the ANO5432 and aug-cc-pVTZ values is of 190–210, 57–68, and 200–220 a.u., i.e. approximately 27, 85–100, and 24% of the total γ^{nr} for the EOKE, ESHG, and DFWM processes, respectively. Particularly interesting are the results for the $\gamma^{nr}(-2\omega;\omega,\omega,0)_{\omega\to\infty}$ corresponding to the ESHG response, where the substantial part of the value can be attributed to the electron correlation effects. Our final estimates for the longitudinal EOKE, ESHG, and DFWM vibrational non-linear responses in the infinite-frequency approximation are 737 $\pm$ 10, 67 $\pm$ 6, and 899 $\pm$ 25 a.u., respectively.

As aforementioned, the standard coupled-cluster theory behind the CCSD(T) method relies on a single determinant reference wave function and it is desirable to a priori assess its reliability, particularly when no experimental data and/or other theoretical results based on multireference methods are available. Table 7 contains the information about the largest t_1 and t_2 amplitudes as well as the T_1 diagnostics [20] values evaluated in the field free and selected field dependent optimum geometries obtained with the ANO5432 basis set. It can be seen that for all the checked geometries the T_1 amplitudes are well below the critical 0.02 value and both T_1 and the largest excitation amplitudes are quite stable with respect to the changes of the geometry required for evaluating the NR contributions.

4. Conclusions

Difluoroacetylene as a model molecule containing a triple bond and two highly polar ends has been selected to investigate the electron correlation and the vibrational motion effects on the linear and non-linear

Table 7
Largest excitation amplitudes and T_1 diagnostics evaluated for the field-free and selected field dependent optimum geometries obtained at the CCSD and CCSD(T) levels

Field	CCSD			CCSD(T)		
(in a.u.)	$t_1^{\max}$	$t_2^{\max}$	T_1	$t_1^{\max}$	$t_2^{\max}$	T_1
0	0.02679	0.06281	0.0114	0.02703	0.06478	0.0115
0.0064	0.02542	0.06265	0.0114	0.02565	0.06463	0.0115
0.0128	0.02420	0.06191	0.0115	0.02443	0.06298	0.0116

molecular properties. Namely, the electronic and nuclear relaxation contributions to the longitudinal component of the linear polarizability and the second hyperpolarizability have been calculated. While the electron correlation has been treated by the CC methods including singly and double excitations (CCSD) and non-iterative triple excitations (CCSD(T)), treatment of the vibrational effects was based on the NR finite field method formulated by Kirtman and co-workers. Particular attention has been paid to the selection of the appropriate basis sets.

All properties have been evaluated in the geometries optimized with the respective basis set and the quantum chemistry method. Since the experimental geometry parameters were not available, the rotational constant was calculated from the computed data and compared with the experimental value. It was found out that within the ANO-L type basis set series very reasonable parameters – compared to the converged value – can be obtained with the [5s4p3d2f] contraction. While the aug-cc-pVTZ basis set has also provided an acceptable result, the POL and HyPOL basis sets have led to an overestimation of both the C-C and C-F bond lengths resulting in overestimation of the polarizability.

The electronic contributions to the (hyper)polarizabilities have been found very sensitive to the selection of a basis set. Though the inclusion of polarization functions of d- and f-type is without any doubt necessary, the proper description of the diffuse part of the wave function by appropriate and balanced diffuse s and p functions appears to be more difficult. This is particularly relevant for the evaluation of the second hyperpolarizability, where within the ANO-L series we have not approached the convergence limit until going to the ANO7632 basis set. A good compromise between the size of the basis set and the accuracy of the result has been found in the use of the aug-cc-pVTZ basis set. The electronic polarizability converges faster than γ^e and the converged value has been obtained with the ANO5432 basis set. The aug-cc-pVTZ basis set slightly overestimates the converged ANO-L value, which is probably related to a slight overestimation of the C-C bond length. As expected, inclusion of the electron correlation effects have been found essential particularly in the case of the hyperpolarizability, where it ranged between 47–49% of the total value. Our best estimates for the electronic linear polarizability and the second hyperpolarizability are 33.55 ± 0.10 and 1562 ± 55 a.u., respectively.

In general, nuclear relaxation (NR) contributions to dipole (hyper)polarizabilities have been found less sensitive to the basis set choice than the electronic contributions. Already with the ANO4321 basis set we have obtained the results very close to those provided by both the ANO5432 and aug-cc-pVTZ basis sets. Our final estimates for the NR contributions to the longitudinal $\alpha(0;0), \gamma(0;0,0,0), \gamma(-\omega;\omega,0,0)_{\omega\to\infty}, \gamma(-2\omega;\omega,\omega,0)_{\omega\to\infty}$ and $\gamma(-\omega;\omega,-\omega,\omega)_{\omega\to\infty}$ are 4.48 ± 0.02, 2538 ± 50, 737 ± 10, 67 ± 6, and 899 ± 25 a.u., respectively. Unlike for the electron correlation effects on the NR contributions (the NR/EC term) in the case of linear polarizability, the electron correlation effects are essential for the hyperpolarizability, where they represent $\sim$ 15, 27, 85–100, and 24% of the total NR contributions for the static, EOKE, ESHG, and DFWM non-linear responses, respectively.

Work to follow will consist in comparing various correlation methods (MP2, DFT, CC) in the calculations of NLO properties, while the presented CC results are expected to serve as benchmark for evaluation of other correlation methods as well as a tool for the selection of suitable basis sets.

After finishing this study we became aware of a very recent work by Maroulis and Xenides [70] on the evaluation of the electric multipole moments and (hyper)polarizability of dihalogenated acetylene. It appears that our results on the effects of electron correlation on the electronic longitudinal (hyper)polarizability are generally consistent with those obtained by Maroulis and Xenides at the MP2 level, though their conclusions on the role of d- and f-type functions in the basis sets differ from ours.

Acknowledgements

The authors acknowledge the financial assistance of the Commissariat Général aux Relations Internationales (CGRI, Belgium) and the Slovak Academy of Science (SAS, Slovakia), and the Belgian National Interuniversity Research Program on "Supramolecular Chemistry and Supramolecular Catalysis" (PAI/IUAP No. P5-03). MM acknowledges the Belgian Office for Scientific, Technical and Cultural Affairs (Research fellowships for Central and Eastern Europe) for the financial support enabling his research stay in the FUNDP, Namur. MM also thanks the Faculty of Natural Sciences, Matej Bel University (Project No. 8/03-Ch) for a partial support of this research. This work has also been supported by the Scientific Grant Agency of the Slovak Republic (VEGA, Projects No. 1/0115/03 and 3103/03). BC and EAP thank the Belgian National Fund for Scientific Research (FNRS) for their Senior Research Associate and Research Associate positions, respectively. Essential part of the calculations has been performed on a cluster of PC acquired thanks to a "credits aux chercheurs" of the FNRS.

References

[1] D.M. Bishop, in: *Advances in Quantum Chemistry*, (Vol. 25), P.-O. Löwdin, ed., Academic Press, San Diego, 1994, pp. 1–45.
[2] B. Champagne and B. Kirtman, in: *Handbook of Advanced Electronic and Photonic Materials and Devices*, (Vol. 9), H.S. Nalwa, ed., Nonlinear Optical Properties, Academic Press, San Diego, 2001, pp. 63–126.
[3] D.M. Bishop and P. Norman, in: *Handbook of Advanced Electronic and Photonic Materials and Devices*, (Vol. 9), H.S. Nalwa, ed., Nonlinear Optical Properties, Academic Press, San Diego, 2001, pp. 1–62.
[4] M. Torrent-Sucarrat, M. Solà, M. Duran, J.M. Luis and B. Kirtman, *J. Chem. Phys.* **116** (2002), 5363.
[5] M. Torrent-Sucarrat, M. Solà, M. Duran, J. Luis and B. Kirtman, *J. Chem. Phys.* **118** (2003), 711.
[6] D. Jacquemin, B. Champagne and J.-M. André, *Int. J. Quantum Chem.* **65** (1997), 679.
[7] O. Quinet and B. Champagne, *J. Chem. Phys.* **109** (1998), 10594.
[8] J. Paldus and X. Li, *Adv. Chem. Phys.* **110** (1999), 1.
[9] R.J. Bartlett, in: *Modern Electronic Structure Theory*, D.R. Yarkony, ed., Part II, World Scientific, Singapore, 1995, pp. 1047–1131.
[10] M. Urban, I. Černušák, V. Kellö and J. Noga, in: *Methods in Comutational Chemistry*, S. Wilson, ed., Plenum Press, New York, 1987, pp. 117–250.
[11] T.J. Lee and G. Scuseria, in: *Quantum Mechanical Electronic Structure Calculations with Chemical Accuracy*, S.R. Langhoff, ed., Kluwer Academic Publishers, Dordrecht, 1995, pp. 47–108.
[12] J. Čížek, *Adv. Chem. Phys.* **14** (1969), 35.
[13] J. Čížek, *J. Chem. Phys.* **45** (1966), 4256.
[14] R.J. Bartlett, ed., *Recent Advances in Computational Chemistry, Recent Advances in Coupled-Cluster Methods*, Vol. 3, World Scientific, Singapore, 1997.
[15] K. Raghavachari, G.W. Trucks, J.A. Pople and M. Head-Gordon, *Chem. Phys. Lett.* **157** (1989), 479.
[16] R.J. Bartlett, J.D. Watts, S.A. Kucharski and J. Noga, *Chem. Phys. Lett.* **165** (1990), 513.
[17] M. Urban, J. Noga, S.J. Cole and R.J. Bartlett, *J. Chem. Phys.* **83** (1985), 4041.
[18] G.D. Purvis and R.J. Bartlett, *J. Chem. Phys.* **76** (1985), 110.
[19] T.J. Lee, J.E. Rice, G. Scuseria and H.F. Schaefer Jr., *Theor. Chim. Acta* **75** (1989), 81.
[20] T.J. Lee and P.R. Taylor, *Int. J. Quantum Chem. Symp.* **23** (1989), 199.
[21] D. Jaytilaka and T.J. Lee, *J. Chem. Phys.* **98** (1993), 9734.

[22] J. Watts, M. Urban and R.J. Bartlett, *Theor. Chim. Acta* **90** (1995), 341.
[23] V. Parasuk, P. Neogrády, H. Lischka and M. Urban, *J. Phys. Chem.* **100** (1996), 6325.
[24] M. Medveď, M. Urban, V. Kellö and G.H.F. Diercksen, *J. Molec. Struct. (Theochem)* **547** (2001), 219.
[25] J. Breidung, W. Schneider, W. Thiel and T.J. Lee, *J. Chem. Phys.* **97** (1992), 3498.
[26] H. Bürger and S. Sommer, *J. Chem. Soc. Chem. Commun.* (1991), 456.
[27] H. Bürger, W. Schneider, S. Sommer, W. Thiel and H. Willner, *J. Chem. Phys.* **95** (1991), 5660.
[28] R.A. King, N.D. Pettigrew and H.F. Schaefer III., *J. Chem. Phys.* **107** (1997), 8536.
[29] B.S. Jursic, *Int. J. Quantum Chem.* **62** (1997), 515.
[30] A.D. Becke, *J. Chem. Phys.* **98** (1993), 5648.
[31] C. Lee, W. Yang and R.G. Parr, *Phys. Rev. B* **37** (1988), 785; B. Mielich, A. Savin, H. Stoll and H. Preuss, *Chem. Phys. Lett.* **157** (1989), 200.
[32] J.P. Perdew, *Phys. Rev. B* **33** (1986), 8822; *Ibid.* **34** (1987), 7046.
[33] J.P. Perdew and Y. Wang, *Phys. Rev. B* **45** (1992), 13244.
[34] A.D. Becke, *J. Chem. Phys.* **98** (1993), 1372.
[35] K. Ruud, T. Helgaker, P. Jørgensen and K.L. Bak, *Chem. Phys. Lett.* **223** (1994), 12.
[36] D.M. Bishop, M. Hasan and B. Kirtman, *J. Chem. Phys.* **103** (1995), 4157.
[37] J.M. Luis, M. Duran, J.L. Andrés, B. Champagne and B. Kirtman, *J. Chem. Phys.* **111** (1999), 875.
[38] C. Eckart, *Phys. Rev.* **47** (1935), 552; S.R. Polo, *J. Chem. Phys.* **24** (1956), 1133.
[39] D.M. Bishop, B. Kirtman and B. Champagne, *J. Chem. Phys.* **107** (1997), 5780.
[40] B. Kirtman and D.M. Bishop, *Chem. Phys. Lett.* **175** (1990), 601.
[41] D.M. Bishop and B. Kirtman, *J. Chem. Phys.* **95** (1991), 2646.
[42] D.M. Bishop and B. Kirtman, *J. Chem. Phys.* **97** (1992), 5255.
[43] D.M. Bishop, J.M. Luis and B. Kirtman, *J. Chem. Phys.* **108** (1998), 10013.
[44] J.M. Luis, J. Martí, M. Duran, J.L. Andrés and B. Kirtman, *J. Chem. Phys.* **108** (1998), 4123.
[45] B. Kirtman, J.M. Luis and D.M. Bishop, *J. Chem. Phys.* **108** (1998), 10008.
[46] J.M. Luis, M. Duran and J.L. Andrés, *J. Chem. Phys.* **107** (1997), 1501.
[47] J.M. Luis, M. Duran and B. Kirtman, *J. Chem. Phys.* **115** (2001), 4473.
[48] J. Martí, J.L. Andrés, J. Bertránd and M. Duran, *Mol. Phys.* **80** (1993), 625.
[49] J. Martí and D.M. Bishop, *J. Chem. Phys.* **99** (1993), 3860.
[50] P.J. Davis and P. Rabinowitz, *Numerical Integration*, Blaisdell, London, 1967.
[51] D. Jacquemin, B. Champagne and J.M. André, *Int. J. Quantum. Chem.* **65** (1997), 679.
[52] D. Feller and K.A. Peterson, *J. Chem. Phys.* **108** (1998), 154; A. Halkier, T. Helgaker, P. Jørgensen, W. Klopper, H. Koch, J. Olsen and A.K. Wilson, *Chem. Phys. Lett.* **286** (1998), 243; D.G. Truhlar, *Chem. Phys. Lett.* **294** (1998), 45; A. Halkier, T. Helgaker, W. Klopper, P. Jørgensen and A.G. Császár, *Chem. Phys. Lett.* **310** (1999), 385.
[53] P. Neogrády, M. Medved', I. Černušák and M. Urban, *Mol. Phys.* **100** (2002), 541.
[54] G. Maroulis, *J. Chem. Phys.* **108** (1998), 5432; G. Maroulis, *Chem. Phys. Lett.* **289** (1998), 403; G. Maroulis and C. Pouchan, *Phys. Rev. A* **57** (1998), 2440; G. Maroulis, *Chem. Phys. Lett.* **312** (1999), 255; G. Maroulis, *J. Chem. Phys.* **111** (1999), 6846.
[55] G. Maroulis and A.J. Thakkar, *J. Chem. Phys.* **95** (1991), 9060.
[56] S.Y. Liu and C.E. Dykstra, *J. Phys. Chem.* **91** (1987), 1749; T. Zhou and C.E. Dykstra, *J. Phys. Chem. A* **104** (2000), 2204.
[57] T.H. Dunning, Jr., *J. Chem. Phys.* **90** (1989), 1007; R.A. Kendall, T.H. Dunning, Jr. and R.J. Harrison, *J. Chem. Phys.* **96** (1992), 6769.
[58] T. Pluta and A.J. Sadlej, *Chem. Phys. Lett.* **297** (1998), 391.
[59] R. Krishnan, J.S. Binkley, R. Seeger and J.A. Pople, *J. Chem. Phys.* **72** (1980), 650.
[60] M.J. Frisch, J.A. Pople and J.S. Binkley, *J. Chem. Phys.* **80** (1984), 3265.
[61] P.-O. Widmark, P.-Å. Malmquist and B.O. Roos, *Theor. Chim. Acta* **77** (1990), 291.
[62] P.-O. Widmark, B.J. Persson and B.O. Roos, *Theor. Chim. Acta* **79** (1991), 419.
[63] F. Jensen, *Introduction to Computational Chemistry*, John Wiley & Sons, Chichester, 1999, p. 161.
[64] A.J. Sadlej, *Collect. Czech. Chem. Commun.* **53** (1988), 1995.
[65] A.J. Sadlej, *Theor. Chim. Acta* **79** (1991), 123.
[66] J.F. Stanton, J. Gauss, J.D. Watts, W.J. Lauderdale and R.J. Bartlett, *Advanced Concepts in Electronic Structure (ACES II)*, Quantum Theory Project: University of Florida, Gainesville, FL, 1991.
[67] GAUSSIAN 98, Revision A.11, M.J. Frisch, G.W. Trucks, H.B. Schlegel, G.E. Scuseria, M.A. Robb, J.R. Cheeseman, V.G. Zakrzewski, J.A. Montgomery, R.E. Stratmann, J.C. Burant, S. Dapprich, J.M. Millam, A.D. Daniels, K.N. Kudin, M.C. Strain, O. Farkas, J. Tomasi, V. Barone, B. Mennucci, M. Cossi, R. Cammi, C. Pomelli, C. Adamo, S. Clifford, J.W. Ochterski, G.A. Petersson, P.Y. Ayala, K. Morokuma, Q. Cui, D.K. Malick, A.D. Rabuck, K. Raghavachari, J.B. Foresman, J.V. Ortiz, J. Cioslowski, B.B. Stefanov, G. Liu, A. Liashenko, P. Piskorz, I. Komaromi, R. Gomperts, R.L.

Martin, D.J. Fox, T. Keith, M.A. Al-Laham, C.Y. Peng, A. Nanayakkara, C. Gonzalez, M. Challacombe, P.M.W. Gill, B. Johnson, W. Chen, M.W. Wong, J.L. Andres, M. Head-Gordon, E.S. Replogle and J.A. Pople, *Gaussian, Inc.*, Pittsburgh, PA, 2001.

[68] GAMESS, M.W. Schmidt, K.K. Baldridge, J.A. Boatz, S.T. Elbert, M.S. Gordon, J.J. Jensen, S. Koseki, N. Matsunaga, K.A. Nguyen, S. Su, T.L. Windus, M. Dupuis and J.A. Montgomery, *J. Comput. Chem.* **14** (1993), 1347.

[69] M.M. Gallo and H.F. Schaefer III., *J. Chem. Phys.* **93** (1991), 865.

[70] G. Maroulis and D. Xenides, *Int. J. Mol. Sci.* **4** (2003), 263.

Journal of Computational Methods in Sciences and Engineering 4 (2004) 267–275
IOS Press

PLS prediction of hyperpolarizabilities for donor-bridge-acceptor organic systems

A.E. de A. Machado*, B. de Barros Neto and A.A. de S. da Gama
Departamento de Química Fundamental, Universidade Federal de Pernambuco, Recife, 50670-901, Pernambuco, Brazil

Abstract. Elevated magnitudes of the first and second static hyperpolarizabilities (AM1/TDHF), respectively, of organic molecules having electron-donation (D) and electron-accepting (A) groups can be obtained through of selection of molecular systems with appropriate structural and electronic parameters. The Homo-Lumo energy gap, the total number of π-electrons of the molecule, the Homo energy, and the ground state dipole moment were the parameters considered in this work. The designed D-A organic molecules investigated have polyenic/mesoionic bridge, which afford large β and γ hyperpolarizabilities, associated for different strength of donor-acceptor pairs. Large data sets of these parameters for the molecules, obtained from quantum chemistry semiempirical calculations with AM1 hamiltonian, were submitted to a principal components analysis (PCA). In addition, a model was proposed based in a principal components regression (PCR), having in mind the prevision and selection of organic molecules with potential applications in nonlinear optics.

Keywords: β and γ hyperpolarizabilities, AM1/TDHF methodology, nonlinear organic molecules, semiempirical method, principal components analysis, partial least squares regression (PLS), chemometric techniques

1. Introduction

The search for organic molecules with large nonlinear optical coefficients is fundamental for the development of advanced technologies. For example, novel materials are crucial for a substantial increment in data storage densities and transmission speeds [1–4].

Molecules with large first hyperpolarizability values are used as electrooptic modulators and all-optical switches, among other devices [1,2]. The literature reports that higher first (β) and second (γ) hyperpolarizability magnitudes can be obtained by adjusting of structural and electronic parameters [3–7]. These parameters include the strength of the donor (D) and acceptor (A) groups [6–10], the Homo-Lumo energy gap [9,11–13], the nature and length of the conjugate bridge linking the D/A groups [8,10,13, 14], the ionisation potential [11,15], the number of π-electrons and their degree of delocalization [13,16] and the molecular conformation [10,15,17]. In the case of γ, some parameters have been evaluated, but there is not the same level of confidence as occurs for the first hyperpolarizability, and more theoretical and experimental research involving multidisciplinary groups has been performed [7,18]. The second hyperpolarizability is considered especially important, because the implementation of all-optical process in photonics requires a quite large value of this parameter [5].

The first and second hyperpolarizabilities of polyenes and their derivatives have been calculated at several levels, from the semiempirical to the more sophisticated [8–10,19–22]. Semiempirical methods

*Corresponding author. Tel.: +55 81 21268443; Fax: +55 81 21268442; E-mail: aeam@ufpe.br.

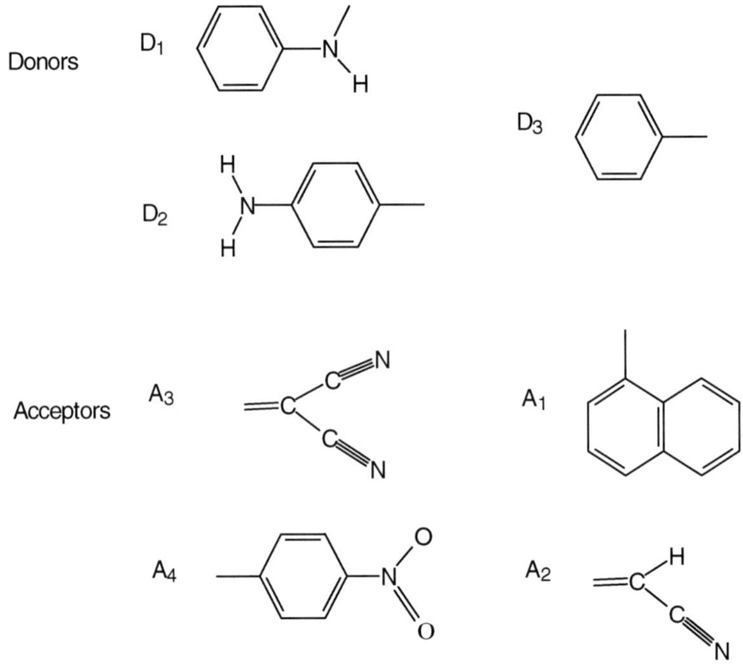

Fig. 1. Structures of donor (D) and acceptor (A) groups investigated in this work.

have the advantage of being computationally less demanding, while ab initio methods are more precise but require more extensive computing support [3–5].

A less expensive procedure for selecting molecules as potential candidates for nonlinear optical applications might be obtained by chemometric modeling of available data. In the present work, to assess the feasibility of this approach, a large data set of electronic and structural parameters obtained from semiempirical calculations is modeled by a multivariate calibration procedure based on projections into latent variables (principal components), which has been successfully used in other fields [23].

A survey of theoretical and experimental reports shows that push-pull polyenic derivatives are an important class of nonlinear material, exhibiting large β and γ hyperpolarizabilities [4,9,24–27]. The introduction, between donor and acceptor groups, of an asymmetric bridge containing a mesoionic ring results in large first hyperpolarizabilities for the designed donor-acceptor systems [28]. Recently, some mesoionic compounds synthesized and fully characterized for γ were found to have large second hyperpolarizability values [29].

The donor-bridge-acceptor molecules investigated in this work present a new bridge model, with polyenic moieties and mesoionic rings. These designed systems exhibit large values for both β and γ, according to the AM1/TDHF calculations [30]. This methodology has been evaluated in relation for β and γ, for a group of molecules fully characterized in the literature [9,15,28–30]. The AM1/TDHF methodology [31–35], therefore, is able to point out donor-acceptor systems with interestingly high nonlinear coefficients.

The strength of different donor-acceptor pairs was investigated. The donor groups used are the phenyl and phenylamine (see Fig. 1). The latter can be attached to the bridge by the phenyl moiety or by the nitrogen atom, and recent results for polyenic derivatives suggest that linking the heteroatom directly to the bridge contributes to enhance the second hyperpolarizability [9]. The acceptor groups are shown in

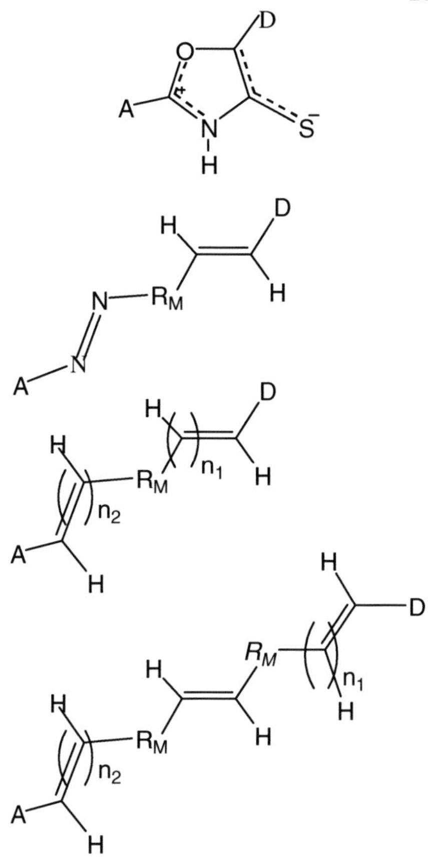

Bridge

I

II

III (n_1=1, n_2=2)

IV (n_1=2, n_2=2)

V (n_1=2, n_2=2)

VI (n_1=2, n_2=2)

Fig. 2. The bridge structures investigated. R_M is the mesoionic ring shown in the bridge I. The mesoionic rings ($\mathrm{R_M}$ and R_M) are in configuration cis in the bridge V and trans in the bridge VI. The $\mathrm{R_M}$ and R_M rings are planar. Groups A_2 and A_3 as acceptors are directly attached to the double bond at end of the bridge (III-VI).

Fig. 1, they are the 1-naphtyl, cyanomethylene, dicyanomethylene and nitrophenyl. The structures of the bridges considered in this work are presented in Fig. 2.

The total number of π-electrons plus AM1 values for the HOMO energy, the Homo-Lumo energy gap and the ground state dipole moment were the theoretical descriptors selected for this investigation. The whole data set was submitted to a partial least squares regression (PLS), in which the scores of the molecules on the first principal components are used as regressors for predicting the values of the static hyperpolarizability [23]. As we shall see, latent variable models can be useful for estimating values that are difficult to measure, or are otherwise unavailable, as well as suggesting possible alternative structures with more desirable properties.

2. Methodology

In a partial least squares regression (PLS) analysis the original multivariate data set is compressed into a small number of uncorrelated axes (PC's) that contain most of the original information [23]. The co-ordinates of these PC axes are linear combinations of the original values, chosen to maximise the

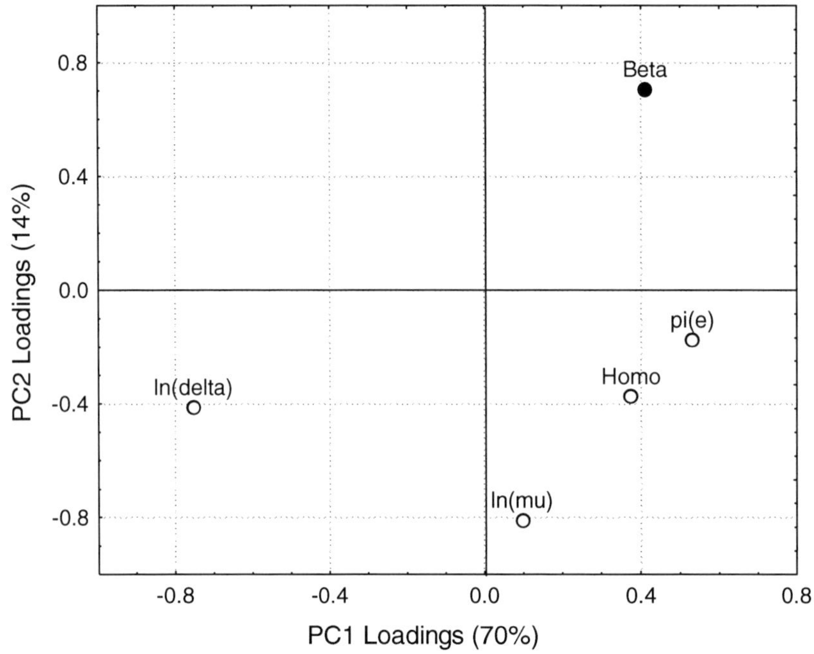

Fig. 3. Loadings plot for the first two PLS principal components for modeling beta. Percentages are explained variances in the regressor matrix by each PC.

covariance matrix between the regressor matrix and the property of interest (the hyperpolarizabilities, in this case). The first PLS component (PC1) is, of all possible axes in the multidimensional space, the one that maximizes at the same time the information in the regressor matrix and its correlation with the variable to be predicted. The second component (PC2) is orthogonal to the first one and explains as much of the remaining information as possible, and so on. Each PC axis is characterised by three quantities: its loadings, which define the contribution of each individual variable for that component; its scores, which locate the molecules along that PC axis; and the percent of the total information reproduced by the component. In the present investigation, as we have seen, the initial variables consist of theoretical data from several organic molecules that are candidates for nonlinear optical applications. After the significant PCs have been identified, the scores of the molecules can be used as regressors for predicting values for other systems of interest.

All multivariate calculations were performed with the Unscrambler 7.6 software [36], with a set of parameters obtained from quantum chemistry calculations, for the designed donor-bridge-acceptor organic molecules. For these D/A molecules, the first and second static optical hyperpolarizabilities were obtained by the time dependent Hartree-Fock methodology (TDHF), as enabled by the MOPAC 93.00 program [31–34]. All parameters – the energies of the highest occupied molecular orbital (HOMO), the ground state dipole moment (mu), the total number of π-electrons in the molecule ($pi(e)$), and the Homo-Lumo energy gap ($delta$) – were evaluated at the semiempirical level using the AM1 Hamiltonian [35].

The parameters delta and mu were chosen because perturbation theory based models show the dependence of the nonlinear parameters β and γ with the inverse of a power of the Homo-Lumo energy gap ($delta$) and directly with a power of the ground state dipole moment (mu) [3,4]. In the case of β, for

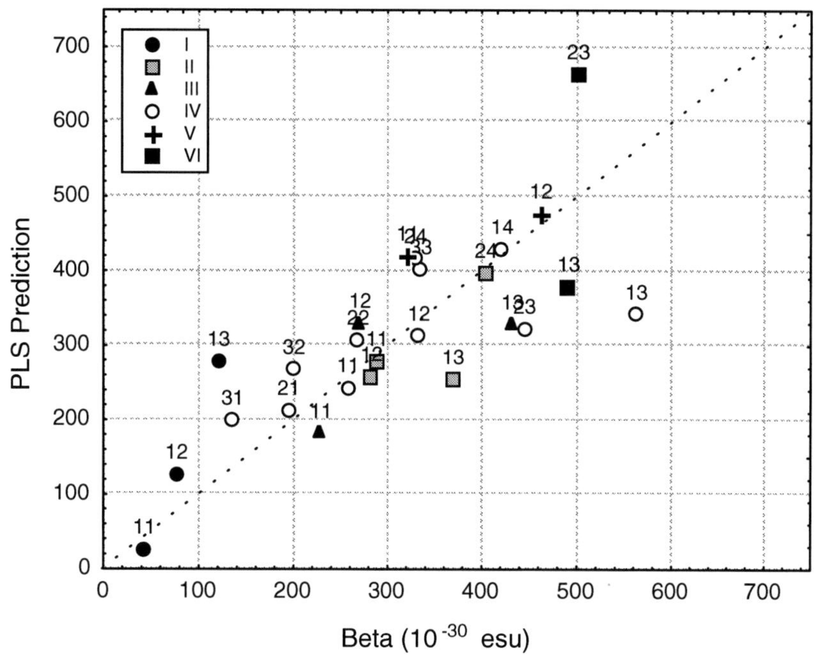

Fig. 4. Comparison between beta values and PLS predicted values, based on a two-component model. Average absolute error of prediction is about 10% of the range of beta values.

example, the following relation for two-level has been used [3]:

$$\beta \propto \frac{f_{gn}\Delta\mu_{gn}}{delta^3}$$

where f_{gn} is the oscillator strength of the transition, $\Delta\mu_{gn}$ is the difference between the ground (g) and excited (n) state dipole moments, and delta is the energy difference between the ground and the excited state. Then, in this study the natural logarithm of the two selected parameters $(mu$ and $delta)$ will be taken instead of the values themselves.

3. Results and discussion

The data were first submitted to a principal component analysis (PCA), in which the original variables are linearly combined to yield new ones describing most of the original information, as described by the variance-covariance matrix [23]. The PCA analysis confirmed that different patterns are associated to different classes of molecules. As already observed from quantum chemical calculations [30], the compounds containing polyenic/mesoionic bridges constitute the class with largest nonlinear optical parameters. Consequently, a PLS model was proposed for the prediction and selection of the members of this class. Prior to the analysis, the values of the Homo-Lumo energy gap and the ground state dipole moment were submitted to a logarithmic transformation, which contributes to linearize the data and leads to better PLS models.

Figure 3 shows the loadings plot of the two first principal components, PC1 and PC2, for the modeling of the first static hyperpolarizability (β). Together, these components account for 84% of the information in the regressor matrix. As the location of the points indicate, higher values of beta tend to be associated

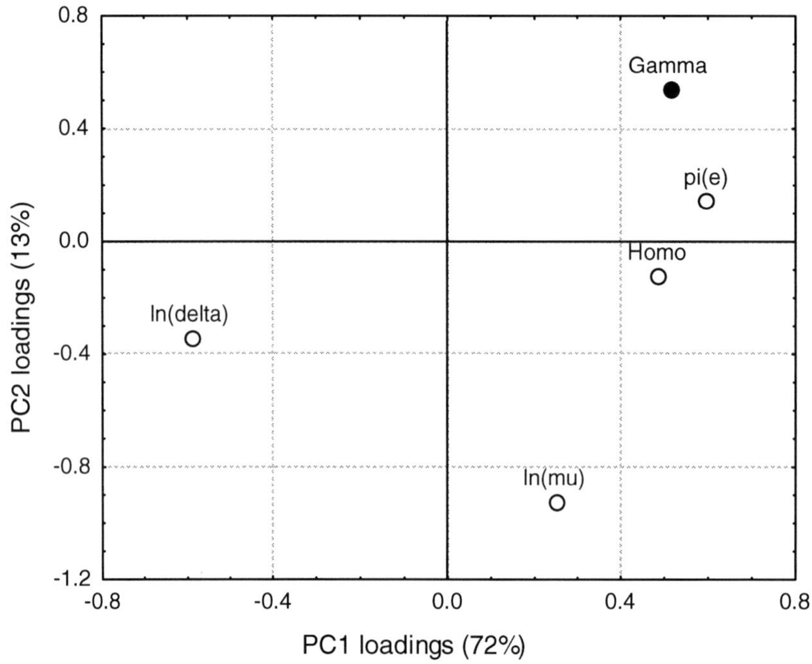

Fig. 5. Loadings plot for the first two PLS principal components for modeling gamma. Percentages are explained variances in the regressor matrix by each PC.

with low values of ln($delta$) and ln(mu), and, to a lesser extent, with large values of $pi(e)$. The dependence of β with delta is assumed from a perturbation theory model [3]. Theoretical and experimental results for some polyenic derivatives show that the increase of the conjugated linker results in strong influence on the magnitude of β in contrast with the value of the ground state dipole moment [37,38]. In addition, the mobility of π-electrons is associated to a higher polarizability value, and consequently a higher value of the number of π-electrons in the molecule results in a high hyperpolarizability magnitude [3]. The *Homo energy*, on the other hand, appears to be quite uncorrelated with beta for this class of compounds. Following Koopmans theorem, the ionisation potential (IP) is made equal the *HOMO energy* [39]. Theoretical investigation of nonplanar aniline oligomers demonstrated that the largest IP values are associated to the smallest β values in contrast with what is observed for the corresponding simulated planar oligomers [15].

Figure 4 shows a fair correlation between the predicted and the original β values. All these compounds have donors and acceptors bonded through polyenic/mesoionic bridges. They are labelled from I to VI according to the different bridges (see Fig. 2) and the way the D and A is linked to them, so the different D and A groups are indicated by numbers 1–4 (see Fig. 1). For example, a compound labelled as III, with donor D_1 and acceptor A_3, is represented on the plot by a graphical symbol (▲) and numbered 13.

The PCA analysis was also carried out for the prediction of the second static hyperpolarizability (γ) from the same set of variables. Figure 5 shows the loadings plot for the two first components, which account for 85% of the information in the regressor matrix. The results are very similar to those obtained for β, and the same conclusions apply: higher values of gamma tend to be associated with low values of ln($delta$) and ln(mu), and also with large values of $pi(e)$. The *Homo energy*, once more, does not appear to play a major role in determining gamma values of this class of nonlinear molecules.

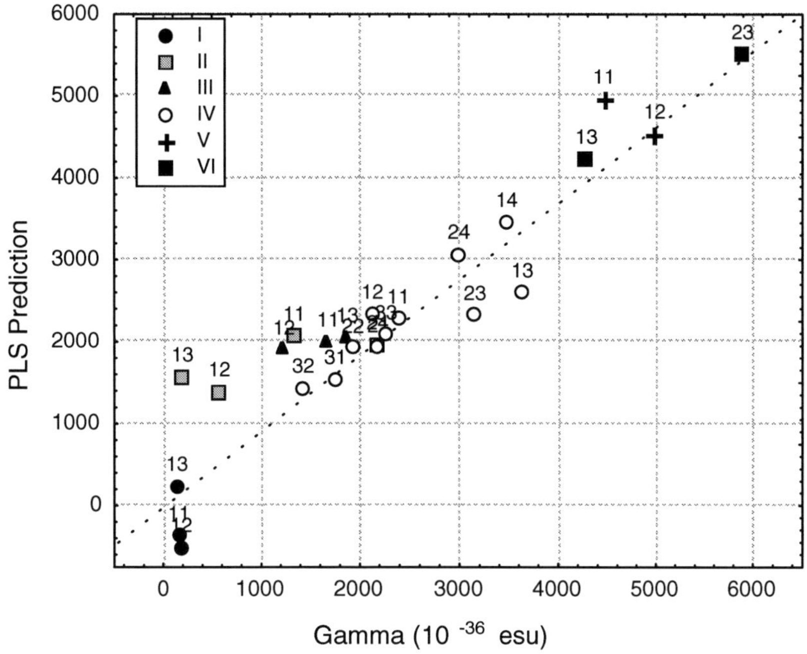

Fig. 6. Comparison between gamma values and PLS predicted values, based on a two-component model. Average absolute error of prediction in about 7% of the range of gamma values.

Figure 6 shows the correlation among the predicted and the initial values of γ. The model for gamma, on the whole, performs a little better than that for beta, as attested by the smaller percent absolute errors. The negative predictions occur for the smaller molecules, with low gamma values (all with type I bridge), which a priori are of less interest for optoelectronic and photonic applications. These systems were considered only to verify the influence of conjugated bridge size on the values of the hyperpolarizabilities when compared to more extended molecules.

The results show that the more extended polyenic/mesoionic bridges (V and VI) are more efficient in promoting large values of the hyperpolarizabilities, associated with the efficient strength of donor-acceptor pairs as the phenylamine-dicyanomethylene and phenylamine-cyanomethylene. This fact can be attributed to an intrinsic charge separation on the mesoionic rings [28,40] and an increased delocalisation of the π-electrons in the more extended linker.

The independent results for both nonlinear optical parameters (β and γ) suggest that they are fairly correlated also, but this conclusion is limited to the class of compounds studied.

4. Conclusions

The models for predicting the β and γ static hyperpolarizabilities from structural-electronic parameters present a good correlation, particularly for the second hyperpolarizability. For the first hyperpolarizability, other parameters can possibly be used to refine the prediction model. As the magnitude of β is null for centrosymmetric molecules, a factor accounting for the symmetry of the molecular system probably would have to be included in the analysis.

The most important result was the demonstration that from a large data set of electronic-structural parameters, obtained at the semiempirical level (it could be also *ab initio*), it is possible to predict and select organic systems with large hyperpolarizability values, especially for the nonlinear response γ. These results demonstrate that this methodology can be useful for cataloging the structural/electronic parameters of nonlinear molecules not yet theoretically studied in the literature in the field of nonlinear optics, and to test the possibility of applications in a short time.

In addition, by including other important classes of nonlinear molecules with interesting nonlinear properties, such as metallic complexes, oligomers of several polymers, octopolar molecules, among others [3–5,7], it might be possible to expand the number of systems in the data set, in order to select the most promising nonlinear molecules for specific applications without the need to carry out a large number of calculations of the hyperpolarizability magnitudes.

This multivariate methodology can also be used to analyze large data sets of experimental parameters available in the literature, in a search for possible patterns that could help to select the more interesting nonlinear systems, prior to the usually laborious, costly and time-demanding measurements of nonlinear phenomena.

Acknowledgments

The authors thank the Conselho Nacional de Desenvolvimento Científico e Tecnológico (CNPQ) for partial financial support.

References

[1] S.R. Marder, J.E. Sohn and G.D. Stucky, Editors, Materials for Nonlinear Optics: Chemical Perspectives, *ACS Symp. Ser.* **455**, American Chemical Society, Washington, 1991.
[2] G.I. Stegeman, J.J. Burke and C.T. Seaton, in: *Optical Engineering: Integrated Optical Components and Circuits*, (Vol. 13), L.D. Huteheson, ed., Marcel Dekker, New York, 1987.
[3] D.R. Kanis, M.A. Ratner and T.J. Marks, *Chem. Rev.* **94** (1994), 195–242.
[4] J.L. Brédas, C. Adant, P. Tackx, A. Persoons and B.M. Pierce, *Chem. Rev.* **94** (1994), 243–278.
[5] B. Champagne and B. Kirtman, in: *Handbook of Advanced Electronic and Photonic Materials and Devices*, (Vol. 9), H.S. Nalwa, ed., Nonlinear Optical Materials, Academic, San Diego, Chapter 2, 2001.
[6] S.M. Risser, D.N. Beratan and S.R. Marder *J. Am. Chem. Soc.* **115** (1993), 7719–7728.
[7] R.R. Tykwinski, U. Gubler, R.E. Martin, F. Diederich, C. Bosshard and P. Günter, *J. Phys. Chem.* **B 102** (1998), 4451–4465.
[8] J.O. Morley, V.J. Docherty and D. Pugh, *J. Chem. Soc. Perkin trans.* **II** (1987), 1351–1355.
[9] A.E. de A. Machado and A.A.S. da Gama, *J. Mol. Struct.* (*THEOCHEM*) **620** (2003), 21–29.
[10] N. Matsuzawa and D.A. Dixon, *Int. J. Quantum Chem.* **44** (1992), 497–515.
[11] S.-L. Lee, K.-C. Yang, J.-H. Sheu and Y.-J. Lu, *Int. J. Quantum Chem.: Quantum Chemistry Symposium* **29** (1995), 509–522.
[12] N. Matsuzawa and D.A. Dixon, *J. Phys. Chem.* **96** (1992), 6232–6247.
[13] H.S. Nalwa, J. Mukai and A. Kakuta, *J. Phys. Chem.* **99** (1995), 10776–10774.
[14] B. Champagne and B. Kirtman, *Chem. Phys.* **245** (1999), 213–226.
[15] A.E. de A. Machado, D.V. Petrov, E.H.L. Falcão, A.A.S. da Gama and W.M. de Azevêdo, *Chem. Phys. Lett.* **356** (2002), 451–456.
[16] T. Geisler, J.C. Petersen, T. Bjørnholm, E. Fisher, J. Larsen, C. Dehu, J.-L. Brédas, G.V. Tormos, P.N. Nugara, M.P. Cava and R.M. Metzger, *J. Phys. Chem.* **98** (1994), 10102–10111.
[17] M. Barzoukas, A. Fort, G. Klein, C. Serbutoviez, L. Oswald and J.F. Nicoud, *Chem. Phys.* **164** (1992), 395–406.
[18] S.R. Marder, W.E. Torruellas, M. Blanchard-Desce, V. Ricci, G.I. Stegeman, S. Gilmour, J.-L. Brédas, J. Li, G.U. Bublitz and S.G. Boxer, *Science* **276** (1997), 1233–1236.
[19] G.J.B. Hurst, M. Dupuis and E. Clementi, *J. Chem. Phys.* **89** (1988), 385–395.
[20] G. Maroulis, *J. Chem. Phys.* **111** (1999), 583–591.
[21] B. Champagne, E. Perpète, *Int. J. Quantum Chem.* **75** (1999), 441–447.
[22] B. Champagne, B. Mennucci, M. Cossi, R. Cammi and J. Tomasi, *Chem. Phys.* **238** (1998), 153–163.

[23] D. Massart et al., *Handbook of Chemometrics and Qualimetrics*, Parts A and B, Elsevier, Amsterdam, 1997, 1998.
[24] M. Blanchard-Desce, R. Wortmann, S. Lebus, J.-M. Lehn and P. Krämer, *Chem. Phys. Lett.* **243** (1995), 526–532.
[25] M. Blanchard-Desce, V. Alain, P.V. Bedworth, S.R. Marder, A. Fort, C. Runser, M. Barzoukas, S. Lebus and R. Wortmann, *Chem. Eur. J.* **3** (1997), 1091–1104.
[26] M. Blanchard-Desce, J.-M. Lehn, M. Barzoukas, C. Runser, A. Fort, G. Puccetti, I. Ledoux and J. Zyss, *Nonlinear Opt.* **10** (1995), 23–36.
[27] G. Puccetti, M. Blanchard-Desce, I. Ledoux, J.-M. Lehn and J. Zyss, *J. Phys. Chem.* **97** (1993), 9385–9391.
[28] G.L.C. Moura and A.M. Simas, *J. Miller, Chem. Phys. Lett.* **257** (1996), 639–646.
[29] A.G. Bezerra, A.S.L. Gomes, P.F. Athayde-Filho, G.B. da Rocha, J. Miller and A.M. Simas, *Chem. Phys. Lett.* **309** (1999), 421–426.
[30] A.E. de A. Machado, *Ph.D. Thesis*, Universidade Federal de Pernambuco, Brazil, 2001.
[31] H.A. Kurtz, J.J.P. Stewart and K.M. Dieter, *J. Comp. Chem.* **11** (1990), 82–87.
[32] S.P. Karna and M. Dupuis, *J. Comp. Chem.* **12** (1991), 487–504.
[33] J.J.P. Stewart, *MOPAC 93.00 Manual*, Fujitsu Limited, Tokyo, Japan, 1993.
[34] J.P. Stewart, *J. Comp. Chem. Aided Mol. Design* **4** (1990), 1–105.
[35] M.J.S. Dewar, E.G. Zoebisch, E.F. Healy and J.J.P. Stewart, *J. Am. Chem. Soc.* **107** (1985), 3902–3909.
[36] The Unscrambler 7.6. CAMO ASA, Nedre Volgate 8, N-0158 OSLO, Norway.
[37] F. Meyers and J.L. Brédas, *J. Zyss, J. Am. Chem. Soc.* **114** (1992), 2914–2921.
[38] M. Blanchard-Desce, V. Alain, L. Midrier, R. Wortmann, S. Lebus, C. Glania, P. Krämer, A. Fort, J. Muller and M. Barzoukas, *J. Photochem. Photobiol. A: Chem.* **105** (1997), 115–121.
[39] J. Sadlej, in: *Semi-empirical Methods of Quantum Chemistry*, I.L. Cooper, ed., Ellis Horwood Limited, Warszawa, 1985.
[40] J. Fabian and B.A. Hess Jr., *Int.J. Quantum Chem.* **90** (2002), 1055–1063.

Journal of Computational Methods in Sciences and Engineering 4 (2004) 277–292
IOS Press

Collisional polarizability correlation functions: A step towards rotational-translational coupling

W. Głaz
Faculty of Physics, A. Mickiewicz University, Umultowska 85, Poznań, Poland
E-mail: glaz@kielich.amu.edu.pl

Abstract. A theory of the spectral distribution of the molecular time correlation function of the collision induced polarizabilities active in the collision induced light scattering has been developed. Within the framework of this approach a new concept of the averaging procedure is proposed, which – unlike the usually applied assumptions of total separation of the angle dependent and the translational contributions – partially includes the coupling between the rotational and translational degrees of freedom due to anisotropic intermolecular interactions. Introduction of the rotational spectral functions, valid for different frequency ranges, in order to produce a numerically treatable method of deriving collisional spectral profiles is discussed. The final results of the theory are formulated in a fashion providing a solid base for numerical calculations capable of interpreting data obtained in various fields of physical, environmental and astrophysical sciences.

PACS: 33.15Kr, 33.70Jg, 34.10+x, 34.150Pi

Mathematics Subject Classification: 42A85, 78A45, 78M25

1. Introduction

Time correlation functions (TCFs) of microscopic molecule quantities and their Fourier-Laplace transforms are one of the basic tools for interpretation and description of processes which occur in molecular media and are affected by their statistical properties. Radiative phenomena like the molecular light scattering and the molecular light absorption are one of the more important among them. Their crucial features may be revealed by the spectral distribution of measured intensities. This spectral distribution considered on the grounds of a theoretical approach can be expressed in terms of a Fourier-Laplace transform of an appropriate TCF function of a relevant quantity, the molecular polarizability/hyperpolarizabilty tensors (scattering) or the molecular dipole moments. In this work we shall focus our considerations on the linear Rayleigh and Raman light scattering phenomena [1–4].

In processes of this kind, when the molecules forming the scattering system may be treated as separate entities, the only contribution to TCF is the permanent polarizability. Yet in general, the properties of molecular systems do not have to be a simple sum of the contributions brought by individual molecules (atoms) of the medium. Important spectroscopic phenomena can also stem from intermolecular interactions and dynamics. Collision-induced light scattering (CILS) refers to radiative transitions, which are forbidden in free molecules, but which occur when components of the molecular polarizabilities are induced or interactively modified. It is obvious that in such effects the so-called collision induced

1472-7978/04/$17.00

polarizabilities must depend both on the individual and intermolecular coordinates, such as rotational or translational degrees of freedom. When this is the case, one has to expect that the averaging over these degrees of freedom can lead to expressions in which a kind of coupling between rotations and translational movement plays a certain role. The main factor responsible for this effect is due to the anisotropic potential of interaction between molecules taking part in a radiative process. This kind of coupling is a source of a number of effects influencing the spectra of scattered or absorbed light, like the so-called line mixing or the state propagation mechanisms [5].

To sum up, in a general case, when calculating the TCF functions we deal with the two types of collisional mechanisms – the CILS effects and the coupling between the degrees of freedom of different kinds. A theoretical treatment, which would take into account both these effects is not a trivial task. The problem becomes particularly severe when, as it happens in the collision induced scattering, the quantities in question (polarizabilities) depend both on the angular and the translational coordinates. As a result, some significant simplifying assumptions must be introduced. One of the most typical approaches is based on neglecting the translational-rotational coupling and performing relevant averagings separately over orientations and the intermolecular vectors. However, as shown e.g. in [5] there are radiative phenomena in which the coupling processes give non-negligent contributions to the experimentally obtained spectra of the scattered light. Moreover, theoretical considerations indicate that the line mixing and the state propagation mechanism might participate visibly in shaping rotational time correlation functions [6,7]. Therefore, developing a theoretical approach accounting for the role of these effects seems to be of essence.

In this work we shall discuss some general properties of the collisional multipolar polarizabilities and their TCF functions as a step towards the main goal, i.e. developing a method suitable for producing collisional spectral profiles capable of reaching behind the limits imposed by the crudest totally factorized procedure. Besides, the new approach should yield expressions describing the collisional spectral profiles which would be fully computationally treatable.

According to the above assumptions the paper is organized as follows. In Section 2 basic formulae for the multipolar polarizabilities expressed in terms of the spherical tensor algebra [8,9] are given. In Section 2.2 we show how the orientational part of these quantities may be determined by the so-called triple-Wigner functions (defined in Appendix B) so that it might be possible to discuss the mathematical properties of the correlation functions in a more general way, without referring to particular physical mechanisms responsible for inducing the collisional properties of the molecules. In Section 3 general expressions for the TCF functions are presented suitable for further analyses. In Section 4 an example of such an approach with totally decoupled TCF parts depending on different degrees of freedom is discussed. Having done this we proceed in Section 5 to the development of a new proposed concept of calculating the CILS spectral profiles with the rotational-translational coupling partially taken into account. There is also Subsection 5.2 devoted to a discussion of the procedures for evaluation of the rotational spectral distribution functions for different regions of the frequency range. The special attention is paid to a new model of the rotational profiles applicable for high frequencies. In concluding remarks we show that the approach is capable of rendering treatable formulae that might be used to analyze the CILS spectra.

2. Pair collisional polarizabilities

The excessive collisional polarizability of a pair of interacting molecules immersed in an external electromagnetic field can be expressed in terms of individual molecular properties, like permanent intrinsic polarizabilities and hyperpolarizabilities of different order, multipole moments and the tensors determining the intermolecular multipole interactions. If other types of interactions are to be taken into account, e.g. overlap and exchange effects, appropriate constants characterizing these forces should be applied accordingly. Fortunately, the structure of the expansions that are of interest to us, considering the main aim of this work, is identical for any of such mechanisms. Therefore, in this section we shall discuss, as an example, just the multipolar contributions to the collisional polarizabilities.

2.1. Pair polarizability tensors of first and second order

The pair polarizability tensor of two interacting molecules p and q is given as

$$\boldsymbol{A}_J = \boldsymbol{A}_J(p) + \boldsymbol{A}_J(q) + \Delta\boldsymbol{A}_J(p) + \Delta\boldsymbol{A}_J(q), \tag{1}$$

where $\Delta\boldsymbol{A}_J(p)$ stands for the excess contribution to the polarizability of the p-th molecule induced by its interactions with another q-th molecule. In this work we limit our discussion to collisional effects of purely linear origin. The induced components of $\Delta\boldsymbol{A}$ can be obtained by having recourse to the well known principles of the perturbation theory [4,10,11]. As suggested in [12] this sort of calculations can be extended also to polarizabilities induced by interactions of dynamical nature. Another way of deriving $\Delta\boldsymbol{A}$, in a more phenomenological fashion, is presented in [13]. The first order quantities of that kind expressed in terms of the irreducible spherical tensor formalism related to the molecular polarizabilities of different order were given some time ago [10]:

$$\begin{aligned}\Delta\boldsymbol{A}_J^{(1)}(p) = &\sum_{J_1J_2J_3}\sum_{l_1l_2}\sum_{q\neq p}(-1)^{J_2+M+1}\left[\frac{2^{l_1+l_2}}{(2l_1)!(2l_2)!}\right]^{1/2}\Pi_{J_1J_2J_3M}\begin{Bmatrix} J_1 & J_2 & J_3 \\ 1 & 1 & J \\ l_1 & l_2 & M\end{Bmatrix}\\ &\times\left\{\boldsymbol{T}_M(pq)\otimes\left\{\boldsymbol{A}_{J_2}^{(1l_2)}(p)\otimes\boldsymbol{A}_{J_1}^{(1l_1)}(q)\right\}_{J_3}\right\}_J, l_1+l_2=M,\end{aligned} \tag{2}$$

where $\boldsymbol{A}_J^{(kl)}(p)$ is the spherical tensor of $2^k - 2^l$ polarizability of the p-th molecule [4,11] and $\boldsymbol{T}_M(pq)$ is the tensor of multipole-multipole intermolecular interaction:

$$\boldsymbol{T}_M(pq) = (-)^M\left(\frac{(2M)!}{2^M}\right)^{(1/2)}\left(\frac{4\pi}{2M+1}\right)^{(1/2)} R_{pq}^{-(M+1)}\boldsymbol{Y}_M(\Omega_{pq}), \tag{3}$$

where $\boldsymbol{Y}_M$ is the spherical harmonic function and Ω_{pq} denotes the orientation of the intermolecular vector $\boldsymbol{R}_{pq}$. In this expression and other appearing hereafter $\otimes$ stands for the irreducible spherical tensor product [8,9] and

$$\begin{Bmatrix} \cdot & \cdot & \cdot \\ \cdot & \cdot & \cdot \\ \cdot & \cdot & \cdot \end{Bmatrix} \tag{4}$$

is the 9-j Wigner symbol; $\Pi_{ab\ldots x} \equiv [(2a+1)(2b+1)\ldots(2x+1)]^{1/2}$. The next possible term in the expansion of the collisional polarizability, $\Delta\boldsymbol{A}_J^{(2)}(p)$ can be obtained by applying the second order perturbation theory [4,10] (this expression is given in Appendix A).

The spherical components of the molecular polarizabilities are given in the laboratory reference frame and they depend on the molecular orientations, Ω_p and Ω_q. According to the known rules of the spherical tensor algebra [8,9] these quantities can be relatively easy transformed to the molecular reference system. Hence, their angular dependence will be totally expressed by appropriate Wigner functions. When we deal with the formulae restricted to bimolecular expressions depending on Ω_p, Ω_q and Ω_{pq}, it is convenient to introduce more general angular functions, so-called *triple Wigner functions*, Φ (Appendix B).

2.2. *Collisional polarizabilities in terms of the triple – Wigner functions*

By applying the functions defined in Appendix B we can express the collective properties of the bimolecular systems in more general and compact form, with angle-dependent part explicitly separated. It should allow us to derive a general form of a theory of the correlation functions and their spectral distribution applicable for physical effects related not only to the light scattering phenomena.

The first step toward this goal is based on transformation of the molecular quantities in Eq. (2) and Eq. (47) to the molecular system of reference:

$$\Delta A^{(1)}_{Jj}(p) = \sum_{kll_1l_2} \sum_{KLM} \sum_{q\neq p} (-)^K \Pi_M \begin{Bmatrix} K & L & M \\ 1 & 1 & J \\ l_1 & l_2 & N \end{Bmatrix} \begin{pmatrix} 2N \\ 2l_1 \end{pmatrix}^{(1/2)} R_{pq}^{-N-1} \times \Phi^{KLN,MJ}_{kl0,j}(\Omega_p, \Omega_q, \Omega_{pq})\, \tilde{A}^{(1l_1)}_{Kk}(p) \tilde{A}^{(1l_2)}_{Ll}(q); \tag{5}$$

$$\begin{aligned}\Delta A^{(2)}_{Jj}(p) = &- \sum_{ll_1l_2n_2n_3} \sum_{KML} \sum_{J_1J_2J_3} \sum_{q\neq p} (-)^{l_2+L+1+J_1} \Pi_{M_1M_2} \left[\frac{(2M_1)!(2M_2)!}{(2l_2)!(2l_1)!(2n_2)!(2n_3)!} \right]^{1/2} \\ &\times \begin{Bmatrix} K & l & 1 & J \\ M_1 & l_2 & 1 & L \\ M & M_2 & n_3 & J_3 \end{Bmatrix} \begin{Bmatrix} J_2 & n_2 & l_2 \\ J_1 & l_1 & 1 \\ K & M_1 & l \end{Bmatrix} \begin{Bmatrix} J_1 & J_2 & K \\ L & J_3 & H \end{Bmatrix} \Pi_{llKKLMJ_1J_3} \\ &\times C^{M0}_{M_10M_20} \Phi^{J_2HM,LJ}_{j_2h0,j}(\Omega_p, \Omega_q, \Omega_{pq}) \tilde{A}^{(n_2l_2)}_{J_2j_2}(q) \left\{ \tilde{\boldsymbol{A}}^{(1l_1)}_{J_1}(p) \otimes \tilde{\boldsymbol{A}}^{(n_31)}_{J_3}(p) \right\}_{Hh} \\ &\times R_{pq}^{-M_1-M_2-2},\end{aligned} \tag{6}$$

where $\tilde{A}_{Jj}$ are spherical tensor components taken in the molecular reference frame and

$$\begin{Bmatrix} \cdot & \cdot & \cdot & \cdot \\ \cdot & \cdot & \cdot & \cdot \\ \cdot & \cdot & \cdot & \cdot \end{Bmatrix} \tag{7}$$

is the 12-j Wigner symbol of the second kind, whereas

$$\begin{Bmatrix} \cdot & \cdot & \cdot \\ \cdot & \cdot & \cdot \end{Bmatrix} \tag{8}$$

denotes the 6-j Wigner symbol; C^{Jj}_{KkLl} are the Clebsch-Gordan coefficients [8,9].

It can be easily seen that these equations have a similar structure which permits us to write both of them in the same generalized fashion as

$$\Delta A_{Jj} = \sum_{KLNM} \sum_{kl} \Pi_J^{-1} B_M(Kk\, Ll\, N, R_{pq}) \Phi^{KLN,MJ}_{kl0,j}(\Omega_p, \Omega_q, \Omega_{pq}), \tag{9}$$

where the coefficient B includes the spherical tensor components related to the physical properties of the molecules (polarizabilities) and depends on the intermolecular separation, R_{pq}, whereas the orientational coordinates are involved through the Φ functions.

Here the generality of Eq. (9) should be stressed – not only is it applicable for the multipolar collisional polarizabilities of the first and the second order considered above, but it can be also used for collisional mechanisms other than the multipolar interactions discussed in this work. The only difference might appear in the inner structure of B resulting from theories of particular physical processes responsible for collisional effects [14–16]. The coefficients can be obtained analytically and/or given as a result of different types of *ab initio* routines [17,18]. Moreover, equations of the shape of Eq. (9) or similar are not only present in the light scattering theories but they also play a significant role in other areas, where the orientational dependence of collective molecular properties is of essence, e.g. in the collision induced absorption [19]. Consequently, the theory of collisional correlation functions that we shall discuss in the next sections of this work will be of a more general nature and importance. Thus, in the section devoted to the problem of the coupling between rotational and translational degrees of freedom in the time correlation functions we shall extend our considerations from just the case of the collisional polarizabilities to a more general analysis of the TCF functions of Φ. However, in this work we concentrate mainly on the CILS phenomena and therefore some introductory remarks considering the time correlation functions of the collisional polarizabilities should be presented.

3. Time correlation functions of the collisional polarizabilities

The basic quantity determining the properties of the scattered light, the pair differential scattering cross section, is proportional to a Fourier transform of the time correlation function of the collisional polarizabilities:

$$I_{JJ}(t) \equiv (-)^J \Pi_J \left\langle \{\Delta \boldsymbol{A}_J(0) \otimes \Delta \boldsymbol{A}_J(t)\}_0 \right\rangle, \tag{10}$$

where $< \ldots >$ denotes appropriate averaging over relevant degrees of freedom. After substituting Eq. (1) in Eq. (10) we have the following contributions to the time correlation function

$$I_{JJ}(t) = I_{JJ}^{(00)}(t) + I_{JJ}^{(10)}(t) + I_{JJ}^{(20)}(t) + I_{jj}^{(11)}(t) + \ldots \tag{11}$$

with

$$I_{JJ}^{(00)}(t) = (-)^J \Pi_J \left\langle \{\boldsymbol{A}_J(0) \otimes \boldsymbol{A}_J(t)\}_0 \right\rangle, \tag{12}$$

$$I_{JJ}^{(10)}(t) = (-)^J \Pi_J \left\langle \left\{\boldsymbol{A}_J(0) \otimes \Delta \boldsymbol{A}_J^{(1)}(t)\right\}_0 \right\rangle + (-)^J \Pi_J \left\langle \left\{\boldsymbol{A}_J(t) \otimes \Delta \boldsymbol{A}_J^{(1)}(0)\right\}_0 \right\rangle, \tag{13}$$

$$I_{JJ}^{(20)}(t) = (-)^J \Pi_J \left\langle \left\{\boldsymbol{A}_J(0) \otimes \Delta \boldsymbol{A}_J^{(2)}(t)\right\}_0 \right\rangle + (-)^J \Pi_J \left\langle \left\{\boldsymbol{A}_J(t) \otimes \Delta \boldsymbol{A}_J^{(2)}(0)\right\}_0 \right\rangle, \tag{14}$$

$$I_{JJ}^{(11)}(t) = (-)^J \Pi_J \left\langle \left\{\Delta \boldsymbol{A}_J^{(1)}(0) \otimes \Delta \boldsymbol{A}_J^{(1)}(t)\right\}_0 \right\rangle. \tag{15}$$

In order to find the influence of specific types of multipolar collisional mechanisms these equations should be transformed applying the components of $\Delta \boldsymbol{A}$ given in Eq. (2):

$$I_{JJ}^{(10)}(t) = -\sum_{l_1 l_2} \sum_{J_1 J_2 J_3 M} \sum_{p \neq q, r} (-)^{J_2+J_3} \left[\frac{2^{l_1} 2^{l_2}}{(2l_1)!(2l_2)!} \right]^{(1/2)} \Pi_{J J_1 J_2 J_3 M} \begin{Bmatrix} J_2 & J_1 & J_3 \\ 1 & 1 & J \\ l_1 & l_2 & M \end{Bmatrix}$$
$$\times \left\langle \left\{ \boldsymbol{T}_M(pq(t)) \otimes \left\{ \left\{ \boldsymbol{A}_{J_2}^{(1l_2)}(q(t)) \otimes \boldsymbol{A}_{j_1}^{(1l_1)}(p(t)) \right\}_{J_3} \otimes \boldsymbol{A}_J^{(11)}(r(0)) \right\}_M \right\}_0 \right\rangle \quad (16)$$
$$l_1 + l_2 = M.$$

The most interesting from the point of view of the present work is the quadratic Eq. (15) term, which after transformations renders

$$I_{JJ}^{(11)}(t) = \sum_{\substack{l_1 L_2 \\ m_1 m_2}} \sum_{\substack{J_1 J_2 J_3 \\ K_1 K_2 K_3}} \sum_{\substack{M_1 M_2 \\ LM}} \sum_{\substack{p \neq q \\ r \neq s}} (-)^{J_1+J_3+M_2+M+K_1+J} \Pi_{J J_1 J_2 J_3 J_3 K_1 K_2 K_3 K_3 M_1 M_2 L M N}$$
$$\times \left[\frac{2^{l_1+l_2+m_1+m_2}}{(2l_1)!(2l_2)!(2m_1)!(2m_2)!} \right]^{(1/2)} \begin{Bmatrix} J_1 & J_2 & J_3 \\ 1 & 1 & J \\ m_1 & m_2 & M_2 \end{Bmatrix}$$
$$\times \begin{Bmatrix} J_1 & J_2 & J_3 \\ L & N & M \\ K_1 & K_2 & K_3 \end{Bmatrix} \begin{Bmatrix} l_1 & l_2 & M_1 \\ 1 & 1 & J \\ K_1 & K_2 & K_3 \end{Bmatrix} \begin{Bmatrix} M_1 & M & M_2 \\ J_3 & J & K_3 \end{Bmatrix} \quad (17)$$
$$\times \left\langle \left\{ \left\{ \left\{ \boldsymbol{A}_{J_1}^{(1m_1)}(p(t)) \otimes \boldsymbol{A}_{K_1}^{(1l_1)}(r(0)) \right\}_L \otimes \left\{ \boldsymbol{A}_{J_2}^{(1m_2)}(q(t)) \otimes \boldsymbol{A}_{K_2}^{(1l_2)}(s(0)) \right\}_N \right\}_M \right. \right.$$
$$\left. \otimes \ \{\boldsymbol{T}_{M_1}(pq(t)) \otimes \boldsymbol{T}_{M_2}(rs(0))\}_M \}_0 \right\rangle, m_1 + m_2 = M_2, \ l_1 + l_2 = M_1.$$

Obviously, a necessary task to perform on the path to the final workable formulae determining the collisional spectra is the averaging procedure involved in the equations in question. In general the averaging should be taken over all the degrees of freedom involved. It is a challenging problem and to the best of our knowledge only a few specific solutions to it, related to more or less restrictive approximating assumptions, have been presented in literature [5]. There are, for instance, relatively few concepts of developing the time correlation functions for the collisional phenomena, in which the molecular quantities, like the dipole moments and the polarizabilities depend both on the translational and rotational variables [20,21]. It comes as no surprise, as from the mathematical point of view, this situation is really involved and because the degrees of freedom of different types may be intertwined not only via the relaxation matrices and/or via nondiagonal matrix elements of the evolution operators, but also through the expressions being averaged.

4. Total rotational-translational decoupling

The crudest form of a possible approximation assumes that the time correlation function can be totally factorized into two parts, of which the first ought to depend solely on the translational degrees of freedom, whereas the other – on the rotational ones. It should be recalled, however, that this kind of treatment is justifiable only when there is no anisotropy in the intermolecular interaction potential or the role of the anisotropy is negligible. Admittedly though, quite often the results obtained on the basis of this approximation are fully satisfactory [13,15]. Hereafter we shall introduce the assumption of a total separation of the rotational and translational degrees of freedom in order to obtain a spectral distribution function for the quadratic term defined in Eq. (15). Usually it gives the most significant component of the collisional spectra especially in the most interesting region of the so-called spectral wings, i.e. for high frequencies. The cross terms, Eqs (13) and (14), (given above for the sake of completeness) contribute mainly to the frequency regions close to the centers of the lines [22,23], where they are often overshadowed by the non-collisional scattering. Their role may be, however, of some importance at the wings through the so-called *correlated terms* in the spectral distribution functions [22]. Though, even then, the averagings in the correlated contributions are to be treated theoretically by means of an approach different from this developed here [22]; therefore they must be a subject of some other discussion.

As far as the quadratic term is considered, the procedure of the separation of the variables leads to

$$\begin{aligned} I_{JJ}^{(11)}(t) = &\sum_{\substack{l_1 l_2\\ m_1 m_2}} \sum_{K_1 K_2} \sum_{p \neq q} (-)^{m_2+l_2+J} \Pi_{K_1 K_2 M J J} \begin{bmatrix} K_1 & J & M & K_2 \\ m_1 & l_1 & 1 & 1 \\ 1 & m_2 & 1 & l_2 \end{bmatrix} \\ &\times \left[\frac{2^{l_1+l_2+m_1+m_2}}{(2l_1)!(2l_2)!(2m_1)!(2m_2)!} \right]^{(1/2)} \left(\left\{ \left\{ \left\langle \left\{ \boldsymbol{A}_{K_1}^{(1m_1)}(p(t)) \otimes \boldsymbol{A}_{K_1}^{(1l_1)}(p(0)) \right\}_0 \right\rangle_p \right. \right. \right. \\ &\otimes \left\langle \left\{ \boldsymbol{A}_{K_2}^{(1m_2)}(q(t)) \otimes \boldsymbol{A}_{K_2}^{(1l_2)}(q(0)) \right\}_0 \right\rangle_q \Big\}_0 + (-)^M \left\{ \left\langle \left\{ \boldsymbol{A}_{K_1}^{(1m_1)}(q(t)) \otimes \boldsymbol{A}_{K_1}^{(1l_1)}(q(0)) \right\}_0 \right\rangle_q \right. \\ &\otimes \left. \left. \left\langle \left\{ \boldsymbol{A}_{K_2}^{(1m_2)}(p(t)) \otimes \boldsymbol{A}_{K_2}^{(1l_2)}(p(0)) \right\}_0 \right\rangle_p \right\}_0 \right) \langle \{ \boldsymbol{T}_M((pq)(t)) \otimes \boldsymbol{T}_M((pq)(0)) \} \rangle_{pq} , \\ &m_1 + m_2 = l_1 + l_2 = M. \end{aligned} \tag{18}$$

where

$$\begin{bmatrix} \cdot & \cdot & \cdot & \cdot \\ \cdot & \cdot & \cdot & \cdot \\ \cdot & \cdot & \cdot & \cdot \end{bmatrix} \tag{19}$$

is the 12-j Wigner symbol of the first kind [8,9]. In this expression $< \ldots >_p$, $< \ldots >_q$ and $< \ldots >_{pq}$ denote averaging over relevant orientations. It can be easily shown that for linear molecules the Fourier transform of $< \ldots >$ for a particular choice of the rotator states denoted by $J_{p/q}$, should read [14,15,24]

$$\begin{aligned} \frac{1}{2\pi} \int \exp(-i\omega t) \left\langle \left\{ \boldsymbol{A}_K^{(1m)}(p(t)) \otimes \boldsymbol{A}_K^{(1l)}(p(0)) \right\}_0 \right\rangle_p dt = \left\{ \tilde{\boldsymbol{A}}_K^{(1m)}(p) \otimes \tilde{\boldsymbol{A}}_K^{(1l)}(p) \right\}_0 \\ g_p g_q \frac{\Pi_{J_p J_q J_p' J_q'}}{\mathcal{Z}_p \mathcal{Z}_q} \exp\left(\frac{E_{J_p} + E_{J_q}}{kT} \right) \delta\left(\omega - \omega_{J_p J_p'} - \omega_{J_q J_q'} \right), \end{aligned} \tag{20}$$

where g_x denotes the nuclear statistic factor, $E_J = J(J+1)B_p$; B_p is the rotational constant, $\mathcal{Z}$ stands for the rotational partition function, and $\omega_{J_p J_p'} = \left[J_p(J_p+1) - J_p'(J_p'+1) \right] B_p$. In order to obtain the final spectral intensity the expression above has to be convoluted with an appropriate function determining the the Fourier transform of the translational correlation function of the $\boldsymbol{T}_M(pq)$ operator. This is possible to achieve by means of a number of procedures of different nature [19,25–27].

5. Partial correlation approximation

In the previous section, when considering the averaging scheme in the expressions for the spectral functions, we followed the crudest possible approximation applied in literature of the subject, i.e. the one in which the total decoupling of the translational and the rotational degrees of freedom is assumed. In this approach the quantities depending on the molecular orientations and on the vector connecting molecular centers of mass are factorized and averaged separately. Quite obviously, it means that certain subtle effects, like e.g. the rotational line mixing and the state propagation, are not included in the calculations and, as a consequence, the spectral profiles might not be precise enough.

Unfortunately, creating the theories that can account fully for the influence of such effects and lead, at the same time, to a reasonably numerically doable formulae is a quite involved problem from the mathematical point of view.

Therefore, in this section we have decided to introduce a new approach based on an approximation, akin to those discussed by some other authors for collision induced absorption [21], which falls half way between the totally decoupling scheme and another fully incorporating interference of translational and rotational degrees of freedom. We propose that the rotational movement of the radiator be affected by the translational degrees of freedom of the perturber, but the opposite is not true. The details of this the so-called *partial coupling approximation* are given in the next sections.

5.1. Rotational-translational correlation function in partial coupling approximation

In the following analysis we restrict the discussion to a model physical situation where the scattering system consists of a linear homo-nuclear active molecule (radiator) interacting with an atomic perturber via an anisotropic potential. Let us consider the time correlation function defined in Eq. (10):

$$I_{JJ}(t) \equiv \langle \Delta\boldsymbol{A}_J(t) \circ \Delta\boldsymbol{A}_J(0)\rangle\,. \tag{21}$$

For such systems the above averaging can be defined as follows:

$$\langle A(t)B(0)\rangle = Tr[A(t)\rho B(0)] = Tr_{\mathcal{R}}\langle A(t)B(0)\rangle_{av}. \tag{22}$$

Here $Tr_{\mathcal{R}}$ denotes the trace taken over the radiator variables, ρ is the density matrix and $<..>_{av}$ stands for the averaging over the translational degrees of freedom of the perturber.

The collisional polarizabilities are defined in a general manner according to Eqs (5), (6) and (9):

$$\Delta A_{Jj} = \frac{1}{\Pi_J}\sum_{\Lambda,L}\sum_{\lambda,l} C^{Jj}_{Ll\Lambda\lambda}\Upsilon^{\Lambda}_{Ll}(R,\Omega_R)\cdot\Phi^{\Lambda 00,\Lambda\Lambda}_{000,\lambda}(\Omega) \tag{23}$$

where Ω stands for the orientation of the radiator ($\Omega \equiv \Omega_p$), Ω_R is the orientation of the intermolecular vector ($\Omega_R \equiv \Omega_{pq}$, $R \equiv R_{pq}$), and the $\boldsymbol{R}$-dependent factor

$$\Upsilon^{\Lambda}_{Ll}(R,\Omega_R) \equiv B_{\Lambda}(\Lambda 0\,00\,L, R)\Phi^{00L,0L}_{000,l}(\Omega_R) \tag{24}$$

is explicitly separated from the part depending on the radiator orientation.

After appropriate substitutions we may write:

$$\begin{aligned} I_{JJ}(t) = \frac{1}{\Pi_{JJ}}\sum_{j}\sum_{l,l',\lambda,\lambda'} (-)^j C^{Jj}_{\Lambda\lambda Ll} C^{J-j}_{\Lambda'\lambda' L'l'} \Big\langle \Upsilon^{\Lambda}_{Ll}(t)\Upsilon^{\Lambda'}_{L'l'}(0)\,\Phi^{\Lambda 00,\Lambda\Lambda}_{000,\lambda}(\Omega(t)) \\ \times\Phi^{\Lambda' 00,\Lambda'\Lambda'}_{000,\lambda'}(\Omega(0))\Big\rangle\,. \end{aligned} \tag{25}$$

For the sake of simplicity we shall assume thereafter that only one type of the collisional mechanism, determined by a single pair of values of Λ and L, is taken into account, so the summation sign over these indices will be omitted.

Here, $< \ldots >$ includes well separated factors depending on translational and/or rotational degrees of freedom. They are, however, involved in the averaging procedure with the density matrix which couples these variables. From the mathematical point of view evaluating such expressions in a reasonably treatable fashion is a rather challenging task. To sort the problem out we may try to introduce some kind of factorization of the procedure. The extremely simple approximation assumes that we totally split the integration over the two types of variables, which is pointless here as it excludes coupling (the line mixing and propagation) effects from our considerations. Therefore we decide to chose a less radical approach, which may be expressed by the following equation:

$$I_{JJ}(t) = \frac{1}{\Pi_{JJ}} \sum_j \sum_{\lambda l} \sum_{\lambda' l'} (-)^j C^{Jj}_{\Lambda\lambda L l} C^{J-j}_{\Lambda'\lambda' L' l'} \left\langle \Upsilon^{\Lambda}_{Ll}(t) \Upsilon^{\Lambda'}_{L'l'}(0) \right\rangle_{tr} \times \left\langle \Phi^{\Lambda 00,\Lambda\Lambda}_{000,\lambda}(\Omega(t)) \Phi^{\Lambda' 00,\Lambda'\Lambda'}_{000,\lambda'}(\Omega(0)) \right\rangle . \tag{26}$$

Here the translational and rotational degrees of freedom are only partially coupled in the latter factors (where full averaging procedure is applied), whereas in the (R, Ω_R)-dependent term Υ only the translational correlation function is considered. On having recourse to Eq. (49) one can express the product $\Upsilon^{\Lambda}_{Ll}(t)\Upsilon^{\Lambda'}_{L'l'}(0)$ in terms of components of a certain irreducible spherical tensor Ψ:

$$\Upsilon^{\Lambda}_{Ll}(t)\Upsilon^{\Lambda'}_{L'l'}(0) = \sum_{A,\alpha} C^{A\alpha}_{LlL'l'} \Psi_{A\alpha}(t), \tag{27}$$

which depends on the relative orientation of $\boldsymbol{R}$ in the moments 0 and t. Assuming that we deal with an isotropic medium, Ψ must be a scalar of rank $A = 0$. Consequently,

$$\left\langle \Upsilon^{\Lambda}_{Ll}(t)\Upsilon^{\Lambda'}_{L'l'}(0) \right\rangle_{tr} = C^{00}_{LlL'l'} \left\langle \Psi^{L}_{\Lambda\Lambda'}(t) \delta_{LL'} \delta_{-ll'} \right\rangle_{tr}, \tag{28}$$

where

$$\Psi^{K}_{\Lambda\Lambda'}(t) = \sum_{k,k'} C^{00}_{KkKk'} \Upsilon^{\Lambda}_{Kk}(t)\Upsilon^{\Lambda'}_{Kk'}(0) = \sum_k \frac{(-)^k}{\Pi_K} \Upsilon^{\Lambda}_{Kk}(t)\Upsilon^{\Lambda'}_{K-k}(0). \tag{29}$$

After substitution Eq. (28) into Eq. (26) and some elementary algebra we have

$$I_{JJ}(t) = \sum_{\lambda} (-)^L \Pi_L^{-1} \left\langle \Psi^{L}_{\Lambda\Lambda}(t) \right\rangle_{tr} \cdot \frac{1}{\Pi_{\Lambda\Lambda}} \left\langle (-)^{\Lambda+\lambda} \Phi^{\Lambda 00,\Lambda\Lambda}_{000,\lambda}(\Omega(t)) \Phi^{\Lambda 00,\Lambda\Lambda}_{000,-\lambda}(\Omega(0)) \right\rangle, \tag{30}$$

which in a more convenient form can be rewritten as:

$$I_{JJ}(t) = I_{tr}(t) I_{rot}(t); \tag{31}$$

$$I_{tr}(t) = \sum_{L,\Lambda} (-)^L \Pi_L^{-1} \left\langle \Psi^{L}_{\Lambda\Lambda}(t) \right\rangle_{tr} = \Bigg\langle B_{\Lambda}(\Lambda 0\, 00\, L,\ R(t))\, B_{\Lambda}(\Lambda 0\, 00\, L,\ R(0)) \times \sum_l \frac{(-)^{L+l}}{\Pi_L} \Phi^{00L,0L}_{000,l}(\Omega_R(t)) \Phi^{00L,0L}_{000,-l}(\Omega_R(0)) \Bigg\rangle_{tr} \tag{32}$$

and

$$I_{rot}(t) = \left\langle \sum_{\lambda} \frac{(-)^{\Lambda+\lambda}}{\Pi_{\Lambda\Lambda}} \Phi^{\Lambda 00,\Lambda\Lambda}_{000,\lambda}(\Omega(t)) \Phi^{\Lambda 00,\Lambda\Lambda}_{000,-\lambda}(\Omega(0)) \right\rangle . \tag{33}$$

Having done the above we may express the intensity of the scattered light (the Fourier transform) in the form of a convolution of two spectral functions:

$$I(\omega) = \frac{1}{2\pi} \int_{-\infty}^{\infty} I_{tr}(\omega') I_{rot}(\omega - \omega') d\omega'. \tag{34}$$

By doing so the problem reduces to the evaluation of two spectral profiles

$$I_{tr/rot}(\omega) = \int_{-\infty}^{\infty} I_{tr/rot}(t) \exp{(i\omega t)} dt. \tag{35}$$

As a result we arrive at an expression, for which obtaining its numerically treatable form seems to be simpler, yet still a tedious, task. We shall discuss some further possible steps of the theory in the following sections concentrating mainly on the rotational component of the equation. The translational part does not include any coupling between orientational and translational degrees of freedom and is not of much interest here considering the main subject of this work. Nevertheless, it is worth noticing that this component can be evaluated numerically in a relatively easy way by means of a number of well known methods The only point that should be carefully looked at, when choosing a particular approach to be applied, is the accuracy of the computing method with regard to the frequency range needed. As we shall discuss it further on, the approach suggested in this work may require using very precise transitional profiles for a wide range of frequencies. Fortunately, such calculating procedures have been developed (e.g. [26,27]).

5.2. The spectral distribution of the rotational correlation function

The main question in the theories describing molecular spectra is the appropriate assessment of the frequency region for which they are to be applied. It can be easily shown on the grounds of the properties of the convolution procedure that the method introduced in the previous section requires that the spectral rotational functions are expressed for a possibly wide range of frequencies. There are unfortunately, very few models that are claimed to give correct results for both the regions near to the center of the line as well as for the far wings [5,28]. Usually both these cases must be treated separately; the question of itself being the problem of the intermediate frequencies.

5.2.1. The rotational profile near the line center

It is the wing region of the total spectrum that is of most interest to us in this work. We must be aware of the fact, however, that within the approach proposed here it is very often inevitable to take into account rotational spectral functions for the frequency regions remote from the points considered. This includes the rotational functions close to the centers of the transition lines. One of the obvious reasons for this necessity is that the convolution used in the calculations depends on the 'overall' shapes of the functions involved. Therefore all cut-offs must be done with utmost care. This is especially the case when the translational profile is a feature of relatively large width, with nonnegligible wings reaching far to the sides. The key function in evaluation of the convolution, Eq. (34), $I_{rot}(t)$, can be obtained by means of several approaches which have been developed in recent decades. Hereafter, we shall discuss general properties of one of the more typical concepts in order to show that the theory presented here indeed should lead to a fully workable procedure of spectral line computing. There are a few theoretical approaches that allow one to calculate the spectral distribution for this function. Some of them are based on the Fano theory and its modifications [29,30]. They refer to the ideas such as the resolvent of the Liouville operator and the relaxation matrix or the Mori-Fano superoperator. As an example we may may write (using the line-space formalism with the Liouville-space vectors, $|pq\rangle\rangle \equiv |q\rangle\langle p|$):

$$\begin{aligned} I_{rot}(\omega) &= \sum_{\lambda} \Re \int_{-\infty}^{\infty} \frac{(-)^{\Lambda+\lambda}}{\Pi_{\Lambda\Lambda}} \left\langle \Phi^{\Lambda 00,\Lambda\Lambda}_{000,\lambda}(\Omega(t)) \Phi^{\Lambda 00,\Lambda\Lambda}_{000,-\lambda}(\Omega(0)) \right\rangle e^{-i\omega t} dt \\ &= -\Im \frac{(-)^{\Lambda+\lambda}}{\Pi_{\Lambda\Lambda}} 4\pi \left\langle Y_{\Lambda\lambda}(\Omega)(\omega - \mathcal{L})^{-1}\, Y_{\Lambda-\lambda}(\Omega)\right\rangle \\ &= Tr_{\mathcal{R}} \sum_{\lambda} \sum_{J_i m_i J_f i m_f} \sum_{J_i' m_i' J_f' m_f'} \frac{(-)^{\Lambda+\lambda}}{\Pi_{\lambda\Lambda}} \langle J_i m_i | \rho_s(0) | J_i m_i \rangle \langle J_f m_f | Y_{\Lambda\lambda}(\Omega) | J_i m_i \rangle \\ &\quad \langle\langle J_f' m_f' J_i' m_i' | [\omega - \mathcal{L}_s - \langle \Gamma(\omega)\rangle_b]^{-1} | J_f m_f J_i m_i \rangle\rangle \langle J_f' m_f' | Y_{\Lambda-\lambda}(\Omega) | J_i' m_i' \rangle. \end{aligned} \tag{36}$$

In order to derive the above equation it has been assumed that the density matrix can be factorized into two parts of which one relates to the active molecule (called *a system*), denoted s and the other part related to the 'bath' (b), in which the system is immersed. Finally, after a number of transformations, we arrive at

$$\begin{aligned} I_{rot}(\omega) = \sum_{J_i J_f} \sum_{J_i' J_f'} \sum_{m_x} \langle J_i | \rho_s(0) | J_i \rangle \; &\times \langle\langle J_f' m_f' J_i' m_i' | [\omega - \mathcal{L}_s - \langle \Gamma(\omega)\rangle_b]^{-1} | J_f m_f J_i m_i \rangle\rangle \\ &\times F[J \ldots J', \Lambda\Lambda'], \end{aligned} \tag{37}$$

where

$$F[J \ldots J', \Lambda\Lambda'] = \frac{\Pi_{\Lambda' J_i J_i' \Lambda}}{\Pi_{J_f J_f'}} C^{J_f m_f}_{J_i i \Lambda\lambda} C^{J_f' m_f'}_{J_i' i' \Lambda-\lambda} C^{J_f' 0}_{J_i' 0 \Lambda 0} C^{J_f 0}_{J_i 0 \Lambda 0}, \tag{38}$$

$$\langle \Gamma(\omega)\rangle_b = Tr_b[\rho_b(0)\Gamma(\omega)]. \tag{39}$$

The basic information on the system evolution due to the intermolecular interactions is given by the relaxation matrix: $\Gamma(\omega)$. This element of the theory is responsible for the possible line mixing or state coupling phenomena. The way in which these two effects are introduced depends on the expression describing the relaxation matrix and the reverse matrix:

$$\langle\langle J_f' m_f' J_i' m_i' | [\omega - \mathcal{L}_s - \langle \Gamma(\omega)\rangle_b]^{-1} | J_f m_f J_i m_i \rangle\rangle. \tag{40}$$

The above problem is usually solved by means of one of the numerous procedures known in literature (e.g. IOS, impact approximation etc.) [5]. The choice of one of them depends on the physical conditions considered and the range of the applicability with regard to the frequency region of the spectra. The methods may be analytical, numerical or both. However, we must stress here once again that within the approach described above the calculating procedure is to be applied only for the small frequency region, and for this region an appropriate theory seems relatively easy to find.

5.2.2. Rotational component of the spectral function in the stationary phase approximation

When the function $I_{rot}(\omega)$ is to be evaluated for the other extreme frequency region, the spectral wings, it has to be treated by means of different methods [6,31,32] Here we shall illustrate one of them based on the theory proposed recently by Glaz in [7]. The reader should refer to this work for the details of the basic assumptions and for the outline of the calculations. Here we shall only remark on the most crucial points.

The theory is based on two major foundations: a diagonalization procedure of the evolution operator of the scattering system [6] and the so-called *stationary phase* approximation [33] used to calculate necessary Laplace transforms.

The starting formula for this approach reads:

$$I_{rot}(\omega) = 2 \lim_{s \to 0} \Re Tr_{\mathcal{R}} \sum_{\lambda} \frac{4\pi}{\Pi_{\Lambda\Lambda}} \left\{ \int_0^\infty d\tau \langle Y^{I\dagger}_{\Lambda\lambda}(\tau) U^I_e(\tau, 0) Y^I_{\Lambda\lambda}(0) U^I_e(0, -\infty) \rangle_{av} e^{i\omega\tau - s\tau} \Sigma_B \right\} \quad (41)$$

Here Σ_B is the Boltzmann distribution function for the radiator states and $<..>_{av}$ stands for averaging over the translational degrees of freedom of the perturber, which in the classical path approximation is given as follows:

$$\langle \ldots \rangle_{av} = \frac{1}{V} \int d\boldsymbol{v}(0) w(v) \int d\boldsymbol{R}(0) \, [\ldots] \, , \quad (42)$$

with $\boldsymbol{v}(0)$ designating the initial perturber velocity and $\boldsymbol{R}(0)$ being the intermolecular vector between the perturbed and the radiator; $w(v)$ is the Maxwell distribution of velocities and V is the volume. The translational averaging is often expressed in a more convenient form in terms of the so-called collisional variables [34,35]:

$$\langle \ldots \rangle_{av} = \frac{1}{V} \int_0^\infty dv 4\pi v^2 w(v) \int_0^\infty db 2\pi b \int_{-\infty}^\infty v \, dt_0 \int_\Omega d\Omega / 8\pi^2 \, [\ldots] \, , \quad (43)$$

where b is the impact parameter, t_0 is the time of the closest approach and $\int d\Omega \ldots$ indicates an angular average [36–38].

The essential terms involved in the procedure here are the operators $U^I_e(t, t')$. They are the one-perturbed collisional propagators, expressed as

$$U^I_e(t, t') = \mathcal{T} \exp \left\{ \int_{t'}^t d\sigma V^{aI}_e(\sigma) \right\} , \quad (44)$$

where $\mathcal{T}$ is the time-ordering operator. In this equation V^{aI}_e is the anisotropic part of the effective interaction potential for the perturber-radiator pair. In fact it is a superoperator acting in the Liouville space transformed to the interaction picture according to:

$$V^{aI}(t) = e^{-L_0 t} V^a(t) e^{L_0 t}; \quad (45)$$

L_0 is the Liouville operator associated with the Hamiltonian of the unperturbed system [22]. Operators $\boldsymbol{Y}_\Lambda$ are defined in the ordinary Hilbert space and consequently $\boldsymbol{Y}^I_\Lambda(t) = e^{-L_0 t} \boldsymbol{Y}_\Lambda(t)$. We shall refrain here from elaborating on the consecutive steps of further development of the theory as the details are given in [6], where also a final numerically workable expressions for I_{rot} are presented. Here we recall only one of them to illustrate the possibility of further calculations:

$$
\begin{aligned}
I_{rot}(\omega) = & \sum_{\{Jm\}} \sum_{\xi\varphi} 2\pi^2 (\mathcal{U}_\xi^{J_c m_c})^* \mathcal{U}_\xi^{J_g m_g} \mathcal{U}_\varphi^{J_i m_i} (\mathcal{U}_\varphi^{J_\alpha m_\alpha})^* \\
& \times \sum_\lambda \frac{\Pi_{\Lambda\Lambda' J_\alpha J_c}}{\Pi_{J_g J_i}} C_{J_\alpha m_\alpha \Lambda\lambda}^{J_g m_g} C_{J_c m_c \Lambda -\lambda}^{J_i m_i} C_{J_c 0 \Lambda 0}^{J_i 0} C_{J_\alpha 0 \Lambda 0}^{J_g 0} \\
& \times \frac{r_2^3}{12\varepsilon_2 |\widetilde{V}_{\xi\varphi}|} \{A(\omega) + B_1(\omega) + B_2(\omega) + B_3(\omega)\} \\
& \times \exp(-\beta < \varphi | V_{aniso} | \varphi >) \exp(-\beta V_{iso}) \Sigma_{\alpha\alpha}(rot).
\end{aligned}
\tag{46}
$$

All the quantities in this equation can be obtained by using appropriate numerical methods – $\mathcal{U}_\xi^{Jm}$ are the coefficients of the diagonalization procedure applied, $\widetilde{V}_{\xi\varphi}$ is a difference between matrix elements of the anisotropic part of the potential taken between states $|\xi >$ and $|\varphi >$, V_{iso} and V_{aniso} are the isotropic and anisotropic part of the potential respectively, whereas $\Sigma_{\alpha\alpha}$ is the Boltzmann distribution function taken for a rotational state denoted by α. Functions A and B are frequency dependent expressions which can be obtained by means of the procedure proposed in [7].

5.2.3. Mid wing region

Unfortunately, both the above mentioned approaches tend to break down in the intermediate region of the frequency range. A satisfying description of the spectra within these limits is a difficult task with very few theoretical treatments suggested. Despite this shortage of purely analytical procedures the situation is not hopeless. Some semi-empirical interpolation methods are available, which offer a numerically treatable formulae that might be included in the convolution. One of such ideas has been developed by Ozanne et al. in [39]. They apply a function determined by means of some fitting parameters that may be calculated so that the continuity of the matrix elements of the relaxation operator within all the three regions of the spectrum should be respected. This method proved to be a successful tool for producing reliable shapes for the collisional absorption processes, yet it seems to be also suitable for the phenomena considered in this work.

All in all, in the above succinct outlook of the possible line calculating routines we have shown that the approach which we suggested in Section 5.1 supplemented with this procedures should turn out to be a suitable and doable method of analyzing the CILS spectral shapes with the rotational-translational coupling taken into account.

6. Conclusion

We have developed a theory that allow us to analyze the spectral distribution of the TCF functions for the CILS processes. The general formulae have been given to express the collisional polarizabilities of different order in terms of the spherical tensor algebra. They were next transform in order to determine their angle dependent parts via the generalized triple-Wigner functions. This approach makes it possible to discuss the TCF functions in more general way to great extent independent of the particular inducing mechanisms involved. The theory of the spectral collisional profiles has been proposed, which, unlike usual previous approaches, does not neglect the coupling between different types of the degrees of freedom. The evidence has been given that the new method of calculating the spectral profiles may be numerically treatable within a wide range of frequencies, providing a useful workable tool for theoretical spectral analyses.

Appendix A

The second order perturbation procedure results in the next possible term in the expansion of the collisional polarizability, $\Delta \boldsymbol{A}_J^{(2)}(p)$:

$$\Delta \boldsymbol{A}_J^{(2)}(p) = - \sum_{ll_1l_2n_2n_3} \sum_{KML} \sum_{J_1J_2J_3} \sum_{q\neq p, r\neq q} (-)^{l_2+K+L+M+J_1+J_3} \left[\frac{2^{l_1+l_2+n_2+n_3}}{(2l_2)!(2l_1)!(2n_2)!(2n_3)!} \right]$$
$$\times \begin{Bmatrix} K & l & 1 & J \\ M_1 & l_2 & 1 & L \\ M & M_2 & n_3 & J_3 \end{Bmatrix} \begin{Bmatrix} J_2 & n_2 & l_2 \\ J_1 & l_1 & 1 \\ K & M_1 & l \end{Bmatrix} \Pi_{llKLMM_1M_2J_1J_2J_3} \tag{47}$$
$$\times \left\{ \{ \boldsymbol{T}_{M_1}(pq) \otimes \boldsymbol{T}_{M_2}(qr) \}_M \otimes \left\{ \left\{ \boldsymbol{A}_{J_1}^{(1l_1)}(p) \otimes \boldsymbol{A}_{J_2}^{(n_2l_2)}(q) \right\}_K \otimes \boldsymbol{A}_{J_3}^{(n_31)}(r) \right\}_L \right\}_J .$$

Appendix B

When quantities under consideration are dependent on orientations of an individual molecule, they are often defined by means of series expansions in terms of the Wigner functions – the matrix elements of the rotation operator. This kind of formalism provide an efficient tool of Racah [40] algebra to be applied in analyses. However, in a case of collective phenomena in which more than one molecule is involved a more general treatment must be applied. To this end, a generalized so-called *triple Wigner* function has been devised in [10] capable of describing quantities depending on three sets of Euler angles – the situation we deal with in the equations defining the collisional polarizabilities of two molecules. In this case we have to consider orientations given by Euler angles Ω_p, Ω_q and Ω_{pq}. The triple-Wigner functions are thus given by

$$\Phi_{kln,j}^{KLN,MJ}(\Omega_p, \Omega_q, \Omega_{pq}) = \sum_{m',n',k'l'} \Pi_{KLN} C_{Mm'Nn'}^{Jj} C_{Kk'Ll'}^{Mm'} D_{k'k}^{K*}(\Omega_p) D_{l'l}^{L*}(\Omega_q) D_{n'n}^{N*}(\Omega_{pq}), \tag{48}$$

where C_{KkLl}^{Jj} are the Clebsch-Gordan coefficients.

Some properties of these functions are discussed thoroughly in [10]. However, it might be of some use for the further course of the analysis in this work to select one of the rules that the Φ functions are to obey. Namely, the product of two such functions can be expanded into a sum in the following manner:

$$\Phi_{kln,j}^{KLN,MJ}(\Omega_p\Omega_q\Omega_{pq}) \cdot \Phi_{k'l'n',j'}^{K'L'N',M'J'}(\Omega_p\Omega_q\Omega_{pq})$$
$$= \sum_{\substack{ABR \\ ST}} \sum_{rstb} C_{JjJ'j'}^{Bb} \Pi_{JKLMNJ'K'L'M'N'} \Pi_A C_{KkK'k'}^{Rr} C_{LlL'l'}^{Ss} C_{NnN'n'}^{Tt} \tag{49}$$
$$\times \begin{Bmatrix} K & L & M \\ K' & L' & M' \\ R & S & A \end{Bmatrix} \begin{Bmatrix} M & M' & A \\ N & N' & T \\ J & J' & B \end{Bmatrix} \Phi_{rst,b}^{RST,AB}(\Omega_p, \Omega_p, \Omega_{pq}).$$

Appendix C

The cross term for the second order collisional polarizability reads

$$I_{JJ}^{(20)}(t) = -\sum_{\substack{l_1 l_2\\ m_1 m_2}} \sum_{\substack{J_1 J_2 J_3\\ M_1 M_2}} \sum_{\substack{KLM\\ XY}} \sum_{p \neq q \neq r,s} (-)^{M+J_1+J} \Pi_{J J_1 J_2 J_3 M_1 M_2 K K M L X X Y}$$
$$\times \left[\frac{2^{l_1+l_2+m_1+m_2}}{(2l_1)!(2l_2)!(2m_1)!(2m_2)!}\right]^{(1/2)} \begin{Bmatrix} K & X & 1 & J \\ M_1 & l_2 & 1 & L \\ M & M_2 & m_2 & J_3 \end{Bmatrix}$$
$$\times \begin{Bmatrix} J_2 & m_1 & l_2 \\ J_1 & l_1 & 1 \\ K & M_1 & X \end{Bmatrix} \begin{Bmatrix} J_2 & J_1 & K \\ J_3 & L & Y \end{Bmatrix} \Big\langle \{\{\boldsymbol{T}_{M_1}(pq(t)) \tag{50}$$
$$\otimes \boldsymbol{T}_{M_2}(qr(t))\}_M \otimes \Big\{\Big\{\boldsymbol{A}_{J_2}^{(m_1 l_2)}(q(t)) \otimes \Big\{\boldsymbol{A}_{J_1}^{(1 l_1)}(p(t)) \otimes \boldsymbol{A}_{J_3}^{(m_2 1)}(r(t))\Big\}_Y\Big\}_L$$
$$\otimes\ \boldsymbol{A}_J^{(11)}(s(0))\Big\}_M\Big\}_0\Big\rangle, l_1 + m_1 = M_1,\ l_2 + m_2 = M_2.$$

References

[1] J.P. McTague and G. Birnbaum, *Phys. Rev. Letters* **21** (1968), 661.
[2] G. Birnabaum, *Phenomena Induced by Intermolecular Interactions*, Plenum, New York, 1985.
[3] G.C. Tabisz and M.N. Neuman, *Collision- and Interaction-Induced Spectroscopy*, Kluwer Academic Publishers, Dordrecht, 1995.
[4] S. Kielich, *Proc. Indian. Acad. Sci.* (*Chem. Sci.*) **94** (1985), 403.
[5] A. Lévy, N. Lacome and I.C.C. Chackerian Jr., in: *Collisional Line Mixing in Spectroscopy of the Earth's Atmosphere and Interstellar Medium*, N. Rao and A. Weber, eds, Academic Press, New York, 1992, p. 261.
[6] W. Głaz and G.C. Tabisz, *Phys. Rev.* **A 54** (1996), 3903.
[7] W. Głaz, *Journal of Physics B* **36** (2003), 2309.
[8] A.R. Edmonds, *Angular Momentum in Quantum Mechanics*, Princeton University, Princeton NJ, 1957.
[9] D.A. Varshalovich, A.N. Moskalev and V.K. Khersonskii, *Quantum theory of angular momentum*, World Scientific, Singapore, 1988.
[10] W. Głaz, *Acta Phys. Pol.* **A 73** (1988), 619.
[11] C.G. Gray and K.E. Gubbins, *Theory of Molecular Fluids*, Vol. 1; Fundamentals, Calderon Press, Oxford, 1984.
[12] W. Głaz, *Physica A* **148 A** (1988), 610.
[13] T. Bancewicz, Y. Le Duff and J.-L. Godet, to be published.
[14] T. Bancewicz, W. Glaz and S. Kielich, *Phys. Lett.* **A 148** (1990), 78.
[15] T. Bancewicz, Y. Le Duff and J.-L. Godet, Multipolar polarizabilities from interaction induced Raman scattering, Modern Nonlinear Optics, Part 1, Second Edition, *Advances in Chemical Physics* **119** (2001), 267.
[16] A. Borysow and M. Moraldi, in: *Collision- and Interaction-Induced Spectroscopy*, G.C. Tabisz and M.N. Neuman, eds, Kluwer Academic Publishers, Dordrecht, 1995, p. 395.
[17] G. Maroulis and A. Haskopoulos, *Chem. Phys. Lett.* **349** (2002), 335.
[18] G. Maroulis, *J. Chem. Phys* **118** (2002), 2673.
[19] L. Frommhold, *Collision-Induced Absorption in Gases*, Cambridge University Press, Cambridge, 1992.
[20] N.N. Filippov and M.V. Tonkov, *J. Chem. Phys.* **108** (1998), 3608.
[21] J.M.M. Roco, A. Medina, A. Calvo Hernandez and S. Velasco, *J. Chem. Phys.* **103** (1995), 9175.
[22] B. Gao, G.C. Tabisz, M. Trippenbach and J. Cooper, *Phys. Rev. A* **A 44** (1991), 7379.
[23] A. Borysow and M. Moraldi, *J.Chem.Phys.* **99** (1993), 8424.
[24] M.S. Brown, S.-K. Wang and L. Frommhold, *Phys. Rev.* **A 40** (1989), 2276.
[25] N. Meinander, *J. Chem. Phys.* **99** (1993), 8654.
[26] W. Glaz, J. Yang, J.D. Poll and C.G. Gray, *Chem. Phys. Lett.* **218** (1994), 183.

[27] W. Glaz and G.C. Tabisz, *Can. J. Phys.* **79** (2001), 801.
[28] J. Szudy and W.E. Baylis, *Phys. Reports* **226** (1996), 127.
[29] U. Fano, *Rev. Mod. Phys.* **29** (1957), 74.
[30] U. Fano, *Phys. Rev.* **131** (1963), 259.
[31] P.W. Rosenkranz, *J. Chem. Phys.* **83** (1985), 6139.
[32] C. Boulet, J. Boissoles and D. Robert, *J. Chem. Phys.* **89** (1988), 625.
[33] M.V. Berry and M. Tabor, *Proc. R. Soc. London Ser. A* **349** (1976), 101.
[34] G. Alber and J. Cooper, *Phys. Rev.* **A 31** (1985), 3644.
[35] G. Alber and J. Cooper, *Phys. Rev.* **A 33** (1986), 3084.
[36] G.C. Maitland, M. Rigby, E.B. Smith and W.A. Wakeham, *Intermolecular Forces; Their Origin and Determination*, Calderon Press, Oxford, 1981.
[37] C.V. Heer, *Statistical Mechanics, Kinetic Theory and Stochastic Processes*, Academic Press, New York and London, 1972.
[38] P. Résibois and M. De Leener, *Classical Kinetic Theory of Fluids*, John Wiley and Sons, New York, 1977.
[39] L. Ozanne, Q. Ma, Nguyen-Van-Thanh, C. Brodbeck, J.P. Bouanich, J.M. Hartmann, C. Boulet and R.H. Tipping, *J. Quant. Spectrosc. Radiat. Transfer* **58** (1997), 261.
[40] G. Racah, *Phys. Rev.* **76** (1949), 1352.

Journal of Computational Methods in Sciences and Engineering 4 (2004) 293–300
IOS Press

Density functional calculations of the frequency-dependent optical rotation: Comparison of theory and experiment for the gas phase

C. Diedrich, S. Causemann and S. Grimme*
Theoretische Organische Chemie, Organisch-Chemisches Institut der Universität Münster, Corrensstraße 40, D-48149 Münster, Germany

Abstract. The optical rotations (OR) of five chiral organic molecules have been calculated by time-dependent density functional response theory (TDDFT) employing four different density functionals. The results are compared with experimental gas phase data in order to explore the inherent accuracy of the functionals. The theoretical results obtained with the BHLYP hybrid functional including 50% "exact" Hartree-Fock exchange compare very well with the experimental data and also better with those derived from the gas-phase than from solution. Non-hybrid functionals show a tendency to overestimate the OR. The anomalous behavior of the methyloxirane molecule which shows sign changes with solvent and excitation frequency is discussed.

Keywords: Time dependent density functional theory, optical rotation, organic molecules

Mathematics Subject Classification: 03-04, 49R50, 37C99, 74C99

1. Introduction

Optical activity is an important research tool routinely used by chemists to solve stereochemical problems of chiral molecules [1,2]. In recent years it has become clear that modern electronic structure methods can provide reliable predictions for electronic circular dichroism (ECD) and optical rotation (OR). The main perspective of such theoretical investigations is to elucidate the absolute stereochemistry of chiral molecules. Although the analysis and interpretation of ECD spectra for this purpose is straightforward and comparisons with experimental data for different absorption bands give a solid basis for such assignments [3], the simple measurement of OR at a single frequency is still the most popular and convenient method of characterization used routinely in almost every chemical laboratory.

Recently, several independent studies have shown that time-dependent density functional theory (TDDFT) [4–6] employing standard functionals such as B3LYP [7,8] can provide accurate ORs for a wide range of organic systems (for a recent review see [9]). The OR is a frequency dependent, second-order property which is related to the electric dipole-magnetic dipole polarizability tensor (see

*Corresponding author. E-mail:grimmes@uni-muenster.de.

Section 2). As such, it is expected to be significantly influenced by the environment in which the experiments are performed. Almost all OR measurements are routinely undertaken in solution and thus, because the calculations usually refer to the gas phase, there remains some uncertainty how good the density functionals (DF) really perform. There are indications [10] that the good agreement between theory and experiment hitherto observed originates at least partially from a lucky error compensation. Although a recent study [11] investigated the OR theoretically by inclusion of solvent effects via electrostatic continuum models, the effect of the van der Waals and Pauli repulsion interactions with a solvent on the chiroptical response properties of complex molecules are completely unknown. Comparison of theoretical and experimental gas phase data is the first step before these contributions (which are most important for many non-polar molecules) can be investigated seriously.

The basic motivation of this study is to explore the inherent accuracy of commonly used density functionals to calculate optical rotations of organic molecules. Somewhat surprisingly, there is only one experimental gas phase study [12] of the OR. The molecules α/β-pinene (1–2), cis-pinane (3), fenchone (4) and methyloxirane (5) (see Fig. 1) have been considered. The limonene molecule, which has also been investigated experimentally, is excluded here because of its conformational flexibility. Because of the special experimental technique used (cavity ring-down polarimetry [12]), the ORs correspond to the excitation wavelength 355 nm which is considerably higher in energy than that used normally in solution (sodium-D-line at 589.3 nm). Although this increases the complexity of the problem (because the measurement frequency more closely approaches the lowest excitation energies of the molecules), our general conclusions are expected to be valid also for lower excitation frequencies.

This paper is organized as follows: in Section 2.1, a brief review of the necessary theory of OR is given. Section 2.2 deals with some technical aspects of the computations. In Section 3, which presents a comparison and discussion of the experimental and theoretical data, also a AO basis set study on the convergence behavior of the calculated property is included which is always an important issue in the treatment of polarizabilities.

2. Theory

In chiral systems, the average dipole moment induced by an electromagnetic field oscillating at frequency ω is given, to first order, by [1]

$$\mu_{\text{ind}}(\omega) = \alpha(\omega)E(\omega) - \tfrac{1}{c}\delta(\omega)B(\omega). \tag{1}$$

E and B are the averaged electric and magnetic fields, $\alpha(\omega)$ is the frequency-dependent isotropic polarizability, and c the light velocity. Equation (1) defines the isotropic OR $\delta(\omega)$ as the response parameter of the induced dipole moment to a frequency-dependent uniform magnetic field. In the density matrix formulation of time-dependent density functional response theory [13], this is conveniently expressed as

$$\delta(\omega) = \frac{c}{3\omega}\,\text{Im}\,\langle\boldsymbol{\mu}(\Lambda - \omega\Delta)^{-1}|\mathbf{m}\rangle. \tag{2}$$

$\boldsymbol{\mu}$ denotes the electric and $\mathbf{m}$ $|\boldsymbol{\mu}\rangle$ the magnetic dipole moment operator, and $\Lambda - \omega\Delta$ the response kernel defined below. In the present notation, a component of $|\boldsymbol{\mu}\rangle$ is a vector

$$|\mu_\alpha\rangle = \begin{pmatrix} P \\ Q \end{pmatrix}, \tag{3}$$

where $P_{ia} = Q_{ai} = \mu_{\alpha\, ia}$ denote the dipole moment operator in the space of occupied times virtual static Kohn-Sham (KS) molecular orbitals (MOs). In case of $|\mathbf{m}\rangle$, which is purely imaginary, $P_{ia} = -Q_{ai} = m_{\alpha\, ia}$. P and Q can be different in general, which necessitates $|P, Q\rangle$. As usual, indices $i, j, \ldots$ are used for occupied and $a, b, \ldots$ for virtual MOs. The expression on the right-hand-side of Eq. (2) should be understood as a scalar product of the x, y and z components of $\boldsymbol{\mu}$ and $\mathbf{m}$. Λ and Δ are 2×2 super-operators,

$$\Lambda = \begin{pmatrix} A & B \\ B & A \end{pmatrix}, \quad \Delta = \begin{pmatrix} 1 & 0 \\ 0 & -1 \end{pmatrix}, \tag{4}$$

where A and B denote the well-known orbital rotation Hessians. For details the reader is referred to ref. [13].

Previous treatments of the OR [10,14–16] have employed the length form of the electric dipole moment,

$$|\boldsymbol{\mu}^{\mathrm{L}}\rangle = -|\mathbf{r}\rangle. \tag{5}$$

This choice is not unique, however, and we consider briefly the implications. By means of a canonical transformation first described by Goeppert-Mayer [17], one obtains the velocity form of $|\boldsymbol{\mu}\rangle$,

$$|\boldsymbol{\mu}^{\mathrm{V}}\rangle = -\Delta\Lambda^{-1}|\boldsymbol{\nabla}\rangle. \tag{6}$$

In a complete basis, $|\boldsymbol{\mu}^{\mathrm{L}}\rangle$ and $|\boldsymbol{\mu}^{\mathrm{V}}\rangle$ are identical. This can be shown by inserting the definitions of Λ and Δ in Eq. (6) and using the completeness of the MO basis as well as the canonical commutator relation for $\mathbf{r}$ and $\boldsymbol{\nabla}$. In a finite basis, the completeness relation does not hold any more, and the length and velocity forms differ.

The magnetic dipole moment operator is given by

$$|\mathbf{m}\rangle = \frac{i}{2c}|(\mathbf{r} - \boldsymbol{R}) \times \boldsymbol{\nabla}\rangle; \tag{7}$$

$\boldsymbol{R}$ is the so-called gauge origin. In the usual atom-centered Gaussian basis sets, the length form of the OR

$$\delta^{\mathrm{L}}(\omega) = \frac{c}{3\omega}\,\mathrm{Im}\,\langle\boldsymbol{\mu}^{\mathrm{L}}(\Lambda - \omega\Delta)^{-1}|\mathbf{m}\rangle \tag{8}$$

exhibits an unphysical gauge-origin dependence. Stephens et al. [16] and Ruud and Helgaker [18] have used the GIAOs to enforce the gauge-origin invariance of δ^{L} in finite basis sets. We pursue a different strategy, which is based on the observation that the velocity form of the OR

$$\delta^{\mathrm{V}}(\omega) = \tfrac{c}{3\omega}\mathrm{Im}\langle\boldsymbol{\mu}^{\mathrm{V}}(\Lambda - \omega\Delta)^{-1}|\mathbf{m}\rangle \tag{9}$$

is gauge-invariant in *any* basis set (the prove of this is outlined in detail in ref. [19]).

2.1. Computational method and technical details

From Eq. (6) it appears that the computation of the vector $|\boldsymbol{\mu}^{\mathrm{V}}\rangle$ requires the solution of an additional static coupled KS equation system with the vector $|\boldsymbol{\nabla}\rangle$ as right-hand side. Since $|\boldsymbol{\nabla}\rangle$ is skew-symmetric, it couples only to the magnetic part of Λ, i.e., eq. (6) takes the form

$$\sum_{jb}(A - B)_{iajb}\,\mu^{\mathrm{V}}_{jb} = -\boldsymbol{\nabla}_{ia}. \tag{10}$$

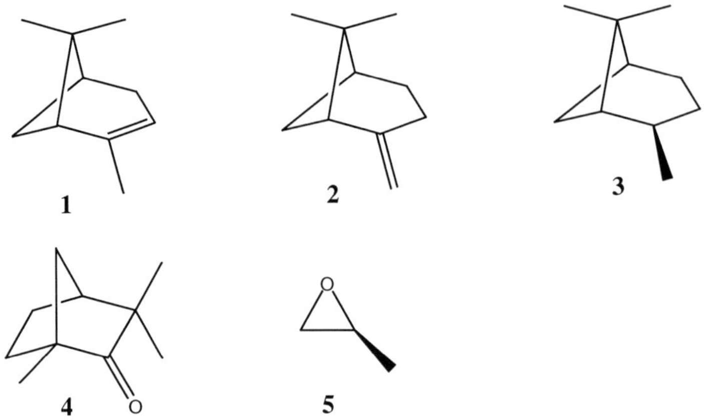

Fig. 1. Schematic structures of the investigated molecules.

For "pure" density functionals without Hartree-Fock exchange, however, the magnetic orbital rotation Hessian $(A-B)$ is diagonal in the basis of KS-MOs, with the differences of virtual and occupied orbital energies as diagonal entries [5]. Thus, δ^{V} can be computed with essentially the same effort as δ^{L}. The time-determining step is the calculation of the vector $(\Lambda - \omega\Delta)^{-1}|\boldsymbol{\mu}\rangle$, which is done by an iterative solution of the linear equation system

$$(\Lambda - \omega\Delta)|X, Y\rangle = |\boldsymbol{\mu}\rangle \tag{11}$$

for $|X, Y\rangle$. The integral direct algorithm we use fully exploits the fact that $(A-B)$ is diagonal; in addition, all spacial components at all desired frequencies can be treated simultaneously, and molecular point group symmetry is fully used for finite groups.

All calculations are carried out with the TURBOMOLE suite of programs [20]. The *escf* response module is used for all OR calculations. The pure, gradient-corrected functionals PBE [21] and BP86 [22–24], the popular B3LYP [7,8] (20% HF exchange) and the BHLYP [25] (50% HF exchange) hybrid functionals are employed in the TDDFT treatments. The ground state structures have been fully optimized at B3LYP level using a split-valence basis set augmented with polarization functions on non-hydrogen atoms (SV(d) [26]). The specific OR at frequency ω is related to δ by

$$[\alpha]_\omega = C \cdot \delta(\omega). \tag{12}$$

where $C = 1.343 \times 10^{-4}\omega^2/M$ with M being the molar mass in g/mol, ω is the frequency in cm^{-1} and $\delta(\omega)$ is given in atomic units. All specific rotations are reported in the usual units deg·[dm·(g/cc)]$^{-1}$ and correspond to the wavelength (355 nm) used in the experiments [12]. The convergence thresholds in the SCF and response calculations and the numerical quadrature in DFT (modified grid `m4` option [27, 28]) are chosen such that at least three significant digits for the OR are obtained. All calculations are performed in center-of-mass atomic coordinates which provides reasonable values for the OR also in the (origin-dependent) length formalism as long as basis sets of at least aug-cc-VDZ quality are used [10].

Table 1
Dependence of calculated optical rotations (length (L) and velocity (V) forms, PBE functional) of the quality of the AO basis set

basis set	$[\alpha]^{L}_{355}$	$[\alpha]^{V}_{355}$
	β-pinene (**2**)	
SV(d)	268	501
aug-SV(d)	387	404
aug-TZV(d,p)	416	413
exp.	70 ± 2	
	fenchone (**4**)	
SV(d)	−46.1	106
aug-SV(d)	−597	−653
aug-TZV(d,p)	−644	−663
exp.	-180 ± 9	

3. Results and discussion

3.1. AO basis set dependence

In order to check the dependence of the results on the quality of the AO basis set we have performed a series of calculations for two of the investigated molecules (β-pinene and fenchone). The main objective here is to monitor the results from the dipole-length (L) and dipole-velocity (V) forms simultaneously. This is of particular importance because (due to technical reasons) the gauge-invariant V-form can currently not be used for the hybrid functionals B3LYP and BHLYP. As a starting point, the non-augmented SV(d) basis, which is not expected to give quantitative results, has been employed. Next, diffuse functions are added to this split-valence basis set [aug-SV(d)]. The largest AO basis set is of augmented valence triple-ζ quality (aug-TZV(d,p) [29]). The results obtained with the pure PBE functional are shown together with the experimental gas phase data in Table 1.

It should be noted here that the quality of a particular basis set can be judged by the deviations of the OR calculated in the length and velocity forms which should be identical in a complete basis. In general, inspection of Table 1 reveals a rather fast convergence to the basis set limit in both cases. Significant differences between the L-and V-forms (in particular a different sign in the case of fenchone) are found only with the non-augmented basis which thus can not be recommended for quantitative work. With the aug-TZV(d,p) basis, the relative differences between L and V forms are less than 3%.

When considering the agreement between calculated and experimental OR, we notice that improving the basis set beyond aug-SV(d) does not improve the results obtained with the non-hybrid functionals. Typically, the absolute magnitude of the OR is overestimated with pure functionals like PBE and BP86 (cf. the following section). Similar as for "normal" polarizabilities, this effect increases as the AO basis set is enlarged. Adding more and more (mainly diffuse) basis functions to an already too diffuse Kohn-Sham solution will contaminate it increasingly with artificial states thereby worsening the agreement between calculated and experimental data. A more extensive basis set study in ref. [19] also revealed that with augmented split-valence basis sets, sometimes favorable error compensation is at work. It should be clear, however, that this behavior results from the inaccuracy of the density functionals used and that better quantum chemical models (like e.g. coupled-cluster theory [30]) should of course yield better results when the AO expansion spaces are improved.

Table 2
Comparison of experimental[a] and calculated[b] specific optical rotations

Molecule	$[\alpha]_{355}$					
	PBE	BP86	B3LYP	BHLYP	exp.(gas)	exp.(solution)[c]
(1S5S)-α-pinene (**1**)	−4.5	−83.6	−104	−171	−189 ± 3	−165
(1S5S)-β-pinene (**2**)	416	349	243	98.2	70 ± 2	22
(1S)-cis-pinane (**3**)	−66.0	−62.0	−51.0	−36.0	−63 ± 6	−88
(1R)-fenchone (**4**)	−644	−564	−377	−259	−180 ± 9	−158
(S)-methyloxirane (**5**)	162	101	52.7	8.6	10 ± 3	−26

[a]Ref. [12]. [b]aug-TZV(d,p) AO basis set using B3LYP/SV(d) optimized geometries. The results refer to the length form for $[\alpha]$ and emplyoing center-of-mass atomic coordinates. [c]In cyclohexane from ref. [12] (extrapolated to 355 nm from measurements at lower frequencies).

3.2. Functional dependence and comparison with experiment

TDDFT calculations employing the aug-TZV(d,p) basis set and the four different functionals PBE, BP86, B3LYP and BHLYP have been carried out for the five molecules shown in Fig. 1. The calculated ORs are compared with experimental gas phase data in Table 2 where also extrapolated solution data are included.

Except for the "outlier" α-pinene with PBE, all calculated OR data show a smooth convergence to the experimental values when the amount of exact HF exchange is increased from zero (PBE/BP86) over 20% (B3LYP) to 50% (BHLYP). In general, the ORs calculated with BHLYP agree very well with those from experiment. For the completely saturated cis-pinane molecule, the BHLYP results are worst although the deviation of about 27 deg·[dm·(g/cc)]$^{-1}$ does not seem alarming because systems without any chromophore can be considered as most difficult even for hybrid functionals. For all systems studied, the ORs calculated with BHLYP agree better with the corresponding gas phase data than with those obtained in cyclohexane solution. The solvent effects on the OR are small for Eq. (1), Eqs (3)–(4) and moderate to large for Eq. (2). A striking example is (S)-methyloxriane which shows different signs for the OR in solution and in the gas phase. Although the precise origin of this effect is not completely clear at present, we note that this system represents a very special (sensitive) case which involves also a sign change with frequency. In Fig. 2 we compare the wavelength dependence of the OR calculated with two functionals with experimental data.

It is seen that both, PBE and B3LYP predict a sign change of the OR at shorter wavelengths (at about 490 and 440 nm, respectively). Although the steepness of the two calculated ORD curves is significantly different, we believe that their qualitative behavior is correct. If this is true we anticipate a negative value for the OR in the gas phase at the sodium-D-line (which is unfortunately not known experimentally). This leads us to the conclusion that the OR for this molecule is strongly influenced by solute-solvent interactions (which seem to vary also with frequency) and that the good agreement between theory (gas phase) and experiment (solution) is pure coincidence.

4. Conclusions

This study has shown that accurate TDDFT calculations of the frequency dependent optical rotations for (mostly saturated) organic molecules require the inclusion of "exact" HF exchange into the functional. Except for cis-pinane, the agreement with experiment improves in the order PBE/BP86, B3LYP and BHLYP. It can thus be concluded, that (opposed to other properties like the energy) a larger fraction (about 50% as in BHLYP) HF exchange is necessary in OR calculations. These findings can be attributed

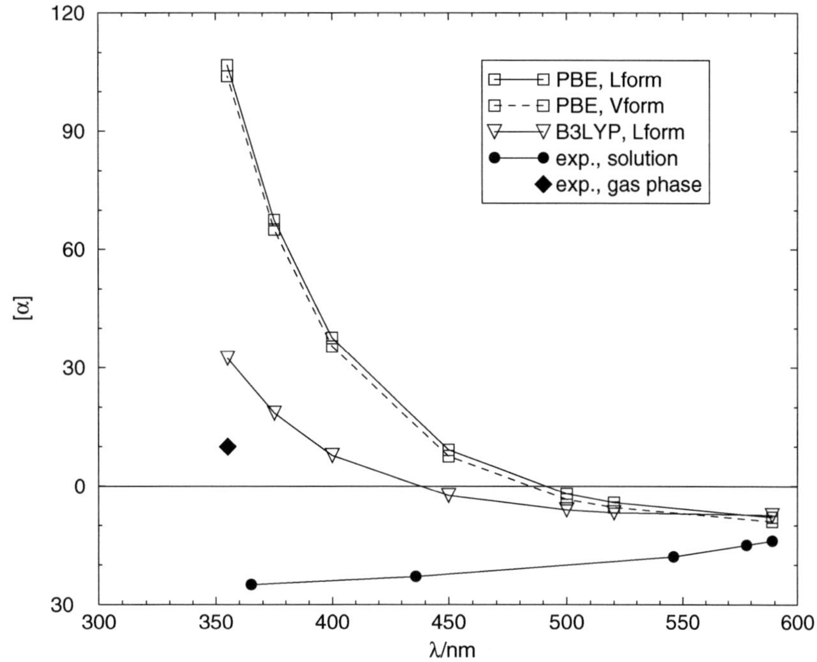

Fig. 2. Optical rotation dispersion for (S)-methyloxirane (aug-cc-pVTZ AO basis [31,32]).

to the wrong asymptotic form of the Becke88/LDA parts in the density functionals at large electron-molecule distances which makes the high-energy part of the electronic spectrum too diffuse. The solvent effects on the OR appear to be moderate for the medium-sized molecules investigated which explains the good accuracy hitherto observed in most theoretical studies. Thus, theoretical OR calculations form a solid basis for the assignment of the absolute configuration of large chiral systems with biological or pharmacological relevance.

Acknowledgment

This work was supported by the Deutsche Forschungsgemeinschaft, SFB 424 ("Molekulare Orientierung als Funktionskriterium in chemischen Systemen").

References

[1] E. Charney, *The Molecular Basis of Optical Activity*, Wiley, New York, 1979.
[2] E.L. Eliel and S.H. Wilen, *Stereochemistry of Organic Molecules*, Wiley, New York, 1994.
[3] S. Grimme, J. Harren, A. Sobanski and F. Vögtle, *Eur. J. Org. Chem.* (1998), 1491.
[4] M.E. Casida, in: *Recent Advances in Density Functional Methods*, (Vol. 1), D.P. Chong, ed., World Scientific, Singapore, 1995.
[5] R. Bauernschmitt and R. Ahlrichs, *Chem. Phys. Lett* **256** (1996), 454.
[6] F. Furche, R. Ahlrichs, A. Sobanski, F. Vögtle, C. Wachsman, E. Weber and S. Grimme, *J. Am. Chem. Soc.* **122** (2000), 1717.
[7] A.D. Becke, *J. Chem. Phys.* **98** (1993), 5648.
[8] P.J. Stephens, F.J. Devlin, C.F. Chabalowski and M.J. Frisch, *J. Phys. Chem* **98** (1994), 11623.

[9] P.L. Polavarapu, *Chirality* **14** (2002), 768.
[10] S. Grimme, *Chem. Phys. Lett.* **339** (2001), 380.
[11] B. Mennucci, J. Tomasi, R. Cammi, J.R. Cheeseman, M.J. Frisch, F.J. Devlin, S. Gabriel and P.J. Stephens, *J. Phys. Chem. A* **106** (2002), 6102.
[12] T. Müller, K.B. Wiberg and P.H. Vaccaro, *J. Phys. Chem. A* **104** (2000), 5959.
[13] F. Furche, *J. Chem. Phys.* **114** (2001), 5982.
[14] R.D. Amos, *Chem. Phys. Lett.* **87** (1982), 23.
[15] T. Helgaker, K. Ruud, K.L. Bak, P. Jørgensen and J. Olsen, *J. Faraday Discuss.* **99** (1994), 165.
[16] P.J. Stephens, F.J. Devlin, J.R. Cheeseman and M.J. Frisch, *J. Phys. Chem.* (*A*) **105** (2001), 5356.
[17] M. Goeppert-Mayer, *Ann. Phys.* **9** (1931), 273.
[18] K. Ruud and T. Helgaker, *Chem. Phys. Lett.* **352** (2002), 533.
[19] S. Grimme, F. Furche and R. Ahlrichs, *Chem. Phys. Lett.* **361** (2002), 321.
[20] R. Ahlrichs, M. Bär, M. Häser, H. Horn and C. Kölmel, *Chem. Phys. Lett.* **162** (1989), 165. current version: see http://www.turbomole.de.
[21] J.P. Perdew, M. Ernzerhof, A. Zupan and K. Burke, *J. Chem. Phys.* **108** (1998), 1522.
[22] J.P. Perdew, *Phys. Rev. B* **33** (1986), 8822–8824.
[23] J.P. Perdew, *Phys. Rev. B* **34** (1986), 7046.
[24] A.D. Becke, *Phys. Rev. A* **38** (1988), 3098–3100.
[25] A.D. Becke, *J. Chem. Phys.* **98** (1993), 1372.
[26] A. Schäfer, H. Horn and R. Ahlrichs, *J. Chem. Phys.* **97** (1992), 2571.
[27] O. Treutler and R. Ahlrichs, *J. Chem. Phys.* **102** (1995), 346.
[28] F. Furche and R. Ahlrichs, *J. Chem. Phys.* **114** 2001, 10362.
[29] A. Schäfer, C. Huber and R. Ahlrichs, *J. Chem. Phys.* **100** (1994), 5829.
[30] K. Ruud, P.J. Stephens, F.J. Devlin, P.R. Taylor, J.R. Cheeseman and M.J. Frisch, *Chem. Phys. Lett.* **373** (2003), 606.
[31] T.H. Dunning, *J. Chem. Phys.* **90** (1989), 1007.
[32] R.A. Kendall, T.H. Dunning, Jr. and R.J. Harrison, *J. Chem. Phys.* **96** (1992), 6796.

Journal of Computational Methods in Sciences and Engineering 4 (2004) 301–306
IOS Press

Ab initio determination of the interaction hyperpolarizability for the H-bond complex NH_3-HF

Wu Di[a], Li Zhi-Ru[a,*], Ding Yi-Hong[a], Zhang Man[a], Zheng Zhi-Ren[b], Wang Bing-Qiang[a] and Hao Xi-Yun[a]
[a]*State Key Laboratory of Theoretical and Computational Chemistry, Institute of Theoretical Chemistry, Jilin University, Changchun 130023, P.R . China*
[b]*Department of Physics, Jilin University, Changchun 130023, P.R. China*

Abstract. The static dipole moment (μ_0), polarizability (α_0) and first hyperpolarizability (β_0) of the hydrogen bond complex NH_3-HF with a equivalent triple π-type hydrogen bond are investigated by means of *ab initio* methods at the MP2 level based on the QCISD potential energy surface. The full counterpoise (CP) method is applied in the studies of the intermolecular interaction contributions to the above properties. The first hyperpolarizability obtained is 31.76 a.u.. The results of the intermolecular interaction contributions to the above properties are 26% for μ_0, -4.2% for α_0 and 10.0% for β_0.

Keywords: *Ab initio*, hydrogen bond, dipole moment, polarizability, hyperpolarizability, counterpoise method

Mathematics Subject Classification: 78A60, 81Q10

1. Introduction

The design of novel materials with large nonlinear optical [NLO] properties (i.e., hyperpolarizabilities) is currently of great interest, mainly due to their potential applications [1–3]. Since the knowledge of atomic and molecular hyperpolarizabilitie is central to understanding of the nonlinear response of matter to light [4], theoretical determination of such properties has been a topic of much interest. A number of reliable *ab initio* studies on the hyperpolarizabilitie of molecules have been reported [3–8]. However, a molecular description of condense phase properties relies on the full characterization of intermolecular interactions. Van der Waals clusters have traditionally served as an source of information on intermolecular interactions to theoreticians and experimentalists alike [9]. So intermolecular interactions among molecules at long range and as close as van der Waals contact form a crucial contemporary frontier for chemical physics, one with many challenges for experimental and theoretical investigation [10].

The effects of intermolecular interactions on the hyperpolarizabilitieare are relatively new and fascinating subjects [11]. Recent experimental work by Donley and Shelton [12] showes that interaction hyperpolarizability is an essential element of the rationalization of nonlinear susceptibility measurements

*Corresponding author: Professor Li Zhi-Ru, Tel.: +86 431 8498964; Fax: +86 431 8945942; E-mail: LZR@mail.jlu.edu.cn.

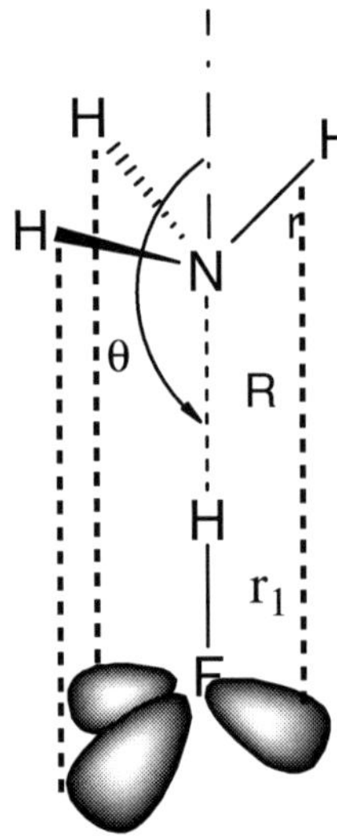

Fig. 1. The structural parameters of the FH-NH_3 with a equivalent triple π-type hydrogen bond

in liquids and solids. Quite recently, using the hyper-Rayleigh scattering method, Yokoyama and coworkers [13] reported that the intermolecular coupling enhanced of the molecular first hyperpolarizability. Theoretical researches of interaction hyperpolarizability have been reported by Hunt[14], Buckingham et al. [15], and recently by Bancewicz [16]. Maroulis has reported, quite recently, computational investigations for a number of small van der Waals systems as $H_2 \ldots H_2$, Ne. . .Ne, Ar. . .Ar, Kr. . .Kr, Ne. . .FH and FH. . .Ne [18], $(H_2O)_2$ [17,18]. The interaction hyperpolarizabilities of Ar-HF have been calculated in our group [19].

In our previous work [20], we have shown that not only a primary hydrogen-bond N. . .H–F (σ– type hydrogen bond) exists but also a equivalent triple π-type hydrogen bond exists in the hydrogen bond dimmer H_3N- HF. The π-type hydrogen-bond interaction is secondary and is an attraction between three H atoms of the H_3N and three lone pairs on the F atom.

In the contemporary frontier of chemical physics, we focus on the calculation of the static first hyperpolarizability (β_0) and the study of their related properties (α_0, μ_0) of the hydrogen bond binary cluster NH_3-HF with a equivalent triple π-type H-bond. Of course, this research for the NH_3-HF, will reveal the contributions of above properties from the intermolecular interaction. The research will also discuss the computational method including the several factors such as atomic basis set and bond function, electron correlation and CP procedure for calculating interaction hyperpolarizability.

2. Computational methods

The equilibrium structure of NH_3-HF is calculated at the QCISD/aug-cc-pVDZ level in the present work. Figure 1 shows the intermolecular coordinates of NH_3-HF. The structural parameters of the clusters are shown in Table 1.

The static response properties of a van der Waals cluster can be defined by expanding the field–dependent energy $E(F)$ as a series in the components of an arbitrary uniform electric field F,

$$E(F) = E(0) - \sum_i \mu_i F_i - \frac{1}{2}\sum_{ij} \alpha_{ij} F_i F_j - \frac{1}{6}\sum_{ijk} \beta_{ijk} F_i F_j F_k - \Lambda, \quad (1)$$

Table 1
The structural parameters of NH_3-HF at the QCISD/aug-cc-pVDZ+BF level

	r(N-H)	R	r_1(F-H)	θ	∠HNH	reference
HF			0.917			[6,34]
NH_3	1.0124				106.68°	[6,34]
NH_3-HF	1.0211	1.7202	0.9521	180°	106.56°	
		1.74				[35]

Distances in Å, angles in deg.

Table 2
Basis set of the bond function[a]

Location	Basis function
At midbond:	
$R_{BF} = \frac{1}{2}R$	3s ($\alpha_s = 0.9, 0.3, 0.1$)
	3p ($\alpha_p = 0.9, 0.3, 0.1$)
	2d ($\alpha_d = 0.6, 0.2$)

[a]Ref. [14,23].

where the tensors μ, α, and β are the dipole moment, polarizability and first hyperpolarizability, respectively.

For the NH_3-HF system, the augmented correlation consistent basis sets (aug-cc-pVNZ(N = D, T)) [21] taken from Gaussian 98 package are used. For an adequate description of the diffuse region of the cluster' wave function, the bond functions (BF) are used for the cluster. The BF are shown in Table 2 [22,23].

It is well known that the reliability of the intermolecular interaction energy calculations depends on the condition of basis set consistency [23–25]. The counterpoise (CP) method [26], which has usually been considered for the correction of the basis set superposition error (BSSE), is then generally employed. For the effects of intermolecular interaction in physical properties, Scheiner et al. [27–29] have applied CP method to the calculations of dipole moments and polarizibilities of the dimers $(HF)_2$ and $(H_2O)_2$, and Maroulis [11,17] has used CP method to the calculations of interaction hyperpolarizabilities of van der Waals systems. Recently, Salvador et al. [30,31] used CP method to study geometries and electron densities of some complexes. In order to determine accurately the contribution (ΔQ) of the intermolecular interaction in the physical property Q, the CP procedure is applied throughout this work and the contribution (ΔQ) of the intermolecular interaction is defined as follows:

$$\Delta Q = Q_{AB}(\chi_{AB}) - Q_A(\chi_{AB}) - Q_B(\chi_{AB}), \quad (2)$$

where $Q = \mu_0, \alpha_0, \beta_0$. In Eq. (2), we use the same basis set, χ_{AB}, for the subsystems as for the van der Waals cluster. When CP is not applied, ΔQ is defined as follows,

$$\Delta Q = Q_{AB}(\chi_{AB}) - Q_A(\chi_A) - Q_B(\chi_B). \quad (3)$$

To calculate the contribution rate (η) of the intermolecular interaction of a physical property Q in a van der Waals cluster, η is defined as follows:

$$\eta = \Delta Q / Q_{AB} \quad Q = \mu_0, \alpha_0, \beta_0 \quad (4)$$

3. Results and discussions

Both with CP and without CP, the static response properties of NH_3-HF at the MP2 level are calculated with different basis sets and their results are listed in Tables 3 and 4, respectively.

Table 3
Calculated values of μ_0, α_0 and β_0 for NH3-HF and its subsystems and by MP2 with CP

	Basis set	aug-cc-pVTZ+BF	aug-cc-pVTZ	aug-cc-pVDZ+BF	aug-cc-pVDZ
	N	206	184	104	82
	NH_3-HF	1.81092	1.80885	1.81724	1.80950
μ_0	NH_3	0.60311	0.60093	0.60730	0.60138
	HF	0.73161	0.73224	0.73313	0.73194
	$\Delta\mu_0$	0.4762	0.47568	0.47681	0.47618
	η	26.3%	26.3%	26.2%	26.3%
	NH_3-HF	19.30732	19.28215	18.99385	18.77118
α_0	NH_3	14.46280	14.45321	14.46454	14.31747
	HF	5.65022	5.63506	5.33946	5.19864
	$\Delta\alpha_0$	−0.8057	−0.80612	−0.81015	−0.74493
	η	−4.2%	−4.2%	−4.3%	−4.0%
	NH_3-HF	31.76	31.57	31.62	30.15
β_0	NH_3	20.50	20.53	18.78	16.38733
	HF	7.95	7.8765	8.97	7.63271
	$\Delta\beta_0$	3.31	3.16	3.87	6.13
	η	10.4%	10.0%	12.2%	20.3%

Firstly, we discuss the dipole moment. In the NH_3-HF cluster, the bond length of the HF subsystem increases 0.035 Å compared with that of the HF molecule. The calculated value of μ_0 of the HF subsystem is 0.023 a.u. larger than the experimental value (0.708 a.u.) of the HF molecule [27]. The contribution ($\Delta\mu_0$) of the intermolecular interaction (hydrogen bond) for the dipole moment is equal to 0.476 a.u., about 26% of μ_0 of the NH_3-HF cluster. So its contribution is very large. From Tables 3 and 4 we know that the application of CP is insignificant to the study of $\Delta\mu_0$ of the NH_3-HF cluster when the basis sets are not less than aug-cc-pVDZ.

Secondly, we are concerned with the polarizability (α_0) of the cluster. Table 1 shows that the structures of the NH_3 subsystem and isolated NH_3 molecule are similar. The α_0 values of the NH_3 subsystem in NH_3-HF calculated by different basis sets, i.e., aug-cc-pVNZ (N = D, T) and/or BF, are closer to the experimental value (14.56 a.u.) of the NH_3 molecule [32]. From Tables 3 and 4, the contributions ($\Delta\alpha_0$) of the intermolecular interaction in α_0 are negative for NH_3-HF. The values of $\Delta\alpha_0$ are about −4.2% of total with CP and −3.6% of total α_0 without CP. It is clear that the CP procedure has an obvious effect on the contribution of the intermolecular interaction in α_0. Lastly, we focus on the static hyperpolarizability, as shown in Tables 3 and 4. The β_0 does not change with the basis sets aug-cc-pVTZ+BF, aug-cc-pVTZ, and aug-cc-pVDZ+BF, as the basis sets are large enough. The β_0 of NH_3-HF is about 31.76 a.u. The contribution ($\Delta\beta_0$) of the hydrogen bond interaction of the β_0 is just about $3 \sim 4$ a.u.. For the application of CP procedure, using aug-cc-pVTZ and aug-cc-pVTZ+BF, the value of η is about 10% with CP, and 6.3% and 11.5% without CP. The discussion about $\Delta\beta_0$ and its η shows that with CP, $\Delta\beta_0$ and η remain almost unchanged with the basis sets above, namely, aug-cc-pVTZ+BF and aug-cc-pVTZ, but without CP, $\Delta\beta_0$ and η are changeable. Therefore, the application of counterpoise procedure to remove the BSSE is significant in the study of the intermolecular interaction contribution of the first hyperpolarizability for van der Waals clusters.

Scheiner [27] pointed out that for the calculations of dipole moments and polarizibilities of the dimers $(HF)_2$ and $(H_2O)_2$, the results of a subsystem depend on the partner orbitals and the effects of secondary BSSE [28,29]. Most of these problems (to correct secondary BSSE) can be avoided if basis set is large enough to begin with [27]. In our study, it is a trivial problem, because the basis sets used are large enough. For example, in NH_3-HF, the β_0 of NH_3 is 20.50 (with CP) and 20.01a.u. (without CP); the β_0 of HF is 7.95 (With CP) and 8.05 a.u. (without CP).

Table 4
Calculated values of μ_0, α_0 and β_0 for NH_3-HF and its subsystems by MP2 without CP

	Basis set	aug-cc-pVTZ+BF	aug-cc-pVTZ	aug-cc-pVDZ+BF	aug-cc-pVDZ
	NH_3-HF	1.81092	1.80885	1.81724	1.80950
μ_0	NH_3	0.603133	0.599431	0.60684	0.60384
	HF	0.73204	0.73188	0.732249	0.73151
	$\Delta\mu_0$	0.4757	0.4775	0.4782	0.4742
	η	26.3%	26.4%	25.4%	26.2%
	NH_3-HF	19.30732	19.28215	18.99385	18.77118
α_0	NH_3	14.42557	14.38428	14.37728	14.06929
	HF	5.60525	5.59745	5.25000	5.13254
	$\Delta\alpha_0$	−0.7235	−0.7002	−0.6334	−0.4300
	η	−3.7%	−3.6%	−3.3%	−2.3%
	NH_3-HF	31.76	31.57	31.62	30.15
β_0	NH_3	20.01	21.48	17.98	14.62
	HF	8.09	8.09	9.15	8.17
	$\Delta\beta_0$	3.66	2.0	4.49	7.36
	η	11.5%	6.3%	14.2%	24.4%

In this work, the BF has shown obvious efficiency to improve the calculated values of μ_0, α_0 and β_0 for NH_3-HF. For instance, the β_0 value (31.62 au) of the aug-cc-pVDZ+BF ($N = 104$) is similar to the value (31.57 au) of the aug-cc-pVTZ ($N = 184$) (see Table 3) . The BF also greatly increases the efficiency of the basis set on the calculations of the effects of intermolecular interaction in physical properties, while CP is used to remove the BSSE (see Table 3, $\Delta\beta_0$). Without CP, the BSSE will disturb the good computational values and the efficiency of BF will disappear (see Table 4, $\Delta\beta_0$).

4. Summary and remarks

For NH_3-HF with a equivalent triple π-type hydrogen bond, the contributions of the intermolecular interaction in the first hyperpolarizability have been studied at the MP2 level with CP by several different basis sets. The obtained results are converging with the size increasing of the basis set. In the NH_3-HF, the contribution of β_0 from the intermolecular interaction (hydrogen bond) is about 10% ($\Delta\beta_0 = 3.16$ a.u.) which should include that from a equivalent triple π-type hydrogen bond . Our calculations show that the counterpoise (CP) procedure is necessary and BF is effectual to improve calculation in the study of interaction NLO property ($\Delta\beta_0$) of the H-bond cluster.

Acknowledgements

This work was supported by the National Natural Science Foundation of China (No.20723024, 20173210, 29873016).

References

[1] J. Eyss and I. Ledoux, *Chem. Rev.* **94** (1994), 77.
[2] D.M. Burland, *Chem. Rev.* **94** (1994), 7.
[3] D.P. Shelton and J.E. Rice, *Chem. Rev.* **94** (1994), 3.
[4] B.L. Hannond and J.E. Rice, *J. Chem. Phys.* **97** (1992), 1138.

[5] J. Guan, P. Duffy, J.T. Carter, D.P. Chong, K.C. Casida, M.E. Casida and H. Wrinn, *J. Chem. Phys.* **98** (1993), 4753.
[6] H. Sekino and R.J. Bartlett, *J. Chem. Phys.* **98** (1993), 3022.
[7] O. Christiansen, C. Hattig and J. Gauss, *J. Chem. Phys.* **109** (1998), 4746.
[8] B. Kirtman, J.M. Luis and B.M. Bishop, *J. Chem. Phys.* **108** (1998), 10008.
[9] G. Chalasinski and M.M. Szczesniak, *Chem. Rev.* **94** (1994), 1723.
[10] C.E. Dykstra and J.M. Lisy, *J. Mol. Struct.* (*Theochem*) **500** (2000), 375.
[11] G. Maroulis, *J. Phys. Chem.* **A 104** (2000), 4772.
[12] E.A. Donley and D.P. Shelton, *Chem. Phys. Lett.* **215** (1993), 156.
[13] S. Yokoyama, T. Nakahama, A. Otomo and S. mashiko, *J. Am. Chem. Soc.* **122** (2000), 3174.
[14] K.L.C. Hunt, *Chem. Phys. Lett.* **70** (1980), 336.
[15] A.D. Buckingham, R.P. Concannon and I.D. Hands, *J. Phys. Chem.* **98** (1994), 10455.
[16] T.J. Bancewicz, *J. Chem. Phys.* **111** (1999), 7440.
[17] G. Maroulis, *J. Chem. Phys.* **113** (2000), 1813.
[18] K.S. Kim, B.J. Mhin, U.-S. Choi and K. Lee, *J. Chem. Phys* **97** (1992), 6649.
[19] B.-Q. Wang, Z.-R. Li, D. Wu and C.-C. Sun, *J. Mol. Struct.* (*Theochem*) **620** (2003), 77–86.
[20] Z.-R. Li, D. Wu, Z.-S. Li, X.-R. Huang, F.-M.- Tao and C.-C. Sun, *J. Phys. Chem. A* **105** (2001), 1163-1168.
[21] S.S. Xantheas, T.H. Dunning, Jr., *J. Chem. Phys.* **99** (1993), 8774.
[22] Z. Li, A. Chou and F.-M. Tao, *Chem. Phys. Lett.* **313** (1999), 313.
[23] F.-M. Tao, *J. Chem. Phys.* **98** (1993), 2481.
[24] F.-M. Tao, *J. Chem. Phys.* **98** (1993), 3049.
[25] F.-M. Tao, *J. Mol. Struct.* **367** (1996), 55.
[26] S.F. Boys and F. Bernardi, *Mol. Phys.* **19** (1970), 553.
[27] S. Scheiner, *Ann. Rev. Phys. Chem.* **45** (1994), 23.
[28] G. Karlstrom and A.J. Sadle, *Theor. Chim. Acta.* **61** (1982), 1.
[29] M.M. Szczesniak and S. Scheiner, *Chem. Phys. Lett.* **131** (1986), 230.
[30] P. Salvador, M. Duran and J.J. Dannenberg, *J. Phys. Chem. A* **106** (2002), 6883.
[31] P. Salvador, M. Duran and X. Fradera, *J. Chem. Phys.* **116** (2002), 6443.
[32] K.P. Huber, G. Herzberg, *Molecular Spectra and Molecular Structure*, Vol. 4, Van Nostrand Reinhold, New York, 1979.
[33] A. Dkhissi, M.E. Alikhani and Y. Bouteiller, *J. Mol. Struct.* **416** (1997), 1.
[34] H. Weber, *J. Mol. Spectrosc.* **107** (1984), 405.

Journal of Computational Methods in Sciences and Engineering 4 (2004) 307–320
IOS Press

Reliable results for the Isotropic Dipole – Dipole and Triple – Dipole Dispersion Energy Coefficients for Interactions involving Formaldehyde, Acetaldehyde, Acetone, and Mono - , Di - , and Tri - Methylamine

Ashok Kumar[a,b] and William J. Meath[b,*]
[a]*Department of Physics, Ch. Charan Singh University, Meerut, 250004, India*
[b]*Department of Chemistry, University of Western Ontario, London, Ontario, N6A 5B7, Canada*

Abstract. Pseudo – spectral dipole oscillator strengths and excitation energies, which are discrete representations of previously developed recommended continuous dipole oscillator strength distributions(DOSDs), are presented for the ground state formaldehyde, acetaldehyde, acetone, and mono - , di - , and tri – methylamine molecules. These pseudo – DOSDs, together with previously published pseudo – DOSDs for other atoms and molecules, are used to evaluate the dipole – dipole and the triple – dipole dispersion energy coefficients for all the two – body and three - body interactions between H_2CO, CH_3CHO, $(CH_3)_2CO$, CH_3NH_2, $(CH_3)_2$ NH, and $(CH_3)_3N$, and between these molecules and forty – four other species, namely Cl_2, SiH_4, SiF_4, CCl_4, H, Li, He, Ne, Ar, Kr, Xe, SF6, HF, HCl, HBr, SO_2, CS_2, OCS, H_2, N_2, O_2, NO, N_2O, H_2O, H_2S, NH_3, CO, CO_2, the normal alkanes CH_4, C_2H_6, C_3H_8, C_4H_{10}, C_5H_{12}, C_6H_{14}, C_7H_{16} and C_8H_{18}, the 1-alkenes C_2H_4, C_3H_6 and C_4H_8, C_2H_2, C_6H_6, and the primary alcohols CH_3OH, C_2H_5OH and C_3H_7OH. Results are presented explicitly for all the dipole – dipole dispersion energy coefficients and for the triple – dipole coefficients for all three -body interactions involving the H_2CO, CH_3CHO, $(CH_3)_2CO$, CH_3NH_2, $(CH_3)_2$ NH, and $(CH_3)_3N$ molecules. The estimated errors in the two - body and three – coefficients are 1% and 1–2%

Keywords: (dipolar) dispersion energies, additive and non – additive interactions, amines, aldehydes, ketones, long range molecular interactions

PACS: 31.15.Md, 31.25.Qm, 31.50.Bc, 32.70.Cs, 34.20.Gj

1. Introduction

The main purpose of this paper is to present discretized representations of previously developed recommended continuous dipole oscillator strength distributions(DOSDs) for the ground state formaldehyde, acetaldehyde, acetone, and mono - , di - , and tri – methylamine molecules and to use these pseudo – DOSDs to evaluate the dipole – dipole dispersion energy coefficients [1,2] for all the two – body interactions between these molecules, and between these molecules and forty – four other species. Results

*Corresponding author. E-mail: wmeath@uwo.ca.

1472-7978/04/$17.00

have also been obtained for the triple – dipole dispersion energy coefficients [1,3,4] for the three – body interactions for all fifty species but these are presented explicitly only for interactions involving the H_2CO, CH_3CHO, $(CH_3)_2CO$, CH_3NH_2, $(CH_3)_2$ NH, and $(CH_3)_3N$ molecules.

The construction [5,6] of the original DOSDs for H_2CO, CH_3CHO, $(CH_3)_2CO$, CH_3NH_2, $(CH_3)_2$ NH, and $(CH_3)_3N$ is summarized in Section 2 which also contains the development of the analogous pseudo – DOSDs. The psuedo – DOSDs are particularly useful [7–9] for evaluating many – fold DOSD integrals over molecular excitation energies such as occur in the expressions for the dispersion energies and in particular for the triple – dipole dispersion energy [8]. The results for the dipole – dipole(C_6) and the triple – dipole(C_9) dispersion energy coefficients are given in Section 3 while Section 4 contains a brief discussion of the importance of these results, together with estimates of their uncertainties, and of the psuedo – DOSDs and the original DOSDs that were used to evaluate them. Generally the results for the dipole – dipole and the triple – dipole dispersion energy coefficients obtained from the DOSD approach used here, are the only values available for the molecules and atoms considered in this paper.

2. Psuedo – DOSDs for Formaldehyde, Acetaldehyde, Acetone, and Mono -, Di -, and Tri - Methylamine

The constrained least squares procedure used to construct the original recommended continuous dipole oscillator strength distributions(DOSDs) for the H_2CO, CH_3CHO, $(CH_3)_2CO$, CH_3NH_2, $(CH_3)_2$ NH, and $(CH_3)_3N$ molecules has been discussed elsewhere in considerable detail [10–12]. The DOSD for a molecule is the molecular differential dipole oscillator strength (df/dE) as a function of excitation energy E from the electronic absorption threshold E_0 for the molecule to very high photon energies.

Generally the dipole properties and dispersion energy coefficients evaluated using DOSDs are isotropic results since only orientationally averaged input dipole oscillator strengths are available over the required wide range of photon energies. In some instances, i.e. CO, H_2, N_2, NO and O_2 [13–19], sufficient anisotropic constraints are available to permit the construction of anisotropic DOSDs and hence the evaluation of anisotropic molecular properties and dispersion enrergies.

In general, and briefly, the available experimental and theoretical dipole oscillator strengths(DOSs) for a given molecule are constrained to satisfy the Thomas – Reiche -Kuhn sum rule [20,21] and to reproduce experimental refractivity data, as a function of wavelength, for the relevant dilute molecular gases. The sum rule states that the area under the plot of (df/dE) versus E is equal to the number of electrons in the molecule of interest. The molar refractivity, which is related to the molecular refractive index, is proportional to the dynamic polarizability of the molecule which is determined, quantum mechanically, by an integral involving the DOSD [1,10,20,22]. These constraints are very important in ensuring that the constructed DOSDs are capable of yielding reliable results for the dispersion energy coefficients of interest in this paper.

For a given molecule the initial DOS data is divided, from the UV absorption threshold to very large values of the photon energy, into N_0 energy intervals as suggested by the structure of the input DOS data and by the photon energy regions associated with the individual sources of the DOS input data. The number of different initial DOSDs that can be obtained by taking all possible combinations of the input DOS data is given by [23,24] $N_D = \Pi_{j=1}^{N_0} N_j$ where N_j is the number of independent sources of DOS data used for the j-th input spectral region. For the molecules considered here N_D is relatively small since the sources of DOS data for these molecules is relatively low [5,6]; in general N_D can be very large (on the order of millions), see for example [25,26]. Each of the initial DOSDs considered is modified by requiring satisfaction of the constraints via application of the constrained least squares technique [10,12]. Briefly, the input initial oscillator strength data base $(df/dE)_{\text{initial}}$ versus E, for each spectral region i, is modified via

$$(df/dE)^i_{constrained} = (1 + a_i)(df/dE)^i_{initial}, \quad i = 1, 2, \ldots\ldots N_0 \tag{1}$$

so that the total constrained DOSD satisfies the imposed constraints through the choice of the a_i. The degree of modification of the initial DOSD data required to satisfy the constraints can be represented by the standard deviation(STD) defined by [11,23,24]

$$\text{STD} = \left[\sum_{i=1}^{N_0} (a_i - \overline{a})^2 / N_0\right]^{1/2} \tag{2}$$

where $\overline{a}$ is the average of all the a_i values. The recommended DOSD for a given molecule generally corresponds to the lowest value of STD. The recommended DOSD is represented by a set of data points E_j and $(df/dE)_j$, for $j = 1, 2, N_p \gg N_0$, and a set of interpolating functions to connect the points. For very high photon energies, $E \geqslant 10^7$ eV, (df/dE) is represented by the Born dipole formula [21] $\text{AE}^{-2.5}$. The dipole properties of molecules are readily evaluated using such representations of DOSDs [10]. Details more specific to the six molecules of relevance to this Section can be found in [5,6].

2.1. *Formaldehyde, acetaldehyde, acetone*

The high and low resolution experimental dipole oscillator strength data of Cooper et al. [27,28] were used to help construct the DOSDs for H_2CO, CH_3CHO and $(CH_3)_2CO$ for $E \leqslant 200$ eV. This data was augmented, particularly for $E \geqslant 200$ eV, by various mixture rules which were based on DOS data associated with previous constrained DOSD work on the H_2 and O_2 [10], CO and CO_2 [29], CH_4 [30] and the C_2H_6 and C_3H_8 [31] molecules. The molar refractivity constraints for acetone and acetaldehyde were developed from the extensive experimental molar refractivity data of Ramaswamy [32], Lowery [33, 34], Prytz [35] and Zahn [36]. For formaldehyde there is apparently no experimental refractivity data and the constraints were developed by using a scaling technique based on the acetone and acetaldehyde results – full details are in [5].

2.2. *Mono - , Di - , and Tri – methylamine*

The low and high resolution experimental DOS data of Burton et al. [6], available for $E \leqslant 250$ eV, were used to help develop constrained DOSDs for CH_3NH_2, $(CH_3)_2$ NH, and $(CH_3)_3N$. This was augmented, particularly for higher photon energies, by using various mixture rules constructed by employing previous constrained DOSDs for the H_2 [10], CH_4 [30], $C_{n+1}H_{2n+4}$ (n = 1,2,3) [31], and NH_3 [37] molecules. The molar refractivity constraints for CH_3NH_2 were provided by the reliable experimental data of Ramaswamy [32]. Scaling techniques [5,6], analogous to those used for formaldehyde, were used to develop the refractivity constraints for $(CH_3)_2$ NH, and $(CH_3)_3N$.

2.3. *Psuedo – DOSDs for H_2CO, CH_3CHO, $(CH_3)_2CO$, CH_3NH_2, $(CH_3)_2$ NH, $(CH_3)_3N$*

Following [7,8] the psuedo – DOSDs for a given atom or molecule are determined from the known values of the dipole sums S_k,

$$S_k = \int_{E_0}^{\infty} (E/E_H)^k (df/dE) dE, \tag{3}$$

Table 1
The values[1] of the pseudo-DOSD excitation energies (in units of E_H) and oscillator strengths for ground state H_2CO, CH_3CHO, $(CH_3)_2CO$

H_2CO		CH_3CHO		$(CH_3)_2CO$	
E_i	f_i	E_i	f_i	E_i	f_i
1.37195(-1)	7.27743(-5)	1.41375(-1)	5.13364(-5)	1.46106(-1)	4.92853(-5)
1.51410(-1)	1.86735(-4)	1.57829(-1)	2.47036(-4)	1.63292(-1)	2.76664(-4)
2.42026(-1)	9.69714(-3)	2.49991(-1)	2.66413(-2)	2.37559(-1)	3.07159(-2)
2.95484(-1)	1.86000(-1)	3.01622(-1)	3.11933(-1)	3.14574(-1)	4.53423(-1)
4.13010(-1)	8.35225(-1)	4.24490(-1)	1.44487	4.34772(-1)	2.22793
6.16709(-1)	2.95961	6.29619(-1)	4.86097	6.42202(-1)	7.14847
1.05708	4.20819	1.06728	6.10894	1.07960	8.31133
2.48478	3.45087	2.53001	4.60345	2.51713	5.59662
1.84503(1)	4.13892	1.77178(1)	6.35515	1.73431(1)	7.90016
2.42916(2)	2.11225(-1)	2.33459(2)	2.87738(-1)	2.32232(2)	3.31021(-1)

[1]Number in parentheses indicate powers of 10.

Table 2
The values[1] of the pseudo-DOSD excitation energies (in units of E_H) and oscillator strengths for ground state CH_3NH_2, $(CH_3)_2NH$ and $(CH_3)_3N$

CH_3NH_2		$(CH_3)_2NH$		$(CH_3)_3N$	
E_i	f_i	E_i	f_i	E_i	f_i
1.95777(-1)	2.88155(-3)	1.92222(-1)	8.52896(-4)	1.83632(-1)	3.16411(-3)
2.14935(-1)	1.42215(-2)	2.23120(-1)	3.37547(-2)	2.05665(-1)	3.15206(-2)
2.60088(-1)	7.26187(-2)	2.56132(-1)	9.35809(-2)	2.40604(-1)	1.36726(-1)
3.57326(-1)	4.28264(-1)	3.51772(-1)	5.94824(-1)	3.26752(-1)	5.24864(-1)
4.80741(-1)	2.12175	4.80410(-1)	3.00602	4.55557(-1)	3.04622
7.02189(-1)	4.50960	6.98044(-1)	6.70805	6.65913(-1)	8.65833
1.18165	4.23660	1.16552	6.14808	1.11106	8.46368
2.74460	2.37369	2.72701	3.22470	2.56891	4.59669
1.86953(1)	4.09247	1.80813(1)	5.98209	1.73019(1)	8.25323
2.35629(2)	1.47893(-1)	2.29964(2)	2.08035(-1)	2.19671(2)	2.85576(-1)

[1]Number in parentheses indicate powers of 10.

which have been evaluated accurately using the original recommended continuous DOSD for the species of interest; here $E_H \approx 4.35975 \times 10^{-18}$ J ≈ 27.21 eV is the Hartree of energy. For the molecules of interest here, the S_k have been determined [5,6] for $k = 2, 1, 0, -1, -2, \ldots 3 - 2n$, with $n \leqslant 20$. The $2n$ sums can then be used to generate n pseudo – dipole (excitation energy – oscillator strength) pairs (E_i, f_i) such that

$$S_k = \sum_{i=1}^{n} (E_i/E_H)^k f_i. \tag{4}$$

It has been convincingly demonstrated [7–9] that ten pseudo states (as defined by a (E_i, f_i) pair) are sufficient to evaluate the dipole – dipole and the triple – dipole dispersion energy coefficients to well within the accuracy inherent in using the original recommended DOSDs. Tables 1 and 2 contain the $n = 10$ pseudo – states for H_2CO, CH_3CHO and $(CH_3)_2CO$, and CH_3NH_2, $(CH_3)_2$ NH, and $(CH_3)_3N$, respectively.

3. Dipole – Dipole and Triple – Dipole Dispersion Energy Coefficients

The dipole – dipole dispersion energy is the dominant interaction energy at long range for interactions of spherically symmetric systems including freely tumbling molecules. The orientation averaged dipole – dipole dispersion energy [1,2] for the interaction of molecules A and B is given by $-C_6(\mathrm{A,B})R_{ab}^{-6}$ where R_{ab} is the distance between species A and B and $C_6(\mathrm{A, B})$ is the dipole – dipole dispersion energy coefficient for the interaction. In terms of the original continuous DOSDs for the interacting species, the expression for the dispersion energy coefficient is [1,2,22],

$$C_6 = E_H a_0^6 (3/2) \int\limits_{E_0(A)}^{\infty} \int\limits_{E_0(B)}^{\infty} dE(A) dE(B) \frac{[df(A)/dE(A)][df(B)/dE(B)](E_H)^3}{E(A)E(B)[E(A)+E(B)]} \tag{5}$$

where $E(A)$, $df(A)/dE(A)$, and $E_0(A)$ are the excitation energy, the differential dipole oscillator strength, and the electronic absorption threshold for molecule A respectively. The discretized version of this result, useable for the evaluation of C_6 by employing the pseudo – DOSDs of the interacting species, is given by [1,7]

$$C_6(A,B) = E_H a_0^6 (3/2) \sum_{i=1}^{n_A} \sum_{j=1}^{n_B} \frac{f_i(A) f_i(B) (E_H)^3}{E_i(A) E_j(B) [E_i(A) + E_j(B)]}. \tag{6}$$

Analogously, the most important non – additive long range interaction energy for the interaction of S – state atoms or orientationally averaged(spherically symmetric) molecules A, B, and C is the triple – dipole dispersion energy [1,3,4] which is given by

$$W^{(3)} = (3\cos\theta_a \cos\theta_b \cos\theta_c + 1) C_9(A,B,C) R_{ab}^{-3} R_{bc}^{-3} R_{ac}^{-3} \tag{7}$$

where θ_a is the angle between R_{ab} and R_{ac}. In terms of the original continuous DOSDs for the interacting molecules A, B and C, the triple – dipole dispersion energy coefficient C_9 is given by [1,3,4]

$$C_9(A,B,C) = E_H a_0^9 (3/2) \int\limits_{E_0(A)}^{\infty} \int\limits_{E_0(B)}^{\infty} \int\limits_{E_0(C)}^{\infty} dE(A) dE(B) dE(C) \times \frac{[df(A)/dE(A)][df(B)/dE(B)][df(C)/dE(C)][E(A)+E(B)+E(C)](E_H)^5}{E(A)E(B)E(C)[E(A)+E(B)][E(B)+E(C)][E(C)+E(A)]} \tag{8}$$

The corresponding pseudo – DOSD expression for this coefficient is [1,8]

$$C_9(A,B,C) = E_H a_0^9 (3/2) \sum_{i=1}^{n_A} \sum_{j=1}^{n_B} \sum_{k=1}^{n_C} f_i(A) f_j(B) f_k(C) \times \frac{[E_i(A) + E_j(B) + E_k(C)](E_H)^5}{E_i(A)E_j(B)E_k(C)[E_i(A)+E_j(B)][E_j(B)+E_k(C)][E_k(C)+E_i(A)]}. \tag{9}$$

While it is possible to evaluate the dipole – dipole and the triple – dipole dispersion energy coefficients directly from their defining expressions, Eqs (5) and (8) respectively, by using the original recommended continuous molecular DOSDs, this can be extremely time consuming especially for the triple – dipole

Table 3
Recommended values for the dipole-dipole dispersion energy coefficients $C_6(H_2CO,B)$ for the interaction of H_2CO with various species B (in units of $E_H a_0^6$). (For organic molecules the results are for normal alkanes, 1-alkenes and primary alcohols.)

B	$C_6(H_2CO,B)$	B	$C_6(H_2CO,B)$	B	$C_6(H_2CO,B)$
H_2CO	165.2	HCl	146.7	C_2H_6	251.0
CH_3CHO	257.6	HBr	188.6	C_3H_8	356.0
$(CH_3)_2CO$	362.2	SO_2	220.3	C_4H_{10}	457.5
Cl_2	253.3	CS_2	372.7	C_5H_{12}	560.8
SiH_4	235.5	OCS	256.6	C_6H_{14}	661.4
SiF_4	229.9	H_2	44.57	C_7H_{16}	762.9
CCl_4	577.8	N_2	109.8	C_8H_{18}	863.9
H	32.19	O_2	100.2	C_2H_4	222.4
Li	296.2	NO	106.9	C_3H_6	330.3
He	15.12	N_2O	174.6	C_4H_8	431.6
Ne	30.80	H_2O	86.39	C_2H_2	183.2
Ar	102.9	H_2S	187.8	C_6H_6	532.3
Kr	146.3	NH_3	121.3	CH_3OH	191.1
Xe	216.8	CO	115.9	C_2H_5OH	296.3
SF_6	303.5	CO_2	161.5	C_3H_7OH	401.1
HF	55.32	CH_4	146.3		

Table 4
Recommended values for the dipole-dipole dispersion energy coefficients $C_6(CH_3CHO,B)$ for the interaction of CH_3CHO with various species B (in units of $E_H a_0^6$). (For organic molecules the results are for normal alkanes, 1-alkenes and primary alcohols.)

B	$C_6(CH_3CHO,B)$	B	$C_6(CH_3CHO,B)$	B	$C_6(CH_3CHO,B)$
CH_3CHO	401.7	HBr	294.2	C_3H_8	555.3
$(CH_3)_2CO$	564.9	SO_2	343.5	C_4H_{10}	713.6
Cl_2	395.1	CS_2	581.8	C_5H_{12}	874.6
SiH_4	367.6	OCS	400.3	C_6H_{14}	1032
SiF_4	358.0	H_2	69.54	C_7H_{16}	1190
CCl_4	901.2	N_2	171.0	C_8H_{18}	1347
H	50.25	O_2	156.0	C_2H_4	347.0
Li	463.5	NO	166.6	C_3H_6	515.2
He	23.54	N_2O	272.2	C_4H_8	673.3
Ne	47.91	H_2O	134.6	C_2H_2	285.8
Ar	160.4	H_2S	293.1	C_6H_6	830.4
Kr	228.1	NH_3	189.1	CH_3OH	298.0
Xe	338.2	CO	180.6	C_2H_5OH	462.1
SF_6	472.6	CO_2	251.7	C_3H_7OH	625.4
HF	86.15	CH_4	228.1		
HCl	228.8	C_2H_6	391.5		

coefficient which involves a triple integral over the relevant DOSDs [7,8]. These difficulties can be avoided, without loss of accuracy [7–9], by using the molecular pseudo – DOSDs and the discrete pseudo – spectral expressions for C_6 and C_9 given by Eqs (6) and (9) respectively.

The results for the dipole – dipole dispersion energy coefficients for all two – body interactions between H_2CO, CH_3CHO, $(CH_3)_2CO$, CH_3NH_2, $(CH_3)_2$ NH, and $(CH_3)_3N$, and between these molecules and forty – four other atoms and molecules, are given in Tables 3–8. The triple – dipole coefficients for all three – body interactions involving the H_2CO, CH_3CHO, $(CH_3)_2CO$, CH_3NH_2, $(CH_3)_2$ NH, and $(CH_3)_3N$ molecules are listed in Table 9. The relevant pseudo – states(pseudo – DOSDs) for H_2CO, CH_3CHO, $(CH_3)_2CO$ and CH_3NH_2, $(CH_3)_2$ NH, and $(CH_3)_3N$ are given in Tables 1 and 2 respectively.

Table 5
Recommended values for the dipole-dipole dispersion energy coefficients $C_6((CH_3)_2CO,B)$ for the interaction of $(CH_3)_2CO$ with various species B (in units of $E_H a_0^6$). (For organic molecules the results are for normal alkanes, 1-alkenes and primary alcohols.)

B	$C_6((CH_3)_2CO,B)$	B	$C_6((CH_3)_2CO,B)$	B	$C_6((CH_3)_2CO,B)$
$(CH_3)_2CO$	794.3	HBr	413.7	C_2H_6	550.5
Cl_2	555.5	SO_2	483.0	C_3H_8	780.8
SiH_4	516.7	CS_2	817.7	C_4H_{10}	1003
SiF_4	503.4	OCS	562.8	C_5H_{12}	1230
CCl_4	1267	H_2	97.79	C_6H_{14}	1450
H	70.64	N_2	240.5	C_7H_{16}	1673
Li	649.4	O_2	219.4	C_8H_{18}	1895
He	33.10	NO	234.3	C_2H_4	487.8
Ne	67.36	N_2O	382.7	C_3H_6	724.3
Ar	225.5	H_2O	189.3	C_4H_8	946.6
Kr	320.7	H_2S	412.1	C_2H_2	401.9
Xe	475.4	NH_3	265.9	C_6H_6	1167
SF_6	664.6	CO	254.0	CH_3OH	419.0
HF	121.1	CO_2	353.9	C_2H_5OH	649.8
HCl	321.7	CH_4	320.8	C_3H_7OH	879.4

Table 6
Recommended values for the dipole-dipole dispersion energy coefficients $C_6(CH_3NH_2,B)$ for the interaction of CH_3NH_2 with various species B (in units of $E_H a_0^6$). (For organic molecules the results are for normal alkanes, 1-alkenes and primary alcohols.)

B	$C_6(CH_3NH_2,B)$	B	$C_6(CH_3NH_2,B)$	B	$C_6(CH_3NH_2,B)$
CH_3NH_2	303.7	SF_6	409.8	CH_4	198.4
$(CH_3)_2NH$	443.6	HF	74.73	C_2H_6	340.5
$(CH_3)_3N$	568.2	HCl	199.0	C_3H_8	483.0
H_2CO	224.0	HBr	256.0	C_4H_{10}	620.6
CH_3CHO	349.3	SO_2	298.6	C_5H_{12}	760.6
$(CH_3)_2CO$	491.1	CS_2	506.8	C_6H_{14}	897.1
Cl_2	343.7	OCS	348.4	C_7H_{16}	1035
SiH_4	320.2	H_2	60.52	C_8H_{18}	1172
SiF_4	310.5	N_2	148.6	C_2H_4	301.9
CCl_4	783.9	O_2	135.4	C_3H_6	448.2
H	43.79	NO	144.7	C_4H_8	585.7
Li	405.5	N_2O	236.5	C_2H_2	248.7
He	20.40	H_2O	117.0	C_6H_6	722.4
Ne	41.46	H_2S	255.2	CH_3OH	259.0
Ar	139.3	NH_3	164.4	C_2H_5OH	401.8
Kr	198.3	CO	157.0	C_3H_7OH	543.8
Xe	294.2	CO_2	218.6		

Those for the other species involved in Tables 3–9 are taken from the literature [7,9,11,12,25,26,29, 38–44].

Results for the dipole – dipole dispersion energy coefficients for the interaction of He with H_2CO, CH_3CHO, $(CH_3)_2CO$, CH_3NH_2, (CH3)2NH, and $(CH_3)_3N$, and for the interaction of Ne with H_2CO and $(CH_3)_3N$, have been obtained by Olney et al. [45] – their results differ from our recommended results for these interactions given in Tables 2 to 8 by -3.2%, 2.5%, -0.4%, 0.1%, 3.1%, 4%, -4% and 3% respectively. The error estimates associated with our calculations of dipole – dipole dispersion energy coefficients is generally $\leqslant 1\%$, see Section 4 and references [7,9,22,29].The reasons for the discrepancies are relevant and illustrate the influence of the choice of constraints on the construction of a DOSD and on properties calculated from the DOSD.

Table 7

Recommended values for the dipole-dipole dispersion energy coefficients $C_6((CH_3)_2NH,B)$ for the interaction of $(CH_3)_2NH$ with various species B (in units of $E_H a_0^6$). (For organic molecules the results are for normal alkanes, 1-alkenes and primary alcohols.)

B	$C_6((CH_3)_2NH,B)$	B	$C_6((CH_3)_2NH,B)$	B	$C_6((CH_3)_2NH,B)$
$(CH_3)_2NH$	647.8	HF	109.1	C_2H_6	497.3
$(CH_3)_3N$	829.9	HCl	290.6	C_3H_8	705.3
H_2CO	327.1	HBr	373.9	C_4H_{10}	906.4
CH_3CHO	510.1	SO_2	436.0	C_5H_{12}	1111
$(CH_3)_2CO$	717.2	CS_2	740.1	C_6H_{14}	1310
Cl_2	501.9	OCS	508.8	C_7H_{16}	1511
SiH_4	467.6	H_2	88.39	C_8H_{18}	1711
SiF_4	453.4	N_2	216.9	C_2H_4	440.9
CCl_4	1145	O_2	197.8	C_3H_6	654.5
H	63.96	NO	211.3	C_4H_8	855.3
Li	592.3	N_2O	345.3	C_2H_2	363.2
He	29.79	H_2O	170.8	C_6H_6	1055
Ne	60.52	H_2S	372.8	CH_3OH	378.3
Ar	203.5	NH_3	240.1	C_2H_5OH	586.7
Kr	289.6	CO	229.2	C_3H_7OH	794.2
Xe	429.6	CO_2	319.2		
SF_6	598.4	CH_4	289.8		

Table 8

Recommended values for the dipole-dipole dispersion energy coefficients $C_6((CH_3)_3N,B)$ for the interaction of $(CH_3)_3N$ with various species B (in units of $E_H a_0^6$). (For organic molecules the results are for normal alkanes, 1-alkenes and primary alcohols.)

B	$C_6((CH_3)_3N,B)$	B	$C_6((CH_3)_3N,B)$	B	$C_6((CH_3)_3N,B)$
$(CH_3)_3N$	1063	HF	139.8	CH_4	371.2
H_2CO	419.0	HCl	372.3	C_2H_6	637.0
CH_3CHO	653.4	HBr	479.0	C_3H_8	903.5
$(CH_3)_2CO$	918.8	SO_2	558.6	C_4H_{10}	1161
Cl_2	643.0	CS_2	948.5	C_5H_{12}	1423
SiH_4	599.2	OCS	651.9	C_6H_{14}	1678
SiF_4	580.7	H_2	113.2	C_7H_{16}	1936
CCl_4	1467	N_2	277.9	C_8H_{18}	2192
H	81.95	O_2	253.3	C_2H_4	564.8
Li	761.6	NO	270.6	C_3H_6	838.5
He	38.15	N_2O	442.3	C_4H_8	1096
Ne	77.52	H_2O	218.8	C_2H_2	465.3
Ar	260.6	H_2S	477.6	C_6H_6	1352
Kr	370.9	NH_3	307.6	CH_3OH	484.6
Xe	550.5	CO	293.6	C_2H_5OH	751.6
SF6	766.4	CO_2	408.9	C_3H_7OH	1017

For the six molecules considered here the initial experimental source of DOS input data is the same for excitation energies $E \leqslant 200$ eV for H_2CO, CH_3CHO, $(CH_3)_2CO$ [27,28] and$E \leqslant 250$ eV for the amines [6]; for higher photon energies this data is extended to very high photon energies by mixture rule DOS data in Kumar and Meath [5] and Burton et al. [6], see Section 2, whereas no such extension is used in Olney et al. [45]. To obtain absolute values for the DOSs for the molecules, the initial raw experimental DOS results were normalized [6,27,28] by using the valence shell TRK(VTRK) sum rule [46]. Olney et al. [45] then renormalized the resulting DOS data to reproduce literature experimental (static) polarizability values for the molecules except in the case of formaldehyde where VTRK normalization was retained because of a lack of reliable experimental polarizability results.

Table 9
Recommended values for the triple-dipole dispersion energy coefficients for all the three – body interactions involving A = H_2CO, B = CH_3CHO, C = $(CH_3)_2CO$, D = CH_3NH_2, E = $(CH_3)_2NH$, and F = $(CH_3)_3N$ molecules (in units of $E_H a_0^9$).

Interaction	C_9	Interaction	C_9	Interaction	C_9
A – A – A	2274	B – B – D	7584	D – D – A	4237
A – A – B	3554	B – B – E	11077	D – D – B	6623
A – A – C	4995	B – B – F	14199	D – D – C	9308
A – A – D	3104	B – C – D	10658	D – D – E	8449
A – A – E	4533	B – C – E	15567	D – D – F	10830
A – A – F	5811	B – C – F	19952	D – E – F	15819
A – B – C	7807	B – D – E	9674	E – E – E	18026
A – B – D	4852	B – D – F	12400	E – E – A	9039
A – B – E	7086	B – E – F	18112	E – E – B	14130
A – B – F	9083	C – C – C	24101	E – E – C	19857
A – C – D	6818	C – C – A	10972	E – E – D	12341
A – C – E	9958	C – C – B	17150	E – E – F	23105
A – C – F	12764	C – C – D	14977	F – F – F	37963
A – D – E	6189	C – C – E	21875	F – F – A	14851
A – D – F	7933	C – C – F	28038	F – F – B	23216
A – E – F	11586	C – D – E	13595	F – F – C	32623
B – B – B	8684	C – D – F	17425	F – F – D	20277
B – B – A	5556	C – E – F	25452	F – F – E	29616
B – B – C	12204	D – D – D	5785		

It should be noted that the static polarizabilty, which in atomic units is just the dipole sum defined by Eq. (3) with $k = -2$, is insensitive to the high energy part of the DOSD [23,24,31,47,48] and so, in principle, the lack of high energy DOS data is not crucial for its evaluation assuming the raw DOS data requires only a simple energy dependent scaling – this is not always the case [26]. If the assumption is correct, then the reliability of the scaled DOS results depends on that of the static polarizability constraint.

For the interactions involving He, the cause of the discrepancies in the C_6's is due to the molecular DOSDs – for example the results of Olney et al. [45] for the dipole properties of He that are sensitive to the same parts [23,24,31,47,48] of the DOSD as the dispersion energy coefficient, namely forS_{-1}, S_{-2}, L_{-1}, and L_{-2}, and for C_6 (He – He) itself, are generally in very good agreement with well established values [40,49,50]. The logarithmic dipole sum L_k is defined by [10,20,51]

$$L_k = \int_{E_0}^{\infty} (E/E_H)^k (df/dE) \ln{(E/E_H)} dE \tag{10}$$

and L_{-2} actually occurs in a very reliable approximate expression [7,22,52,53] for C_6 through the average energy [10,20,51] $I_{-2} = E_H \exp[L_{-2}/S_{-2}]$.

For $(CH_3)_2CO$ and CH_3NH_2 the values of S_{-2} corresponding to the constraints used by Olney et al. [45] are just 0.2% lower and the same, respectively, as that used earlier by Kumar and Meath [5] and Burton et al. [6]; the corresponding C_6 values are in close agreement as pointed out earlier. On the other hand for three of the other molecules the polarizability constraints used by Olney et al. [45] are significantly different than the recommended S_{-2} values of references [5,6] and in the case of formaldehyde their use of the VTRK sum rule as a constraint yields a value of S_{-2} that 3.3% lower than the result of Kumar and Meath [5].

For CH_3CHO Olney et al. [45] based their constraint on the experimental polarizability result of 31.0 quoted by Maryott and Buckley [54], which is due to Zahn [36]. This result is about 2.5% higher than the result of $S_{-2} = 30.25$ obtained in [5]. However the Zahn result is not the static polarizability – rather it corresponds to a wavelength of 5893 Å. In our earlier work [5] we used Zahn's value at 5893 Å as one of the molar refractivity constraints in the construction of our recommended DOSD for this molecule which yielded the value of 30.25 for S_{-2}. For the $(CH_3)_2NH$ molecule the average of the polarizabilities quoted by Stuart [55], due to Fuch and Wolf [56], and by Marryot and Buckley [54] is used as a constraint to normalize the DOS data in reference [45]. The result given in [54] is obtained using atomic additivity and furthermore corresponds to a wavelength of 5893 Å and therefore should not be averaged with the static polarizabilty of Fuch and Wolf [56]. Also the polarizabilies of Fuch and Wolf [56] for NH_3 and CH_3NH_2 have been shown [6,37] to be too high by about 1% – their result for $(CH_3)_2NH$ apparently is too (their result of 39.06 is 0.9% higher than the recommended value of 38.7 due to Burton et al. [6]). Finally the constraint used in [45] for $(CH_3)_3N$ is the polarizability at a wavelength of 5893 Å (obtained using atomic additivity) [54] and therefore is not the static polarizability– it is 4% higher than the recommended result of $S_{-2} = 49.9$ due to Burton et al. [6].

For the two interactions involving Ne, the discrepancies in the C_6 values of this work and Olney et al. [45] also involve the Ne DOSDs used in the calculations. The results for $S_{-1}, S_{-2}, L_{-1}, L_{-2}$ and I_{-2} from reference [45] versus those calculated from the recommended constrained DOSD of Kumar and Meath [23], which lead to the pseudo - DOSD for Ne used in the present calculations, are 3.60, 2.66, 1.707, 0.399 and 31.6 eV versus 3.80, 2.67 ,2.16, 0.474 and 32.49 eV. These dipole properties probe the regions of the DOSD that are particularly important for the evaluation of C_6 [23,24,31,47,48]. While the results for the static polarizability agree well, the discrepancies in the other properties suggest there are probably significant differences in the DOSs of Olney et al. [45] and Kumar and Meath [23] regarding the calculation of C_6; the result C_6 (Ne – Ne) $= 6.18\ E_H a_0^6$ of Olney et al. [45] is 3.2% lower than the recommended value of 6.38 $E_H a_0^6$ [38]. The discrepancies given earlier between our results for the dipole – dipole dispersion energy coefficients and those of Olney et al. [45] are due to the combined effects of the differences in the DOSDs used in the calculations for the relevant excitation energy regions. In general the use of both refractivity data, which also ensures reproduction of the static polarizability, and Thomas- Reiche – Kuhn sum rule as constraints in the construction of DOSDs, as for example in the case of references [5,6,10–12,23–26], furnishes energy dependent scaling of input DOS data that is not available if only the static polarizability is used as a constraint as in the case of reference [45].

Spackman [57] has carried out time – dependent Hartree – Fock (TDHF) calculations of C_6 and C_9 for the dimer and trimer of formaldehyde using a 6-31G(+ sd + sp) basis set. His results of 121.3 and 1407 are ≈26% and 38% lower than our recommended values of 165.2 and 2274 respectively. Spackman [57] has investigated the behaviour of TDHF calculations for C_6 and C_9 relative to reliable results evaluated using our constrained DOSD approach(references cited in [57]) for some twenty-five molecules and has proposed a scaling procedure to correct the TDHF calculations which are systematically considerably lower that the DOSD results. His scaled values of 141 and 1780 for $C_6(H_2CO-H_2CO)$ and $C_9((H_2CO)_3)$, respectively, are about 15% and 22% lower than our results. Analogous TDHF calculations have been carried out for acetone [57]. However in this case the initial TDHF/6-31G(+ sd + sp) results for C_6 and C_9, namely 686.0 and 19,060, are about 14% and 21% lower than our recommended values of 794.3 and 24,101 respectively and the scaled results, 797 and 24,190 respectively, agree extremely well, to within 0.4%, with the recommended DOSD results.

The discrepancies between Spackman's predicted results for C_6 and C_9 for formaldehyde and our recommended results are due to the lack of a reliable "experimental" value of the static dipole polarizability at the time of his work, see previous discussions and, for example, references [5,58]. The Spackman's equations for the predicted values are [57]

$$C_{6,pre} = 0.9435[\alpha(0)_{\exp}/\alpha(0)_{TDHF}]^2 C_{6,TDHF} \tag{11}$$

and

$$C_{9,pre} = 0.9283[\alpha(0)_{\exp}/\alpha(0)_{TDHF}]^3 C_{9,TDHF} \tag{12}$$

Repeating his calculations for formaldehyde, except for replacing $\alpha(0)_{\exp} = 17.8$, a value estimated from an MP2 calculation [58], by our recommended value [5] of 19.3, gives $C_6 = 165.8$ and $C_9 = 2278$ which are in excellent agreement with our recommended DOSD values for these coefficients.

4. Discussion and conclusions

The results for the dipole – dipole and the triple – dipole dispersion energies for the interactions considered in this paper are often the only values available for these molecules. They represent prototype interactions involving small prototypical molecules containing the carbonyl (aldehyde, ketone) and amine (primary, secondary, tertiary) groups and therefore should be of considerable general interest.

The estimated uncertainties in our recommended results for the dipole – dipole dispersion energy coefficients of Tables 3 to 8 are 1% , those for the triple - dipole dispersion coefficients of Table 9 are 1 to 2%. These error estimates are established by well tested methods [7–10,22,29–31]. Alternative molecular DOSDs, satisfying the same constraints as the recommended molecular DOSD, have been constructed for a variety of molecules by using different DOS data in spectral regions where there is a significant disagreement between DOS data obtained from different sources. Comparison of the dipole dispersion energy coefficients obtained from the original and the alternative DOSDs establishes the uncertainties in the results. As pointed out earlier various tests of psuedo – DOSD results for the dispersion energy coefficients show that these results agree with those obtained from the recommended molecular DOSDs to well within their uncertainties. As emphasized in all our work in this area, the error estimates in the calculated dispersion energy coefficients, or other molecular properties, are valid only if the molar refractivity constraints use to construct the original DOSDs are reliable(errors within a few tenths of a percent).

Reliable results for dipole – dipole dispersion energies are useful in many areas of research. The dipole – dipole dispersion energy is the dominant interaction energy at long range for interactions of "spherically symmetric" molecules. Its importance lies in its use for representing long range interactions between two atoms or molecules, in constructing potential energy models that are valid for all intermolecular distances through the use of dispersion energy damping functions [59–63], and in scaling relevant portions of ab initio calculations of potential energy surfaces [64–67]. There are several relatively recent successful potential energy models based on these ideas, see for example the review contained in [61], references [59,60,68–71], and references therein.

For interactions involving molecules, long range interaction energies other than dispersion are also important, for example second order induction and first order electrostatic energies. These, like the long range dispersion energies [72,73], are valid only for sufficiently large R and should be modified, as R decreases, by including charge overlap effects ("damping functions") [72,74,75]. If the Hartree Fock dimer energy is used to help construct the model potential energy surface, the first order electrostatic and second order induction energies, including charge overlap corrections, are included in this energy (at an SCF level) and a "complete" model can be obtained by adding in the damped dispersion energy. For other potential energy models, the first order electrostatic energies and their charge overlap corrections are included in the first order Coulomb energy , or the Heitler London interaction energy, and dispersion energies and (if significant) induction en ergies, including the incorporation of charge overlap effects through damping functions, will be needed to obtain reasonable results.

The triple – dipole dispersion energy gives the dominant non-additive interaction energy for an assembly of well separated "tumbling" or spherically symmetric species and can be used to help model non-additive effects for other intermolecular configurations as well [61,76,77]. It plays a role in the non-additive part of a many – body potential energy analogous to that played by the dipole – dipole dispersion energy in the two – body or additive part of the many body potential [61,77]. The dependence of the triple – dipole dispersion energy on the relative orientation of the interacting molecules is also important for interactions involving non – spherical species [78], see also [18] and references therein, and the orientationally averaged C_9 can be used in this context as well [78,79], however see also references [18, 79,80]. It is relevant to mention that the triple – dipole energy need not be the only important long – range non – additive energy for interactions involving molecules, see for example [61,81,82] and references therein. Also all multipolar results for the non – additive energies, including the triple – dipole energy, are representations of portions of the exact non – additive energy valid only for large intermolecular separations. As these distances decrease there are non – additive charge overlap and electron exchange energies which modify the multipolar results in significant ways, sometimes for surprisingly "moderately small" values of the inter species distances, see for example [61,77,81–83].

Accurate DOSD values of C_6 for important prototype interactions furnish useful checks on other methods for evaluating dispersion energy coefficients, for example empirical, non – empirical, ab initio quantum mechanical, and Padé approximant approaches [7,8,22,29,64–66,79,84–87]. In this context it is relevant to point out that in the potential energy models discussed briefly previously, it is necessary to include the dispersion energy to at least terms that vary as R^{-10} at long range. The higher order terms, that is the terms varying as powers of R^{-1} higher than R^{-6}, are not accessible by DOSD techniques due to the lack of (experimental) higher (than dipole) oscillator strengths. The method of choice for determining these higher order dispersion energies is via ab initio quantum mechanical methods analogous to those used in the dipole – dipole case. The validity, or the lack thereof, of these methods for these energies can be checked indirectly via the checks on the dipole –dipole dispersion energy referred to earlier – the ab initio calculation of the higher order terms is not easier than for the dipole – dipole dispersion energy. In any event the dipole – dipole dispersion energy is the lead or dominant dispersion energy and the success of the potential energy models discussed previously depends on a reliable value of it's coefficient C_6 [47].

The pseudo – DOSDs employed in this work to calculate the dipole – dipole and the triple – dipole dispersion energy coefficients form a concise representation of the original recommended DOSDs, which are continuous functions of the excitation energy from the UV absorption threshold to many thousands of eV, for the molecules under consideration. They give reliable results for the dispersion energy coefficients, and a variety of other molecular properties, through the use of discretized analogues of the usual expressions for these properties [7–9]. It is relevant to point out that more than ten pseudo states are needed to reliably evaluate some of these properties, for example the logarithmic dipole sums L_k for "larger" values of k, which have significant negative and positive contributions [23,24,26]. The original recommended DOSDs are constructed from extensive experimental and theoretical input information and are not developed only to give reliable results for the dispersion energy coefficients which depend on the relatively low energy portions of the DOSDs. The recommended DOSDs, and the related dipole properties, are of use in a variety of applications ranging from intermolecular forces to medical physics.

Acknowledgements

The authors wish to thank Professor Spackman for helpful discussions concerning his calculations of the dispersion energy coefficients for formaldehyde.This research was supported by grants from the

Natural Sciences and Engineering Research Council of Canada.

References

[1] J.O. Hirschfelder, C.F. Curtiss and R.B. Bird, *Molecular Theory of Gases and Liquids*, Wiley, 1954.
[2] R. Eisenschitz and R.F. London, *Z. Phys.* **60** (1930), 491.
[3] P.M. Axilrod and E. Teller, *J. Chem. Phys.* **11** (1943), 299.
[4] Y. Muto, *Proc. Phys. Math. Soc. Jap.* **17** (1943), 629.
[5] A. Kumar and W.J. Meath, *Molec. Phys.* **90** (1997), 389.
[6] G.R. Burton, W.F. Chan, G. Cooper, C.E. Brion, A. Kumar and W.J. Meath, *Can. J. Chem.* **72** (1994), 529.
[7] D.J. Margoliash and W.J. Meath, *J. Chem. Phys.* **68** (1978), 1426.
[8] D.J. Margoliash, T.R. Proctor and W.J. Meath, *Molec. Phys.* **35** (1978), 747.
[9] B.L. Jhanwar and W.J. Meath, *Molec. Phys.* **41** (1980), 1061.
[10] G.D. Zeiss, W.J. Meath, J.C.F. MacDonald and D.J. Dawson, *Can. J. Phys.* **55** (1977), 2080.
[11] A. Kumar, G.R.G. Fairly and W.J. Meath, *J. Chem. Phys.* **83** (1985), 70.
[12] R.J. Pazur, A. Kumar, R.A. Thuraisingham and W.J. Meath, *Can. J. Chem.* **66** (1988), 615.
[13] W.J. Meath and A. Kumar, *Int. J. Quantum Chem.* **S24** (1990), 501.
[14] A. Kumar and W.J. Meath, *Theoret. Chim. Acta.* **82** (1992), 131.
[15] A. Kumar and W.J. Meath, *Chem. Phys.* **189** (1994), 467.
[16] A. Kumar, W.J. Meath, P. Bundgen and A.J. Thakkar, *J. Chem. Phys.* **105** (1996), 4927.
[17] P. Bundgen, A.J. Thakkar, A. Kumar and W.J. Meath, *Molec. Phys.* **90** (1997), 721.
[18] S.A.C. McDowell, A. Kumar and W.J. Meath, *Molec. Phys.* **87** (1996), 845.
[19] S.A.C. McDowell, A. Kumar and W.J. Meath, *Can. J. Chem.* **74** (1996), 1180.
[20] J.O. Hirschfelder, W. Byers Brown and S.T. Epstein, *Adv. Quant. Chem.* **1** (1964), 255.
[21] H.A. Bethe and E.E. Salpeter, *Quantum Mechanics of One – and Two – Electron Atoms*, Academic Press, 1957.
[22] G.D. Zeiss and W.J. Meath, *Molec. Phys.* **33** (1977), 1155.
[23] A. Kumar and W.J. Meath, *Can. J. Chem.* **63** (1985), 1616.
[24] A. Kumar and W.J. Meath, *Can. J. Phys.* **63** (1985), 417.
[25] A. Kumar, M. Kumar and W.J. Meath, *Chem. Phys.* **286** (2003), 227.
[26] A. Kumar, M. Kumar and W.J. Meath, *Molec. Phys.* **101** (2003), 1535.
[27] G. Cooper, J.E. Anderson and C.E. Brion, *Chem. Phys.* **209** (1996), 61.
[28] G. Cooper, J.E. Anderson and C.E. Brion, *private communication.*
[29] B.L. Jhanwar and W.J. Meath, *Chem. Phys.* **67** (1982), 185.
[30] G.F. Thomas and W.J. Meath, *Molec. Phys.* **34** (1977), 113.
[31] B.L. Jhanwar, W.J. Meath and J.C.F. MacDonald, *Can. J. Phys.* **59** (1981), 185.
[32] K.L. Ramaswamy, *Proc. Indian Acad. Sci.* **A 4** (1936), 675.
[33] H. Lowery, *Proc. R. Soc. London* **A 133** (1931), 188.
[34] H. Lowery, *Phil. Mag.* **14** (1932), 743.
[35] K. Prytz, *Wed. Ann.* **11** (1980), 104.
[36] C.T. Zahn, *Physik Z.* **33** (1932), 686.
[37] G.R. Burton, W.F. Chan, G. Cooper, C.E. Brion, A. Kumar and W.J. Meath, *Can. J. Chem.* **71** (1993), 341.
[38] A. Kumar and W.J. Meath, *Molec. Phys.* **54** (1985), 823.
[39] A. Kumar and W.J. Meath, *Chem. Phys.* **91** (1984), 411.
[40] R.E. Johnson, S.T. Epstein and W.J. Meath, *J. Chem. Phys.* **47** (1967), 1271.
[41] A. Kumar and W.J. Meath, *Molec. Phys.* **75** (1992), 311.
[42] A. Kumar, *J. Molecular Structure (Theochem)*, **591** (2002), 91.
[43] M. Kumar, A. Kumar and W.J. Meath, *Molec. Phys.* **100** (2002), 3271.
[44] B.L. Jhanwar and W.J. Meath, unpublished work.
[45] T.N. Olney, N.M. Cann, G. Cooper and C.E. Brion, *Chem.Phys.* **223** (1997), 59.
[46] G.R. Burton, W.F. Chan, G. Cooper and C.E. Brion, *Chem. Phys.* **177** (1993), 217.
[47] W.J. Meath, D.J. Margoliash, B.L. Jhanwar, A. Koide and G.D. Zeiss, in: *Intermolecular Forces*, B. Pullman, ed., Dordrecht – Reidel, 1981, p. 101.
[48] G.D. Zeiss, W.J. Meath, J.C.F. MacDonald and D.J. Dawson, *Molec. Phys.* **39** (1980), 1055.
[49] A. Koide, W.J. Meath and A.R. Allnatt, *J. Phys. Chem.* **86** (1982), 1222.
[50] A.J. Thakkar, *J. Chem. Phys.* **75** (1981), 4496.
[51] U. Fano and J.W. Cooper, *Rev. Mod. Phys.* **40** (1968), 441 and **41** (1969), 724.
[52] H.L. Kramer, *J. Chem. Phys.* **53** (1970), 2783.

[53] H.L. Kramer and D.R. Herschbach, *J. Chem. Phys.* **53** (1970), 2792.
[54] A.A. Maryott and F. Buckley, *US National Bureau of Standards Circular*, 1953, No. 537.
[55] H.A. Stuart, Landolt-Bornstein, *Atom und Molekular Physik* **1**(part 3) (1951), 509.
[56] Fuchs u. Wolf, Hand – und – Jahrbuch der Chemischen Physik, *Bd.6, Abschn.* **I** (1935), s262.
[57] M.A. Spackman, *J. Chem. Phys.* **94** (1991), 1295.
[58] M.A. Spackman, *J. Phys. Chem.* **93** (1989), 7594.
[59] C. Douketis, G. Scoles, S. Marchetti, M. Zen and A.J. Thakkar, *J. Chem. Phys.* **76** (1982), 3057.
[60] K.T. Tang and J.P. Toennies, *J. Chem. Phys.* **80** (1984), 3726.
[61] W.J. Meath and M. Koulis, *J. Molec. Struct.* **226** (1991), 1.
[62] A.J.C. Varandas, *Molec. Phys.* **60** (1987), 527.
[63] A.K. Dham, A.R. Allnatt, A. Koide and W.J. Meath, *Molec. Phys.* **196** (1995), 81.
[64] M.P. Hodges and R.J. Wheatley, *J. Chem. Phys.* **114** (2001), 8836.
[65] M.P. Hodges, R.J. Wheatley and A.H. Harvey, *J. Chem. Phys.* **116** (2002), 1397.
[66] M.P. Hodges and R.J. Wheatley, *J. Molec. Struct.* (*Theochem*) **591** (2002), 67.
[67] K. Patkowski, T. Korona, R. Moszynski, B. Jeziorski and K. Szalewicz, *J. Molec. Struct.* (*Theochem*) **591** (2002), 231.
[68] M.S. Bowers, K.T. Tang and J.P. Toennies, *J. Chem. Phys.* **88** (1988), 5465.
[69] J. Dias da Silva, J. Brandao and A.J.C. Varandas, *J. Chem. Soc. Faraday Trans. II* **85** (1998), 1851.
[70] A.K. Dham, F.R. McCourt and W.J. Meath, *J. Chem. Phys.* **103** (1995), 8477.
[71] A.K. Dham and W.J. Meath, *Molec. Phys.* **99** (2001), 991.
[72] H. Kreek and W.J. Meath, *J. Chem. Phys.* **50** (1969), 2289.
[73] A. Koide, W.J. Meath and A.R. Allnatt, *Chem. Phys.* **58** (1981), 105.
[74] K.-C. Ng, W.J. Meath and A.R. Allnatt, *Molec. Phys.* **38** (1979), 449.
[75] R.J. Wheatley and W.J. Meath, *Chem. Phys.* **179** (1994), 341, and **203** (1996), 209.
[76] J.A. Barker, *Molec. Phys.* **57** (1986), 755.
[77] W.J. Meath and R.A. Aziz, *Molec. Phys.* **52** (1984), 225.
[78] D.E. Stogryn, *Phys. Rev. Lett.* **24** (1970), 971.
[79] S.A.C. McDowell and W.J. Meath, *Can. J. Chem.* **76** (1998), 483.
[80] S.A.C. McDowell and W.J. Meath, *Chem. Phys.* **223** (1997), 23.
[81] G. Chalasinski, S.M. Cybulski, M.M. Szczesniak and S. Scheiner, *J. Chem. Phys.* **91** (1989), 7048.
[82] G. Chalasinski, M.M. Szczesniak, P. Cieplak and S. Scheiner, *J. Chem. Phys.* **94** (1991), 2873.
[83] S.F. O'Shea and W.J. Meath, *Molec. Phys.* **28** (1974), 1431 and *Molec. Phys.* **31** (1976), 515.
[84] F. Mulder, G.F. Thomas and W.J. Meath, *Molec. Phys.* **41** (1980), 249.
[85] F. Mulder, G. van Dijk and A. van der Avoird, *Molec. Phys.* **39** (1980), 407.
[86] R.M. Berns and A. van der Avoird, *J. Chem. Phys.* **71** (1980), 6107.
[87] H. Hettema, P.E.S. Wormer and A.J. Thakkar, *Molec. Phys.* **80** (1993), 533.

Journal of Computational Methods in Sciences and Engineering 4 (2004) 321–332
IOS Press

First principle calculations of dipole-dipole dispersion coefficients for the ground and first $\pi \rightarrow \pi^*$ excited states of some azabenzenes

Patrick Norman*, Auayporn Jiemchooroj and Bo E. Sernelius
Department of Physics and Measurement Technology, Linköping University, SE-581 83 Linköping, Sweden

Abstract. The complex polarization propagator method has been applied to the calculation of dipole-dipole dispersion coefficients (also known as C_6 coefficients) of pyridine, pyrazine, and *s*-tetrazine. These calculations refer to the electronic ground states as well as the first excited states of $\pi \rightarrow \pi^*$ character. It is argued that accurate ground state dispersion coefficients are obtained with density functional theory using the B3LYP exchange-correlation functional. The proposed values for the C_6 coefficients of pyridine, pyrazine, and *s*-tetrazine in their ground states are 1543 a.u., 1398 a.u., and 1014 a.u., respectively. Multi-configurational complete active space calculations are performed on these compounds in their respective $\pi \rightarrow \pi^*$ excited state. The isotropic averages of the frequency-dependent polarizabilities are smaller in the excited states, but the effective frequencies – defined in the London – van der Waals dispersion relation – are on the other hand larger.

1. Introduction

Intermolecular potentials and forces play an important role for the properties of molecular complexes, liquids, and solids as well as for site interactions in extended proteins. For neutral species the forces are dispersive in nature, and the molecular property of interest, which governs these interactions, is the polarizability. A number of polarization propagator methods in quantum chemistry have been used in this context, and there exist several publications in the literature concerned with the calculation of dipole-dipole as well as higher order multipole dispersion coefficients [1–13]. In a recent publication [13], we showed that the complex polarization propagator [14] provides an efficient and straightforward way of calculating the polarizability on the imaginary frequency axis also for larger molecules. Without exceptions, the calculations in that work as well as others have referred to the interactions between molecules in their respective electronic ground states.

It is clear, however, that interactions between electronically excited species will be different both in magnitude and nature. This aspect has been investigated recently: on the one hand, experimentally by Failache et al. [15] in the determination of the van der Waals interaction between excited Cs atoms and a sapphire surface, and, on the other hand, theoretically by Boström et al. [16]. In the work by Boström et al. [16] the resonant interaction between a molecule in its ground state and another of the same kind in its excited state is considered. Such resonant interactions can be much stronger than the regular $1/R^6$

*Corresponding author. E-mail: panor@ifm.liu.se.

energy dependence which is observed in the van der Waals region, and they may serve as a mechanism for energy transport. The non-retarded interaction energy is shown to have a $1/R^3$ dependence in this case [16].

In this work we will explore the possibility of calculating dynamic polarizabilities of molecules in their excited states. With a real frequency argument, these excited state electric dipole polarizabilities $\alpha(\omega^R)$ can be obtained with quite different techniques. First, and if using a multi-determinant based reference state, the electronic structure of the excited state can be optimized followed by a regular linear polarization propagator calculation. In this case, one or more resonances of the propagator will correspond to negative excitation energies, but the convergence of the linear response equations turns out not to be a critical factor. Second, and applicable also to single determinant based methods, the excited state polarizability can be identified from the double residue of the second-order nonlinear propagator of the electronic ground state. Implementations of this approach are available for a number of different electronic structure methods [17–19], and several applications have been presented in the literature [17–23].

Due to the larger density of states in the vicinity of the excited state, however, these two alternative methods will be applicable only in a quite limited frequency region due to the approximation of infinite lifetimes for the excited states. The response functions will be divergent at resonant frequencies when damping is excluded in the theory. If we turn to the situation of interest in the present work, namely intermolecular dispersion interactions, the pertinent molecular property is the polarizability evaluated with a purely imaginary frequency argument $\alpha(i\omega^I)$. This quantity is clearly well behaved at all frequencies also in the infinite lifetime approximation, but calculations using standard propagator methods are not straightforward due to the imaginary frequency. One way, which has been pursued in ground state applications, is to determine the so-called Cauchy moments from calculations of $\alpha(\omega^R)$ and then retrieve $\alpha(i\omega^I)$ by summing the moments to a finite order [4,7,8,24]. In the case of excited state polarizabilities, this may prove difficult because of the mentioned divergences in the propagator. A second approach to this problem has been presented by us in an earlier publication [13] and which is based on the complex polarization propagator method [14] that allows for a direct evaluation of $\alpha(i\omega^I)$. The same method should, at least in principle, be applicable to properties of excited states, an issue which we will investigate in the present work.

As a set of sample systems we have chosen three of the members of the azabenzenes, namely pyridine, pyrazine, and s-tetrazine. A good reason for this choice is that there are results available in the literature for the static polarizabilities of these species in their lowest excited states [19,21,25]. We will here address the lowest excited states of $\pi \rightarrow \pi^*$ character.

2. Theory and methodology

The non-retarded long-range interaction energy between two randomly oriented molecules A and B is given by

$$\Delta E(R) = -\frac{C_6}{R^6} - \frac{C_8}{R^8} - \frac{C_{10}}{R^{10}} - \cdots, \tag{1}$$

where R is the intermolecular separation in the van der Waals region. The coefficient of the leading term equals

$$C_6 = \frac{3\hbar}{\pi}\int_0^\infty \overline{\alpha}_A(i\omega^I)\overline{\alpha}_B(i\omega^I)d\omega^I, \tag{2}$$

where $\overline{\alpha}_A(i\omega^I)$ is the isotropic average of the electric dipole polarizability tensor of molecule A evaluated at a purely imaginary frequency. The molecules are here considered to be polarizable entities in a given electronic state. Additional contributions to the energy will arise from higher-order electric multipole interactions [26] as well as magnetic interactions [27].

In this paper we will carry out quantum chemical calculations of the dipole-dipole interaction energy as given by Eqs (1) and (2) between two like species in their respective ground or excited electronic state. The molecular property needed in this case is the polarizability on the imaginary axis for the electronic state in question. If we consider $\omega = \omega^R + i\omega^I$ to be a complex frequency argument, we may express the electric dipole polarizability for a molecule in state $|n\rangle$ in terms of a sum-over-states (SOS) formula

$$\alpha_{\alpha\beta}^{|n\rangle}(\omega) = \hbar^{-1}\sum_{k\neq n}\left[\frac{\langle n|\hat{\mu}_\alpha|k\rangle\langle k|\hat{\mu}_\beta|n\rangle}{\omega_{kn}-\omega} + \frac{\langle n|\hat{\mu}_\beta|k\rangle\langle k|\hat{\mu}_\alpha|n\rangle}{\omega_{kn}+\omega}\right], \tag{3}$$

where $\hbar\omega_{kn}$ are the transitions energies between the reference electronic state $|n\rangle$ and state $|k\rangle$.

It has recently been shown, although in a different context, that a direct evaluation of Eq. (3) by means of the linear polarization propagator techniques is feasible [14]. The complex linear polarization propagator was developed and implemented in the multi-configurational self-consistent field (MCSCF) approach [14] for the purpose of a determination of the electronic polarizability in the near-resonant region of the spectra where the incorporation of lifetime broadening effects is imperative. The lifetime broadening enters in the expression for the polarizability as a small and constant imaginary number corresponding to the inverse lifetime of the excited states and which adds to the real frequency of the perturbing electric field. In order to utilize this development in the present context, we merely need to interpret the lifetime broadening quantity, not as a constant term but as an imaginary dynamic frequency $i\omega^I$. Since, in essence, it is Eq. (3) which is being implemented in the complex linear polarization propagator approach, the propagator is well-behaved in the entire complex frequency plane (apart from the resonances at the excitation energies on the real axis of course). Its applicability to the calculation of $\alpha^{|0\rangle}(i\omega^I)$, i.e., to the polarizability of species in their ground electronic state, has been demonstrated by us [13], and it is now our intention to use the same technique for the calculation of excited state polarizabilities.

The linear response function, or the linear polarization propagator, equals minus the linear polarizability

$$\langle\langle\hat{\mu}_\alpha;\hat{\mu}_\beta\rangle\rangle_\omega = -\alpha_{\alpha\beta}^{|n\rangle}(\omega), \tag{4}$$

where $\hat{\mu}_\alpha$ is the electric dipole operator along the Cartesian axis α. The calculation thus involves an optimization of the electronic structure of the unperturbed molecule in state $|n\rangle$, and a subsequent evaluation of the linear response function using time-dependent perturbation theory. The computationally tractable formulas for the response functions differ depending on the electronic structure method at hand and a true spectral representation as in Eq. (3) is valid only in the limit of full-configuration interaction (FCI). With methods where electron correlation is only partially included we meet matrix equations instead. In the MCSCF approach, and indeed also in time-dependent density functional theory methods, the linear response function takes the form [14,28]

$$\langle\langle A;B\rangle\rangle_\omega = -A^{[1]\dagger}\left[E^{[2]} - (\omega^R + i\omega^I)S^{[2]}\right]^{-1} B^{[1]}, \tag{5}$$

where $E^{[2]}$ and $S^{[2]}$ are the so-called Hessian and overlap matrices, and $A^{[1]}$ and $B^{[1]}$ are the property gradients corresponding to the components of the polarizability. The evaluation of Eq. (5) is not achieved by explicit matrix inversion due to the large dimension of the involved matrices; instead one rather finds an iterative solution to the complex linear response equation [14]

$$\left[E^{[2]} - (\omega^R + i\omega^I)S^{[2]}\right] N(\omega) = B^{[1]}, \tag{6}$$

where, in the computational basis, all matrix quantities except for the linear response vector $N(\omega)$ are real. We can thus re-write Eq. (6) as a set of two coupled equations

$$\left[E^{[2]} - \omega^R S^{[2]}\right] N^R(\omega) = B^{[1]} - \omega^I S^{[2]} N^I(\omega), \tag{7}$$

$$\left[E^{[2]} - \omega^R S^{[2]}\right] N^I(\omega) = \omega^I S^{[2]} N^R(\omega), \tag{8}$$

where the response vector has been written as $N(\omega) = N^R(\omega) + iN^I(\omega)$. The structure of the real matrices are [28]

$$E^{[2]} = \begin{bmatrix} A & B \\ B & A \end{bmatrix}, \quad S^{[2]} = \begin{bmatrix} \Sigma & \Delta \\ -\Delta & -\Sigma \end{bmatrix}, \quad B^{[1]} = \begin{bmatrix} g_B \\ -g_B \end{bmatrix}, \tag{9}$$

whereas the structure of the response vector will be

$$N^R(\omega) = \begin{bmatrix} Z^R \\ -Y^R \end{bmatrix}, \quad N^I(\omega) = \begin{bmatrix} Z^I \\ Y^I \end{bmatrix}. \tag{10}$$

In the present application to intermolecular interaction potentials, we have $\omega^R = 0$ and therefore we find $Z^R = Y^R$ as well as $Z^I = Y^I$. This will further imply that the final value of the linear response function will be real since

$$\langle\langle A; B\rangle\rangle_{i\omega^I} = -A^{[1]\dagger} N(i\omega^I) \;=\; -[g_A \;\; -g_A]\left(\begin{bmatrix} Z^R \\ -Z^R \end{bmatrix} + i\begin{bmatrix} Z^I \\ Z^I \end{bmatrix}\right) \;=\; -2g_A Z^R, \tag{11}$$

which of course is in agreement with the fact that the polarizability on the imaginary axis is real. So it is only by the coupling between Eqs (7) and (8) that the imaginary part of the response vectors enters, and not in Eq. (11), due to the fact that the A operator is real.

3. Computational details

The property calculations are performed using the time-dependent formulations of Hartree-Fock, complete active space, and density functional (with the hybrid B3LYP exchange-correlation functional) theory. In the subsequent sections, the three methods will be referred to as SCF, CAS, and DFT/B3LYP, respectively.

Due to their single-determinant character the SCF and DFT/B3LYP methods have only been applied to the electronic ground state calculations, whereas CAS results are presented also for the excited states. We have considered the three azabenzenes pyridine, pyrazine, and s-tetrazine. The multi-configurational CAS states are formed with an inactive σ-space and with six active electrons in a π–π^* space of molecular orbitals. This wave function parameterization is identical to that employed by Jonsson et al. [21]. The polarization basis sets of Sadlej [29] have been used at all instances.

Table 1
Averaged static polarizabilities $\overline{\alpha}(0)$, dipole-dipole dispersion coefficients C_6, and effective frequency ω_1 for the lowest singlet states of pyridine. All quantities are given in atomic units

State	Method	α_{xx}	α_{yy}	α_{zz}	$\overline{\alpha}$	C_6	ω_1
X^1A_1	SCF	75.0	70.5	41.0	62.2	1498	0.516
	SCF[a]					1444	
	DFT/B3LYP	78.2	73.9	40.6	64.2	1543	0.499
	CAS	70.4	66.0	38.8	58.4	1374	0.537
	SDQ-MP4[b]	76.6	72.0	38.8	62.5		
	Exp[c]	80.2	73.2	39.0	64.1		
1^1B_2	SCF[d]	86.7	78.1	54.7	73.1		
	CAS[e]	66.7	63.8	39.5	56.6		
	CAS	63.1	61.3	40.2	54.9	1278	0.565

[a]SCF/6-31G($+sd+sp$) result by Spackman [11].
[b]SDQ-MP4/6-31G($+sd+sp$) result by Archibong and Thakkar [34].
[c]Experimental value quoted from Ref [33].
[d]SCF/ANO($+pd$) cubic response calculation by Jonsson et al. [21].
[e]CAS/ANO($+pd$) cubic response calculation by Jonsson et al. [21].

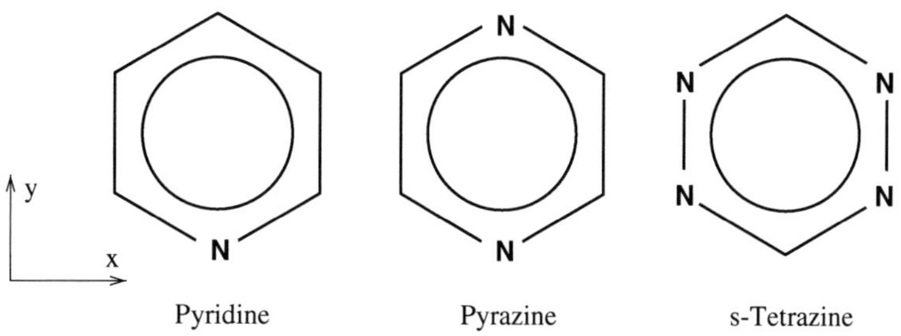

Fig. 1. Coordinate system for the azabenzenes.

Experimental ground state structures [30] have been employed in the present study, and the orientation of molecules is chosen in accordance with Fig. 1. In C_{2v} (pyridine) and D_{2h} (pyrazine and s-tetrazine), the x and y components of the electric field will span the irreducible representations B_2 and A_1 and B_{3u} and B_{2u}, respectively. Hence, the lowest $\pi \longrightarrow \pi^*$ excited states of singlet spin symmetry will be 1^1B_2, 1^1B_{3u}, and 1^1B_{3u} for pyridine, pyrazine and s-tetrazine, respectively.

The polarizability $\alpha(i\omega^I)$ has been evaluated for a dense grid of frequencies and we have therefore not needed to pay particular attention to the scheme for the numerical integration of the dispersion integral [Eq. (2)] on the imaginary frequency axis. The grid density is 0.0911 a.u.

All calculations have been carried out with a locally modified version of the Dalton program [31] with DFT capability added by Helgaker et al. [32].

4. Results and discussion

4.1. Static polarizabilities

The results for the polarizabilities for pyridine, pyrazine, and s-tetrazine are found in Tables 1, 2 and 3, respectively. As far as ground state static polarizabilities are concerned, we consider the best reference

Table 2
Averaged static polarizabilities $\overline{\alpha}(0)$, dipole-dipole dispersion coefficients C_6, and effective frequency ω_1 for the lowest singlet states of pyrazine. All quantities are given in atomic units

State	Method	α_{xx}	α_{yy}	α_{zz}	$\overline{\alpha}$	C_6	ω_1
X^1A_g	SCF	74.8	64.2	37.4	58.8	1355	0.523
	DFT/B3LYP	76.9	67.9	37.4	60.7	1398	0.506
	CAS	69.6	60.0	35.8	55.1	1245	0.547
	SDQ-MP4[a]	72.7	63.2	35.4	57.1		
	Exp[b]				60.6		
1^1B_{3u}	SCF[c]	77.0	66.0	53.0	65.4		
	CAS[d]	61.8	58.5	37.8	52.7		
	CAS	59.0	56.2	38.3	51.2	1147	0.583

[a]SDQ-MP4/6-31G($+sd+sp$) result by Archibong and Thakkar [34].
[b]Experimental value quoted from Ref. 33.
[c]SCF/ANO($+pd$) cubic response calculation by Jonsson et al. [21].
[d]CAS/ANO($+pd$) cubic response calculation by Jonsson et al. [21].

Table 3
Averaged static polarizabilities $\overline{\alpha}(0)$, dipole-dipole dispersion coefficients C_6, and effective frequency ω_1 for the lowest singlet states of s-tetrazine. All quantities are given in atomic units

State	Method	α_{xx}	α_{yy}	α_{zz}	$\overline{\alpha}$	C_6	ω_1
X^1A_g	SCF	53.6	58.3	31.4	47.7	976.5	0.572
	DFT/B3LYP	55.6	61.4	32.2	49.8	1014	0.545
	CAS	52.9	54.4	31.4	46.2	919.6	0.575
	SDQ-MP4[a]	53.0	57.9	31.3	47.4		
	CCSD[b]	56.0	60.7	32.7	49.8		
	CASPT2[c]	55.1	60.0	32.6	49.2		
1^1B_{3u}	SCF[d]	50.2	61.3	31.8	47.8		
	CAS	44.1	50.1	32.8	42.3	835.0	0.622

[a]SDQ-MP4/6-31G($+sd+sp$) result by Archibong and Thakkar [34].
[b]CCSD/Sadlej results by Hättig et al. [19].
[c]CASPT2/Sadlej results by Schütz et al. [25].
[d]SCF/ANO($+pd$) cubic response calculation by Jonsson et al. [21].

values for pyridine and pyrazine to be a set of experimental results quoted from Ref. 33. The local field and dispersion effects present in the experimental data are believed to be small [33]. For s-tetrazine there exist highly accurate CASPT2 [25] as well as coupled cluster singles and doubles (CCSD) [19] results. For all three molecules there also exist highly correlated SDQ-MP4 calculations by Archibong and Thakkar [34]. However, the 6-31G($+sd+sp$) basis set employed in this work appears insufficient to match the level of correlation, a fact which is also discussed by the authors. When compared against the more recent CASPT2 [25] and CCSD [19] results for s-tetrazine, it is seen that the SDQ-MP4 values underestimates $\overline{\alpha}(0)$ by respectively 1.8 and 2.4 a.u., see Table 3. A similar discrepancy is observed for the ground state polarizabilities of pyridine and pyrazine when the SDQ-MP4 values are compares against the experimental results – the differences are in these cases 1.6 and 3.5 a.u., respectively. With an improved basis set we believe this discrepancy should be reduced, and we have therefore chosen the experimental results for the polarizabilities of pyridine and pyrazine as reference values.

With the mentioned values taken as reference, we see that the contribution of electron correlation to the ground state $\overline{\alpha}(0)$ amounts to no more than 3% for any of the molecules. The SCF values slightly underestimate the reference results, and it is apparent that the addition of static correlation in terms of a π–π^* active space does not improve this situation for the polarizabilities of the azabenzenes. On the contrary, the CAS results are between 6–10% below the reference values, and dynamic correlation is needed in order to remedy this discrepancy.

The use of density functional theory in molecular property calculations is still not to be performed routinely [35], as for instance in structure optimizations. In our previous work, we showed that the hybrid B3LYP functional performed very well in the calculation of polarizabilities and dispersion parameters for the n-alkanes but that it suffered noticeable discrepancies when applied to the noble gas atoms [13]. In the present study on the azabenzenes, we note the excellent agreement between the DFT/B3LYP results and the reference values for the polarizabilities but are careful not to consider this as a general finding. However, with a method such as Kohn–Sham density functional theory, it is important to gain trust by experience, and, in that respect, the calculations on the n-alkanes [14] as well as the azabenzenes are useful.

As mentioned above there are two fundamentally different ways to determine the excited-to-ground state polarizability difference – either by state-specific optimization or as a double residue of the cubic response function – and the two techniques provide identical results only in the limit of full configuration interaction (FCI). Since the latter method avoids the explicit optimization of the electronic structure of the excited state it can be used in conjunction with wave functions based on single determinants. With an implementation of the cubic response function in the DFT/B3LYP approximation, it would be rewarding to address the excited state polarizability of the azabenzenes considering the quality of results for the ground state counterpart. At present, however, the polarizabilities of the $\pi \rightarrow \pi^*$ states in pyridine, pyrazine, and s-tetrazine are available at the SCF and CAS levels only [21]. In Tables 1–3, we see that the excited-to-ground state polarizability difference is small in general. We also note that our CAS results, which are obtained by optimization of the electronic structure of the excited states, are in close agreement with the CAS results of Jonsson et al. [21] and which are obtained with the residue technique. The average polarizabilities as determined with the two methods agree to within 4%, and we therefore believe that our π–π^* active spaces provide sufficient flexibility for an optimization of the electronic structure of the excited states. In neither of these calculations has the molecular structure of the excited states been optimized as motivated by the their short lifetimes.

4.2. Dynamic polarizabilities with purely imaginary frequencies and dispersion parameters

The dynamic polarizabilities $\alpha(i\omega^I)$ are important molecular properties and do often appear in expressions for long-range intermolecular interactions. The most familiar example is perhaps the van der Waals interaction as given by a combination of Eqs. (1) and (2). The predominant reason to calculate excited state polarizabilities is, however, not to determine the van der Waals interaction between electronically separated species in their respective excited states. Photon-induced excited states in organic compounds are short-lived by nature, and it is therefore not relevant to consider the van der Waals interaction. Far more important is instead the resonant interaction between a species in its ground state and another species in its excited state. This interaction involve an exchange of energy and thus constitutes a mechanism for energy transport. The theory for this interaction has recently been revisited by Boström et al. [16] and some erroneous results from the past were pointed out and corrected. In their analysis, however, Boström et al. [16] assumed the polarizability of the ground and excited states to be the same. It is our intention to relieve this constraint in a future work, but, to begin with and in the present work, we will investigate the possibility to carry out first principle calculations of excited state polarizabilities.

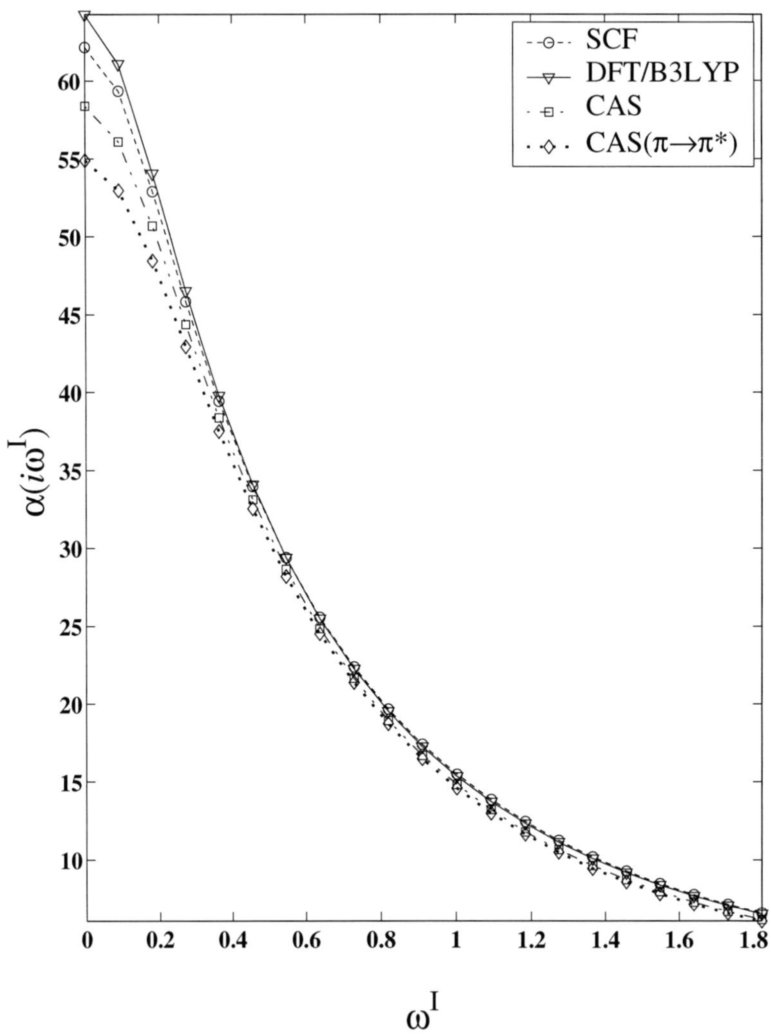

Fig. 2. Averaged polarizabilities $\overline{\alpha}(i\omega^I)$ for pyridine in the ground and lowest $\pi \rightarrow \pi^*$ excited state. All quantities are given in a.u.

For convenience and computational ease, the dispersion of the polarizability is sometimes assumed to be of the form

$$\overline{\alpha}(i\omega^I) = \frac{\overline{\alpha}(0)}{1 + (\omega^I/\omega_1)^2}, \tag{12}$$

where the effective frequency ω_1 has been introduced. With this assumption, Eq. (2) can be evaluated analytically and we arrive at

$$C_6 = \frac{3}{4}\overline{\alpha}(0)^2\omega_1. \tag{13}$$

We have included results for the effective frequency in the tables. It is clear that ω_1 is a measure of the speed at which the polarizability tend to zero, and the form of Eq. (12) indicates that the dispersion of the polarizability is smooth and monotonic, c.f. Figs 2–4.

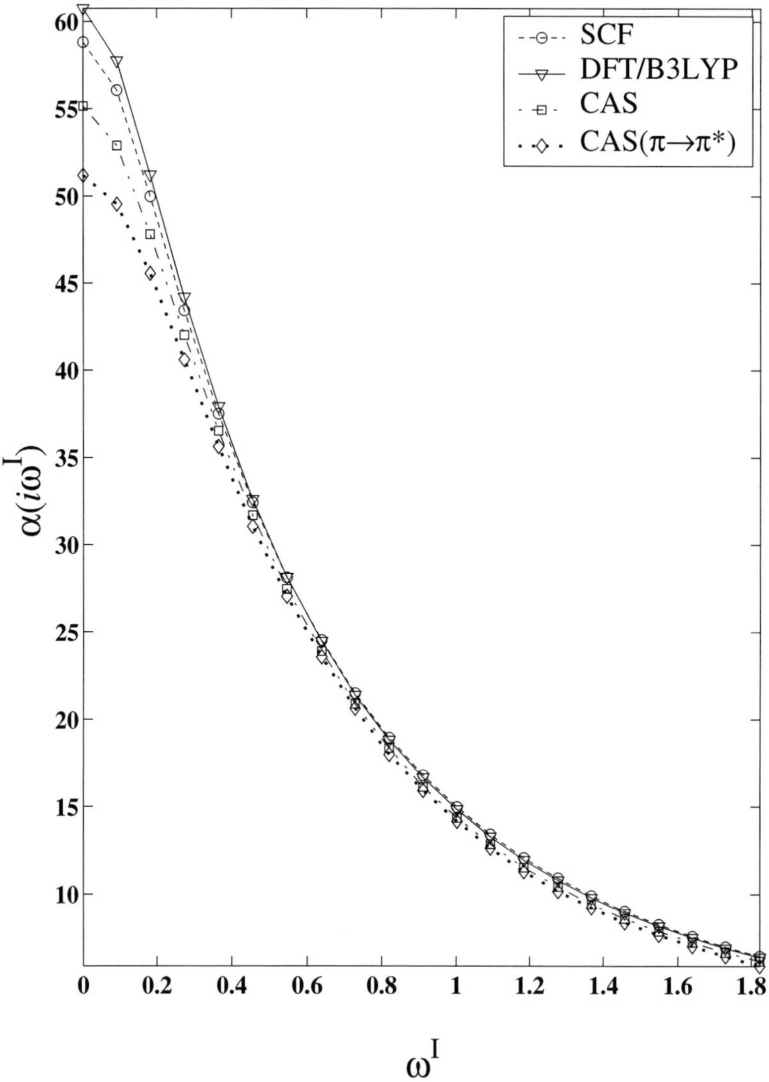

Fig. 3. Averaged polarizabilities $\overline{\alpha}(i\omega^I)$ for pyrazine in the ground and lowest $\pi \rightarrow \pi^*$ excited state. All quantities are given in a.u.

We are aware of two previous first principle calculation of the ground state dispersion coefficient for these systems, namely the SCF calculation on pyridine by Spackman [11] and the SCF/Unsöld calculations on the azabenzenes by Mulder et al. [33]. The present value of $C_6 = 1498$ a.u. is in fair agreement with the value of 1444 a.u. obtained by Spackman. Our slightly larger value is most likely an effect of the use of a basis set optimized for this property (the polarizing basis set of Sadlej [29]). The older work by Mulder et al. [33] and in which the Unsöld approximation has been adopted in the calculations of the polarizabilities, report a dipole-dipole dispersion coefficient for pyridine of 2379 a.u. It is clear that this approximation is not suitable to introduce in this case.

The DFT/B3LYP result of 1543 a.u. reported in the present work must be taken as a very accurate estimate of the dispersion parameter considering the high quality of the static polarizability at this level of theory. We have already concluded that it is difficult to retain the effect of electron correlation on the

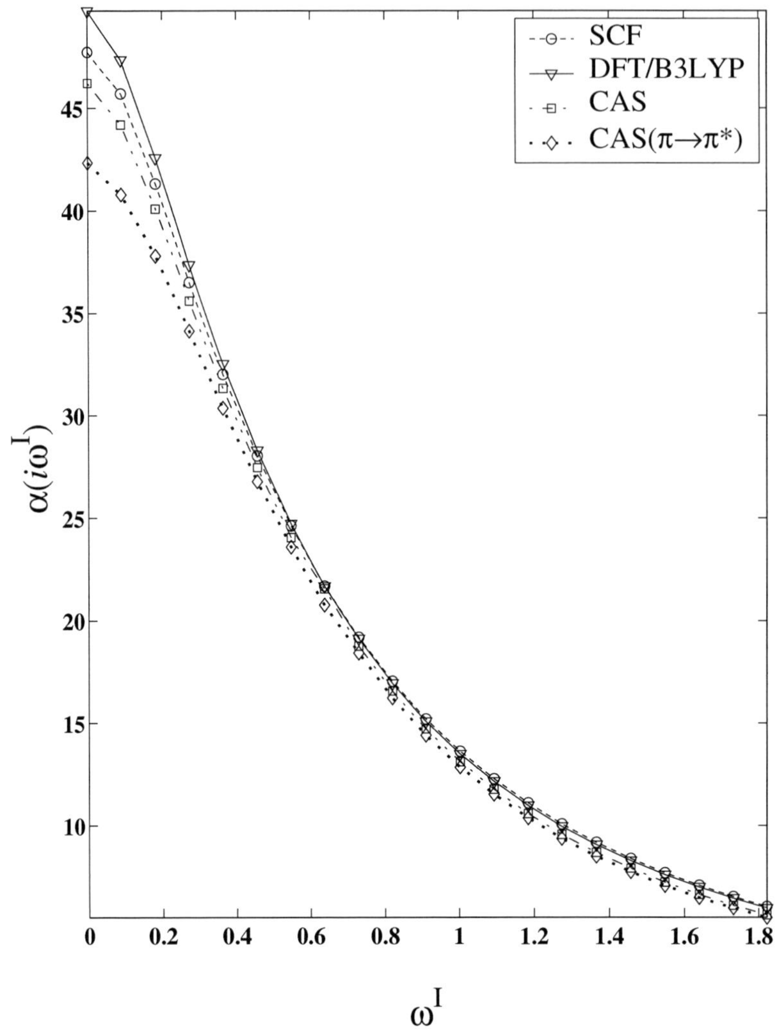

Fig. 4. Averaged polarizabilities $\overline{\alpha}(i\omega^I)$ for s-tetrazine in the ground and lowest $\pi \rightarrow \pi^*$ excited state. All quantities are given in a.u.

polarizability with use of a solely static correlated method such as CAS, and the dispersion coefficient of 1374 a.u. for pyridine underestimates the DFT/B3LYP result with about 11%. Considering the London – van der Waals dependence between the dispersion coefficient and the static polarizability in Eq. (13) one should expect the discrepancy in the polarizability to be doubled for the dispersion coefficient. So with the observed 10% error in the CAS polarizability we would expect a 20% error for the dispersion coefficient. However, this effect is counterbalanced by a larger CAS value for the effective frequency ω_1, which is seen in Fig. 2 as a result of a diminishing sensitivity to correlation with increasing frequency. In other words, the CAS and DFT/B3LYP polarizability curves start off with a 10% difference, but at $\omega^I = 0.5$ a.u. the difference is reduced to being less than 1%. We offer no explanation as to why this seemingly general observation should occur.

The dispersion interaction is reduced along with the aromaticity of the azabenzenes, which follows as a result of a competition between a decreasing polarizability with the number of nitrogen atoms in the system and an enhanced effective frequency. For pyrazine and s-tetrazine our best estimates of the ground state dispersion coefficients are 1398 and 1014 a.u., respectively, and both are obtained with the DFT/B3LYP method. Similar discrepancies as for pyridine are observed for the CAS ground state C_6 results for these molecules; for pyridine the discrepancy was 11% and for pyrazine and s-tetrazine it amounts to 12% and 10%, respectively.

The possibility of making a formal identification of the excited state $\alpha(i\omega^I)$ with the residue of a higher-order ground state response function has not been explored in the present work, and we are thus left with no alternative but to optimize the electronic structure of the excited state in order to determine its $\alpha(i\omega^I)$. That also leaves us with no choice but to employ the CAS method even though we have seen that, with the affordable active spaces, it provides ground state polarizabilities and dispersion coefficients that are about 10% lower than the best estimates. The accuracy for the corresponding excited state properties are not likely to be better.

The average polarizabilities $\overline{\alpha}(i\omega^I)$ of the $\pi \rightarrow \pi^*$ excited states is below that of the corresponding ground states for pyridine, pyrazine, and s-tetrazine, see Figs 2–4. The effective frequencies, on the other hand, are larger, which, as discussed above, can be seen in the figures as a diminishing ground-to-excited state polarizability difference with increasing frequency ω^I. The effect on the excited state dispersion coefficient is a modest reduction of 7–9% for the molecules under investigation.

5. Conclusions

To our knowledge, we have performed the first calculations of the dynamic polarizabilities (with an imaginary frequency argument) of species in their electronically excited states. We have achieved this with use of the complex polarization propagator approach [14] in the multi-configurational self-consistent field approximation. This accomplishment should be viewed in light of the possibility to study intermolecular – as well as intramolecular in applications involving extended systems – energy transport mechanisms.

We have also presented highly accurate density functional theory results (with the B3LYP exchange-correlation functional) for the ground state dispersion coefficients of three members of the azabenzenes.

References

[1] A.J. Thakkar, *J. Chem. Phys.* **75** (1981), 4496.
[2] P.W. Fowler and N.C. Pyper, *Mol. Phys.* **59** (1986), 317.
[3] W. Rijks and P.E.S. Wormer, *J. Chem. Phys.* **88** (1988), 5704.
[4] P.W. Fowler, P. Jørgensen and J. Olsen, *J. Chem. Phys.* **93** (1990), 7256.
[5] A.J. Thakkar, H. Hettema and P.E.S. Wormer, *J. Chem. Phys.* **97** (1992), 3252.
[6] C. Hättig and B.A. Hess, *Chem. Phys. Letters* **233** (1995), 359.
[7] C. Hättig and B.A. Hess, *J. Phys. Chem.* **100** (1996), 6243.
[8] C. Hättig, O. Christiansen and P. Jørgensen, *J. Chem. Phys.* **107** (1997), 10592.
[9] R.D. Amos, N.C. Handy, P.J. Knowles, J.E. Rice and A.J. Stone, *J. Phys. Chem.* **89** (1985), 2186.
[10] P.W. Fowler, P. Lazzeretti and R. Zanasi, *Mol. Phys.* **68** (1989), 853.
[11] M.A. Spackman, *J. Chem. Phys.* **94** (1991), 1295.
[12] P.E.S. Wormer and H. Hettema, *J. Chem. Phys.* **97** (1992), 5592.
[13] P. Norman, A. Jiemchooroj and B.E. Sernelius, *J. Chem. Phys.* **118** (2003), 9167.
[14] P. Norman, D.M. Bishop, H.J.Aa. Jensen and J. Oddershede, *J. Chem. Phys.* **115** (2001), 10323.
[15] H. Failache, S. Saltiel, M. Fichet, D. Bloch and M. Ducloy, *Phys. Rev. Lett.* **83** (1999), 5467.
[16] M. Boström, J.J. Longdell and B.W. Ninham, *Europhys. Lett.* **21** (2002), 59.
[17] P. Norman, D. Jonsson, O. Vahtras and H. Ågren, *Chem. Phys.* **203** (1996), 23.

[18] D. Jonsson, P. Norman and H. Ågren, *J. Chem. Phys.* **105** (1996), 6401.
[19] C. Hättig, O. Christiansen, S. Coriani and P. Jørgensen, *J. Chem. Phys.* **109** (1998), 9237.
[20] D. Jonsson, P. Norman, Y. Luo and H. Ågren, *J. Chem. Phys.* **105** (1996), 581.
[21] D. Jonsson, P. Norman and H. Ågren, *Chem. Phys.* **224** (1997), 201.
[22] D. Jonsson, P. Norman, H. Ågren, K.O. Sylvester-Hvid and K.V. Mikkelsen, *J. Chem. Phys.* **109** (1998), 6351.
[23] P. Norman, D. Jonsson and H. Ågren, *Chem. Phys. Lett.* **268** (1997), 337.
[24] F. Visser and P.E.S. Wormer, *Mol. Phys.* **52** (1984), 923.
[25] M. Schütz, J. Hutter and H.P. Lüthi, *J. Chem. Phys.* **103** (1995), 7048.
[26] A. Salam and T. Thirunamachandran, *J. Chem. Phys.* **104** (1996), 5094.
[27] M. Marinescu and L. You, *Phys. Rev.* **A59** (1999), 1936.
[28] J. Olsen and P. Jørgensen, *J. Chem. Phys.* **82** (1985), 3235.
[29] A.J. Sadlej, *Collect. Czech. Chem. Commun.* **53** (1988), 1995.
[30] Landolt-Börnstein, *New Series II/23*, Springer Verlag, 1995.
[31] T. Helgaker, H.J.Aa. Jensen, P. Jørgensen, J. Olsen, K. Ruud, H. Ågren, A.A. Auer, K.L, Bak, V. Bakken, O. Christiansen, S. Coriani, P. Dahle, E.K. Dalskov, T. Enevoldsen, B. Fernandez, C. Hättig, K. Hald, A. Halkier, H. Heiberg, H. Hettema, D. Jonsson, S. Kirpekar, R. Kobayashi, H. Koch, K.V. Mikkelsen, P. Norman, M.J. Packer, T.B. Pedersen, T.A. Ruden, A. Sanchez, T. Saue, S.P.A. Sauer, B. Schimmelpfennig, K.O. Sylvester-Hvid, P.R. Taylor and O. Vahtras, Dalton, an ab initio electronic structure program, Release 1.2. See http://www.kjemi.uio.no/software/dalton/dalton.html,2001.
[32] T. Helgaker, P.J. Wilson, R.D. Amos and N.C. Handy, *J. Chem. Phys.* **113** (2000), 2983.
[33] F.G. Mulder, G. Van Dijk and C. Huiszoon, *Mol. Phys.* **38** (1979), 577.
[34] E.F. Archibong and A.J. Thakkar, *Mol. Phys.* **81** (1994), 557.
[35] Z.-L. Cai, K. Sendt and J.R. Reimers, *J. Chem. Phys.* **117** (2002), 5543.

Journal of Computational Methods in Sciences and Engineering 4 (2004) 333–344
IOS Press

Electronic and vibrational polarizability and first hyperpolarizability of charge transfer chromophores: Quantum chemistry investigation

Amar Saal and Ourida Ouamerali*
Laboratory of Physico-Chimie Théorique et de Chimie Informatique, Faculty of Chemistry
USTHB-University, Algiers, Algeria

Abstract. Static vibrational and electronic contributions to the longitudinal polarizability (α) and first hyperpolarizability (β) have been calculated, within the double harmonic oscillator approximation, at the HF/6-31G for a set of polyconjugated, D-linker-A, compounds. Effects due to the nature of the linker and of the D/A-pair type, on the geometrical structures and the electrical properties, are investigated. It turns out that, for a given linker, in almost all cases, the static β_L^e (0) and β_L^v (0) vary in the same order, with the type of the acceptor group A (A # NO_2, COH, CF_3). As well, the pair HO/NO_2 was associated the greatest β_L^e (0) and β_L^v (0) values. The analysis upon the SOM expressions, due to Bishop and Kirtman, demonstrates that, more than 50% of β_L^v owes to three, two or even to only one vibrational normal mode contribution. Moreover, the vibrational modes with frequencies at, or more than, 3000 cm^{-1} present a null individual vibrational first hyperpolarizability.

Keywords: Polarizability, hyperpolarizability, vibrational contribution, *ab-initio*

1. Introduction

The field of NonLinear Optics (NLO) is subject of great interest because of its industrial applications. The computation of (hyper)polarizabilities, properties governing the optical responses of matter, has been a purpose of numerous investigations [1–3].

The electrical polarizability P ($P = \alpha, \beta, \gamma$) is a consequence of many contributions: electronic, vibrational, . . .etc. The perturbation of the electronic motion leads to electronic polarizability component and the vibrational contribution arises from the distortion of the vibrational motion upon the applied electric fields. The contributions due to other physical phenomena, with long response times, like rotation, are not discussed in this paper. To get agreement with experiment, it is indispensable to take into account the vibrational polarizability, see references (2) and (3) and references therein.

The support of the perturbation of vibrational motion to electrical properties is originated from two sources [4]: changes of the potential energy surface (leads to the curvature term) and changes of the nuclear equilibrium positions (determines the so-called nuclear relaxation term). Another way to see the

*Corresponding author. E-mail: oouamerali@usthb.dz.

vibrational polarizability is: as the sum of a pure vibrational contribution and the zero point vibrational averaging correction. It has been demonstrated [5] that nuclear relaxation + curvature term = pure vibrational contribution + zero point vibrational averaging correction.

As well as, there are two different techniques to calculate the electronic (hyper)polarizability: the Sum-Over-States methods and finite field theory, also, two major approaches have been used for the evaluation of the vibrational contributions to the (hyper)polarizability: perturbation treatment and finite field approximation.

General Sum-Over-States formulas have been given by Orr et Ward [6]. By adopting the clamped nucleus approximation [7], Bishop and Kirtman [8,9] have given, on the basis of the SOS formulas, compact expressions for the dynamic vibrational (hyper)polarizability for polyatomic molecules [9,10]. The BK-expressions are simple, non-divergent and easy to manipulate. The publication of the BK-perturbation approach has led to numerous determinations of the vibrational support to the electrical properties (polarizabilities) [2,3].

Aside this perturbation theory, the finite field-nuclear relaxation techniques is widely used for the evaluation of static vibrational (hyper)polarizabilities [11,12]. This is based on a finite differentiation, with respect to the electric fields, of the field-dependent energy or of the induced dipole moment. Since the static quantities are experimentally meaningful, a new Finite Field-Nuclear Relaxation (FF-NR) version was developed by Bishop et al. [13,14] within the infinite optical frequency approximation [15]. This version has allowed to obtain the dynamic nuclear relaxation [16] and the curvature terms. Note also that, this FF-NR method was recently been "improved" by using a relatively small number of field-induced coordinates (FICs) instead of the 3N-6 normal coordinates of the molecule to evaluate the vibrational nuclear relaxation (hyper)polarizability [17].

For about three decades, π-conjugated organic compounds have been found to be of interest for nonlinear optic devices [18–28]. Eight different series of oligomers: polyacetylene, polyyne, polydiacetylene, polybutatriene, polycumulene, polysilane, polymethineimine and polypyrrole, were studied by Champagne et al. [29] They have found that vibrational second hyperpolarizability γ^v is substantial in comparison with static electronic γ^e. They also concluded that, for semi-quantitative purposes, in planar π-conjugated polymers, anharmonicity due to nuclear relaxation may be ignored. The polymethineimine oligomers have, also, been of another investigation made by Perpète et al. [30] In this reference, the authors found that for dc-Pockels and Optical Rectification processes, the vibrational component β^v is substantial. It also turns out that, till fifteen unit cells (CH=N) the vibrational contribution per unit cell β^v/N increases nearly linearly with chain length, whereas, its electronic counterpart, β^e/N, tends to saturate. Furthermore, β^v/N is clearly larger than its corresponding β^e/N. Polyene, polyyne and polythiophene have been a subject of an investigation made by Champagne and Kirtman [20]. These same authors have studied [31], the effect of the nature of the linker on the first hyperpolarizabilities of π-conjugated organic molecules at the CHF/6-311G**. They have carried out that, most often, increases in β^e are accompanied by increases in β^v, though they have mentioned counter examples to this observation.

Although, polyenic-like molecules are largely studied in theory they are less investigated experimentally. In fact, these compounds are irresistible to the processing conditions. Conditions which are indispensable to obtain NLO phenomena. The stabilization of the chemical molecular structure of organic materials to these conditions is usually ensured by introduction of aromatic moieties. Furthermore, including π-electron donor and/or acceptor groups, on para positions of the conjugated backbone, increases the values of both contributions to the (hyper)polarizabilities [1,2,26,32–38].

In addition to effects due to the increase of the conjugation length and the strength of the D/A pair, it has been demonstrated that geometrical structure and solvent effects have a great influence on NLO properties [39–44]. Solvent effects on linear polarizability and first hyperpolarizability of push-pull polyenes have been investigated by Cammi et al. [39]. Also, Lipinski et al. [40,41] have studied the influence of geometry/conformation and solvent effects on first and second molecular hyperpolarizabilities of CT chromophores. They have shown that these effects should be considered for a reasonable agreement with experiment.

In this paper, a set of push-pull π-conjugated molecules, with two aromatic rings, has been considered. A study upon the effect of inter-ring linker and the D/A pair on the polarizability and the first hyperpolarizability, as well as on the geometrical structures, was made. In the next section, we give summary of the method used for the calculation of the vibrational properties. The results and their discussion are given in section three, finally, the conclusion this investigation and some outlooks are outlined in the last section.

2. Methodological and computational aspects

The induced dipole moment due to the application of external electric fields on a molecule can be expressed, in a form of a Taylor series

$$\Delta\mu_r(\omega_\sigma) = \sum_s \alpha_{r,s}(-\omega_\sigma;\omega_1)F_s(\omega_1) + \frac{1}{2!}\sum_{s,t}\beta_{rst}(\omega_\sigma;\omega_1,\omega_2)F_s(\omega_1)F_t(\omega_2) + \ldots \tag{1}$$

Where $F_s, F_t, \ldots$ are the Cartesian components of the field, α and β are the polarizability and the first hyperpolarizability tensors, respectively. $\omega_\sigma = \sum_i \omega_i, \omega_i$ is the frequency dispersion of the field F_i, and $r, s, t, \ldots$ are associated to the Cartesian coordinates.

By assuming that the fields act sequentially rather than simultaneously upon the two types of motions (electronic and vibrational), i.e. adopting the clamped nucleus approximation [7,33], a separation may be made between the different contributions of the polarizability and the hyperpolarizabilities, the factors of the Taylor series of the induced dipole moment. Furthermore, the electrical property P ($P = \alpha, \beta, \ldots$) can be expressed as the sum of three contributions: electronic P^e, the zero point vibrational correction ΔP^{ZPVA}, and the pure vibrational one P^v.

The electronic contribution can be evaluated by either the Sum-Over-States (SOS) method, where the sums are made over the electronic states, or the derivative methods, where the electrical property is expressed as a function of the energy or the dipole moment derivatives with respect to the components of the field these derivatives are calculated analytically or numerically. As well, the vibrational contribution can be calculated by two techniques, finite field approach or perturbation theory [5].

Bishop and Kirtman [8,9] wrote handling expressions of the vibrational (hyper)polarizability of polyatomic molecules. Within the Double-Harmonic-Oscillator approximation (DHO) [45], holding on only electrical and mechanical harmonic terms, where the ZPVA corrections are zeros [5], the tensor components of the polarizability α and the first hyperpolarizability β are, by invoking the fact that the vibrational frequencies are much smaller than the optical ones the so-called "enhanced" or the infinite frequency approximation [45,46], expressed as [47]:

$$\alpha^v_{rs}(-w_\sigma;w_1) = [\mu^2]^{0,0} = k^{(1)}(w_\sigma;w_1)\sum_a^{3N-6}\left(\frac{\partial\mu_r}{\partial Q_a}\right)_0\left(\frac{\partial\mu_s}{\partial Q_a}\right)_0\frac{1}{w_a^2}, \tag{2}$$

$$\beta^{v}_{rst}(-w_\sigma;w_1,w_2)=[\mu\alpha]^{0,0}=k^{(2)}(w_\sigma;w_1,w_2)\sum_{a}^{3N-6}\left[\left(\frac{\partial\mu_r}{\partial Q_a}\right)\left(\frac{\partial\alpha_{st}}{\partial Q_a}\right)+\left(\frac{\partial\mu_s}{\partial Q_a}\right)\left(\frac{\partial\alpha_{rt}}{\partial Q_a}\right)+\left(\frac{\partial\mu_t}{\partial Q_a}\right)\left(\frac{\partial\alpha_{rs}}{\partial Q_a}\right)\right]\frac{1}{w_a^2}, \quad (3)$$

where $\sum_a$ indicates the summation over the vibrational normal modes of the molecule, $k^{(i)}$ are factors associated to the phenomena; $k^{(1)} = 1$ for the static polarizability ($w_\sigma = w_1 = 0$), $k^{(1)} = 0$ for dynamic polarizability. The static, dc-Pockels, OR, and SHG are associated to $k^{(2)} = 1$, 1/3, 1/3, and 0, respectively. $w_a = 2\pi\text{v}_\alpha$ is the circular frequency of the vibrational mode Q_a.

In this work, we will evaluate longitudinal components of the two contributions, vibrational and electronic, to the polarizability and the first hyperpolarizability. Once the geometrical structure are optimized using the SCF wave functions, the electrical properties are calculated.

By using the CPHF method implemented in GAUSSIAN 98W [48] set of programs, the electronic quantities as well as the force constants and the dipole moment and the polarizability derivatives wrt the normal coordinates are computed. Since these derivatives are very sensitive to the molecular geometrical structures, a tight convergence threshold on the forces is adapted.

The choice of the basis set is a very delicate question while evaluating the (hyper) polarizabilities. It is well known that for a good determination of these properties, which depend very strongly upon the geometrical structure, it is recommended to use extended basis sets augmented by diffuse and polarization functions. The basis set 6-31G is adapted in this work. Although, it appears as insufficient for the study of small molecules, it becomes more and more adequate when the molecular size increases [49,50]. Indeed, the limitation due to the finite number of Gaussian functions on a given atom is reduced by the Gaussian functions on the neighboring atoms. Furthermore, Perpète et al. [30] have made calculations on polymethineimine oligomer series with different basis sets and compared the results to those obtained at HF/6-31G. Their study have shown that the ratios β^e(6-31G)/$\beta^e(X)$ and β^v(6-31G)/$\beta^v(X)$ (where X = 6-311G**, cc-pVDZ, AUG-cc-pVDZ) tend to a constant when the number of cells increases. Note, also, that in this study, we are more interested by the qualitative and general variation of the electrical properties under consideration, rather than to compute their exact value.

3. Results and discussion

We have proposed to study a set of polyconjugated D/A compounds: HO-C_6H_4-X-Y-C_6H_4-A and HO-C_6H_4-C_6H_4-A where: the segment X-Y may be CH=CH, CH=N, N=CH, N=N, C≡C and A # COH, NO_2, CF_3. We will design, through this paper, for commodity of writing, the molecule HO-C_6H_4-X-Y-C_6H_4-A by HOxyR and: HO-C_6H_4-C_6H_4-A by HOR. For example, HOchchNO_2 refers to the molecule: HO-C_6H_4-CH=CH-C_6H_4-NO_2 and HONO_2 to: HO-C_6H_4-C_6H_4-NO_2. The electrical properties; dipole moments, electronic and vibrational (hyper)polarizabilities are computed. Note that, as it has been mentioned by Bonifassi et al. [35,36], some of these molecules have been synthesized at the Macromolecular Chemistry Laboratory of Saint Etienne University URA CNRS N° 842 [51–54].

All the calculations were performed at the Hartree – Fock level and the 6-31G basis set by using GAUSSIAN98W [48] package. The longitudinal axis is the direction of the permanent dipole moment.

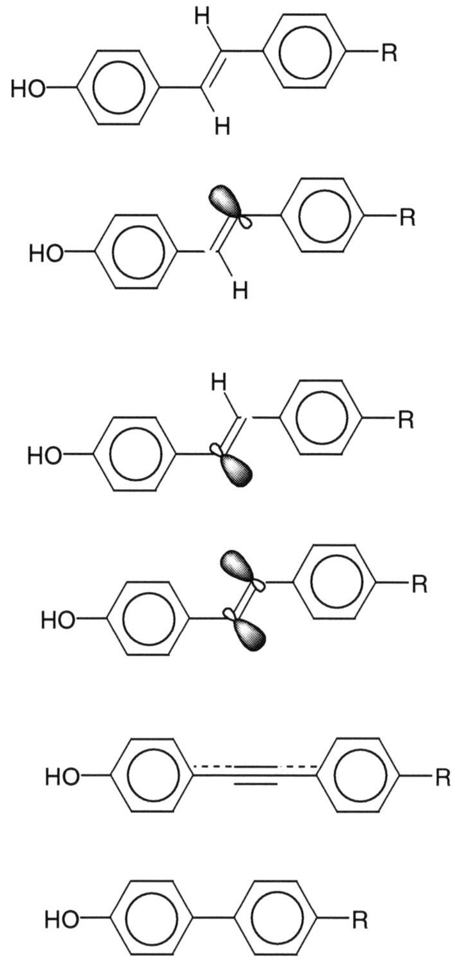

Fig. 1. The molecular geometrical structure of the molecules studied. The bond length, given in Å, are obtained at HF/6-31G.

3.1. Geometrical structure

We have, first, made a complete optimization of the geometrical structure of the molecules under consideration (Fig. 1). It is well known that the length of the conjugation has a significant influence on the molecular hyperpolarizabilities. The geometrical parameters of importance, which affect and perturb the conjugation, here, is the torsion angles between the plans of the two aromatic rings, Figures 1 and 2.

From the data in Table 1, one can see that, the molecules HOchchNO_2, HOchnNO_2, HOnnR, and HOccR (where R # COH, NO_2, CF_3) are "planar", however, an angle of $\sim 30°$ is formed between the plans of the two aromatic cycles in the cases of HOnchR (R# COH, NO_2, CF_3), and a torsion of about 40° between the rings of the molecules HOchchR', HOchnR', and HOR (R'# COH, CF_3 and R # COH, NO_2, CF_3). As well as the torsion angle increases the conjugation decreases. Two important remarks must be noted.

The first is the fact that, in the cases of HOchchR', the linker -CH=CH- is on a plan which subdivides the torsion angle between the two rings to two equal ones, i.e. of $\sim 21°$. This certainly increases the conjugation from one aromatic cycle to the linker -CH=CH- and then to the other cycle.

Table 1
The values of the inter-ring torsion angles, Figure 2, of the different molecules studied, calculated at the HF/6-31G, are given in this table in (°)

Molecules*	R	ϕ_1^{**}	ϕ_2^{**}	ϕ_3^{**}
R1-CH=CH-R2	COH	19.4	0.5	21.1
	NO_2	0.0	0.0	0.0
	CF_3	19.4	0.5	21.8
R1-CH=N-R2	COH	1.4	3.2	40.6
	NO_2	0.0	0.0	0.0
	CF_3	1.5	2.1	40.5
R1-N=CH-R2	COH	29.1	1.5	1.4
	NO_2	26.0	1.4	1.3
	CF_3	28.8	1.4	1.5
R1-N=N-R2	COH	0.0	0.0	−0.0
	NO_2	0.0	0.0	0.0
	CF_3	0.1	0.0	0.1
R1-C≡C-R2	COH	0.0	–	–
	NO_2	0.0	–	–
	CF_3	0.0	–	–
R1-R2	COH	42.8	–	–
	NO_2	42.4	–	–
	CF_3	43.4	–	–

*R1 # HO-C_6H_4-, R2 # -C_6H_4-R.
**See Fig. 2.

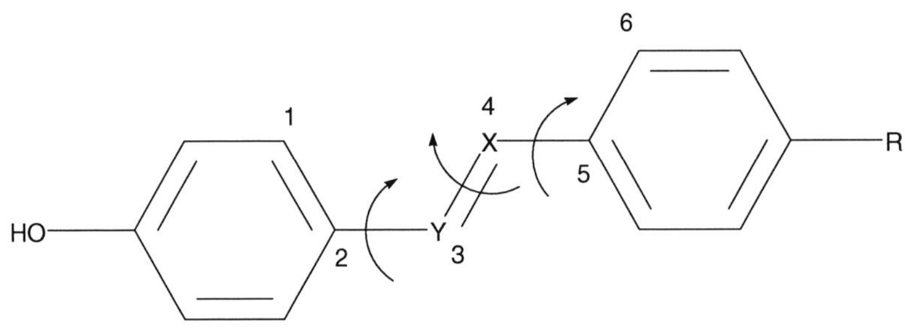

Fig. 2. In this figure, we show the different torsion angles, given in Table 1, existing between the two aromatic rings of the molecules under consideration (R # COH, NO_2, CF_3). $\phi 1$ = angle (1234), $\phi 2$ = angle (2345), $\phi 3$ = angle (3456) and ϕ = angle (1256).

The second remark is: the torsion angle in the case of the segment -CH=N- and -N=CH- is due to the rotation around the Nitrogen – Phenyl bond. This may be a consequence of the interaction between the Hydrogen atoms H1 and H5 (a repulsion) and between H9 and N (an attraction which maintain the part -CN=CH-Ph- in the same plane), Figure 3. In the case of the segment -CN=N-, there are two interactions N ... H (attractions). One of them maintain the part -CN=N-Ph- in the same plane and the other maintain, in the same plane, the part -CPh-N=N-. And because the two nitrogen atoms are hybridized sp2 the molecule -CPh-N=N-Ph- is almost planar.

The results show also that small torsion angles are always associated to the NO_2 acceptor group. This fact may be due to the strength of this group relatively to the other electro -acceptors CF_3 and COH. NO_2 pull the electronic cloud and let the other atoms of the molecule less charged thus the interactions are weaken.

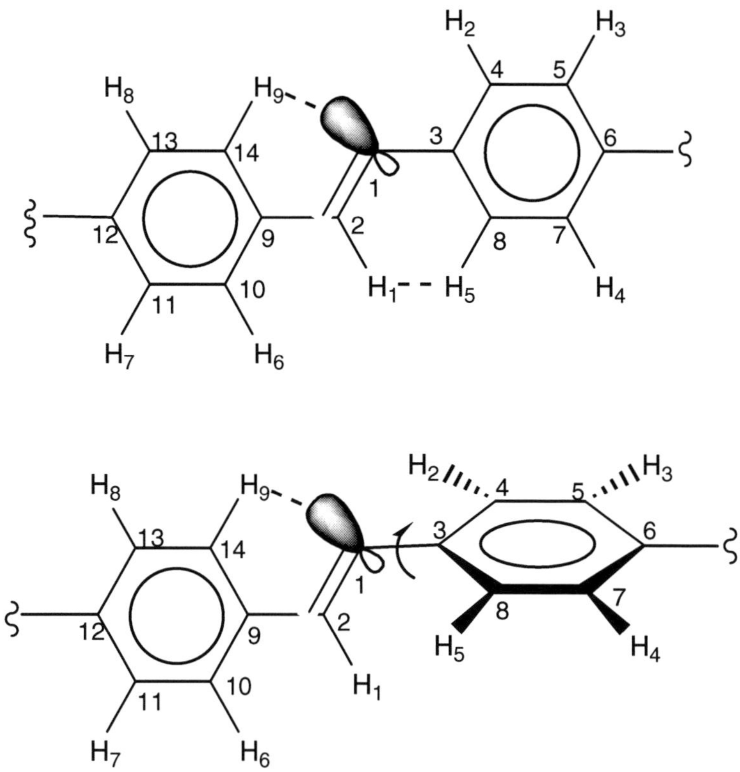

Fig. 3. In this figure, the existing interaction which lead to the inter-ring torsion angle are reported, in the case of the CH=N linker.

3.2. *Dipole moment and polarizability*

In Table 2, we compare the longitudinal dipole moments and the longitudinal components of the static polarizabilities, electronic and vibrational, of the different molecules.

According to the results reported in this table we can see that the dipole moments, in all cases and for a given segment-type, increases in the order μ_L(COH) $<$ μ_L (CF_3) $<$ μ_L (NO_2). The variation of the longitudinal component μ_L with the inter-ring linker, Table 2, show that it varies very strongly because, as we have seen, the segment influences the geometrical structure (the molecular planarity) thus the charge distribution. One can also remark that the molecular dipole moment depends more on the nature of the acceptor groups (NO_2, COH, CF_3) than on the segment moieties (CH=CH, CH=N, N=CH, N=N, C≡C, without segment).

The two contributions to the longitudinal static polarizability, electronic and vibrational, are reported in Table 2. From these data, α^e is more affected by changing the charge-acceptor group than is its vibrational counterpart α^v. The ratio α^v/α^e shows a clear inferiority of α^v relatively to α^e. In all the cases, this ratio is comprised between 12 and 26%. Thus, the classification of the different acceptor groups, for a given segment-type, upon the polarizabilities of their corresponding molecules, is not affected by the vibrational contribution. In the case, for example, of the CH=CH moiety we have: α^e and α^T ($= \alpha^e + \alpha^v$) increase in the order $CF_3 <$ COH $< NO_2$, although, increases in the sense $CF_3 < NO_2 <$ COH.

Table 2

The molecular dipole moments and the two contributions to the linear polarizability are listed in this table in a.u.. We also give the inter-ring torsion angle in (°). All the values are obtained at the HF/6-31G implemented in GAUSSIAN98W

Molecules*	R	μ_L	$\alpha_L^e(0)$	$\alpha_L^v(0)$	α_L^v/α_L^e	Angle**
R1-CH=CH-R2	COH	1.589	272.8	55.4	0.20	41.058
	NO_2	2.804	282.7	46.2	0.16	0.047
	CF_3	1.942	241.6	43.3	0.18	41.681
R1-CH=N-R2	COH	1.827	256.9	49.0	0.19	45.183
	NO_2	2.834	280.4	48.8	0.17	0.048
	CF_3	1.946	255.5	50.1	0.19	44.131
R1-N=CH-R2	COH	1.373	252.9	65.7	0.26	31.983
	NO_2	2.829	239.4	57.4	0.24	28.711
	CF_3	2.096	198.3	51.5	0.26	31.639
R1-N=N-R2	COH	1.460	292.3	51.9	0.18	0.005
	NO_2	2.773	272.8	41.1	0.15	0.000
	CF_3	1.945	241.2	42.4	0.18	0.181
R1-C≡C-R2	COH	1.446	302.0	46.6	0.15	0.001
	NO_2	2.791	298.4	36.2	0.12	0.019
	CF_3	1.943	364.2	38.4	0.14	0.033
R1-R2	COH	1.417	190.2	41.8	0.22	42.811
	NO_2	2.568	208.5	29.5	0.14	42.359
	CF_3	1.795	182.5	30.8	0.17	43.430

*R1 # HO-C_6H_4-, R2 # -C_6H_4-R.
**See Fig. 2.

3.3. *First hyperpolarizability*

In addition to the electronic component of the longitudinal static first hyperpolarizability, the vibrational support as well as the ratio β^v/β^e are listed in Table 3 for all the molecules displayed in Fig. 1. As we have mentioned above, the longitudinal axis is the permanent dipole moment direction.

The data of Table 3 indicate that, generally, β^e and β^v are almost of the same order of magnitude, as it has been pointed out for many push-pull molecules [16,18–28,31] Although, some particular cases have to be mentioned. The molecule HOchchCF_3 shows a negative electronic first hyperpolarizability β^e, thus, the two contributions β^e and β^v will not be complementary (we have $\beta^e < \beta^e + \beta^v < \beta^v$). The values of the ratio β^v/β^e reported in Table 3, show that for certain molecules: HOchchCF_3, HOchnNO_2 and HOnchCF_3 the vibrational β^v is greater than its electronic β^e, at least at the static limit. In the case of HOchchCOH, this ratio is only of 29%. In fact, for this compound and for HOchchNO_2, the other first hyperpolarizability β^v components, i.e. β^v_{xxx} and β^v_{yyy}, are, rather than being negligible in front of $\beta^v_L(\beta^v_L = \beta^v_{zzz})$, more important, as it is the case for the other molecules.

For a given inter-ring linker, apart CH=CH, β^e, β^v and $\beta^T(= \beta^e + \beta^v)$ vary in the same order. When the linker is CH=N this variation is β(COH)< β (CF_3) < β (NO_2). For the other linkers these quantities increase in the sense β(CF_3) < β (COH)< β (NO_2). Remark, finally, that, for same segment the acceptor group NO_2 (inductive and mesomeric attractive effects) gives, always, the larger values of β^e and β^v. A semi-empirical calculation (AM1) [37], on the molecules HO-Ph-CH=N-Ph-R (where R # NO_2, CN, CF_3), have given the classification: β^e(CF_3) < β^e (CN)< β^e (NO_2).

The results of Table 1, show that, when HO/NO_2 is chosen to be the D/A pair, the insertion of a segment between the phenyl rings, i.e. passing from Ph-Ph linker to Ph-CH=CH-Ph, Ph-CH=N-Ph, Ph-N=N-Ph and Ph-C≡C-Ph, decreases the torsion angle formed by the plans of the two aromatic rings

Table 3
Vibrational and electronic longitudinal first hyperpolarizabilities (in a.u.), of the molecules considered, computed at the HF/6-31G level are given in this table. The Bond Length Alternation (BLA) values, in Å, are also listed

Molecules*	R	$\beta_L^e(0)$	$\beta_L^v(0)$	$\beta_L^T(0)^{**}$	β_L^v/β_L^e	BLA
R1-CH=CH-R2	COH	2046.4	590.9	2637.3	0.29	0.0457
	NO_2	3393.8	5817.8	3393.8	1.71	0.0413
	CF3	−99.1	2289.1	2190.0	−23.10	0.0499
R1-CH=N-R2	COH	1439.5	1978.3	3417.8	1.37	0.0466
	NO_2	3112.9	3828.5	6941.4	1.23	0.0406
	CF_3	1491.6	2178.5	3670.1	1.46	0.0498
R1-N=CH-R2	COH	1381.2	3148.7	4529.9	2.28	0.0520
	NO_2	1650.1	3944.2	5594.3	2.39	0.0498
	CF_3	290.5	1807.3	2097.8	6.22	0.0575
R1-N=N-R2	COH	2182.4	2924.9	5107.3	1.34	0.0482
	NO_2	2918.3	3023.9	5942.2	1.04	0.0452
	CF_3	1244.5	1482.6	2727.1	1.19	0.0535
R1-C≡C-R2	COH	2454.8	2647.6	5102.4	1.08	0.0597
	NO_2	3438.2	2792.5	6230.7	0.81	0.0553
	CF_3	1264.7	1242.2	2506.9	0.98	0.0633
R1-R2	COH	899.9	1175.7	2075.6	1.31	0.0299
	NO_2	1554.9	2146.5	3701.4	1.38	0.0287
	CF_3	422.1	1056.4	1478.5	2.50	0.0341

from 42° to 0° and increases both β^e and β^v by a factor between 1.0 and 2.7 (Table 3). This angle is of 28.7° for the molecule HO-Ph-N=CH-Ph-NO_2, its $\beta^e_{28^\circ}$ is higher than $\beta^e_{42^\circ}$ and smaller than $\beta^e_{0^\circ}$, we have: $\beta^e_{42^\circ} < \beta^3_{28^\circ} < \beta^3_{0^\circ}$. A similar conclusion was found by Barzoukas et al. [55] Thus, upon the nature of the segment linker, we have the classification of the static electronic first hyperpolarizability: β^e (without segment) $\ll$ β^e (N=CH) $\ll$ β^e (N=N) $<$ β^e (CH=N) $<$ β^e (CH=CH) $<$ β^e (C≡C). However, we remark that when the torsion angle between the two rings is almost fixe the computed β^e remains unchanged.

The vibrational contribution to the first hyperpolarizability β^v, for molecules with HO/NO_2 pair, increases in the order: (without segment) < C≡C < N=N < N=CH < CH=CH < CH=N. We can see that there is no correlation between the inter-cycle torsion angle and the static vibrational β^v. Relatively to its electronic counterpart, the largest vibrational contribution to the first nonlinear polarizability β is, in the case of the pair HO/NO_2, obtained when N=CH is used as a linker with a ratio β^v/β^e of 2.39.

The correlation between electrical properties and the Bond Length Alternation (BLA), defined as the difference between the average single and double-bond lengths along the conjugated pathway, is largely investigated for several organic compounds. In Table 3, we have reported the values of BLA in the cases of the different molecules studied. It results, from this table, that the BLA increases, for a given inter-ring segment, in the order NO_2 < COH < CF_3. This means that the substituent NO_2 acts more strongly on the bond lengths, so that, it reduces the length of the single-bonds and increases those of the double-bonds, while the CF_3-group amplifies the difference between the lengths of the two bond-types, i.e. the value of BLA. The longitudinal first hyperpolarizabilities, electronic and vibrational, vary, almost, in the inverse order of the BLA, i.e. in the order NO_2 > COH > CF_3.

The vibrational contribution to the first hyperpolarizability, β^v_L, can be written [47] as the sum of individual normal mode vibrational hyperpolarizability, i.e. one can write, for a molecule with N normal

Table 4

In this table, we report the individual contributions(in a.u.) of the three most contributing normal modes to the longitudinal static hyperpolarizability β_L^v of the molecules HO-Linker-NO_2. We have also computed the ratio $\beta_{a,L}^v/\beta_L^v$

Linker	β_L^v	Normal mode contributions			$\Sigma\beta_{a,L}^v/\beta_L^v$
		N°	$\upsilon(\text{cm}^{-1})$	$\beta_{a,L}^v$	
Ph-CH=CH-Ph		1	23.2	2725(47)*	
	5817	57	1471.9	1037(18)	77**
		68	1801.4	672(12)	
Ph-CH=N-Ph		55	1471.3	872(22)	
	3829	66	1797.0	909(24)	63
		68	1878.3	635(17)	
Ph-N=CH-Ph		1	29.7	2672(68)	
	3944	2	37.1	289(07)	87
		56	1479.9	454(12)	
Ph-N=N-Ph		46	1260.4	285(09)	
	3024	47	1291	263(15)	64
		62	1773.2	1214(40)	
Ph-C≡C-Ph		31	918.9	−88(-03)	
	2793	54	1476.2	938(34)	55
		64	1803.1	658(24)	
Ph-Ph		1	51.2	270(13)	
	2147	3	79.5	520(24)	65
		49	1476.6	602(28)	

*$(\beta_{a,L}^v/\beta_L^v)\times 100$.
**(sum of the three normal mode sited/total vibrational β_L^v) $\times$ 100.

modes:

$$\beta_L^e = \sum_a^N \beta_{a,L}^v = \beta_{1,L}^v + \beta_{2,L}^v + \ldots + \beta_{N,K}^v.$$

We have calculated the vibrational first hyperpolarizability of each normal mode for all the molecules studied and we report the first three more contributing modes, in Table 4. The results of this table show that more than 50% of the total vibrational contribution is due, in all the cases, to three, two or even only one vibrational normal modes. For the molecules $HOchchNO_2$ and $HOnchNO_2$, the lowest vibrational state (N°1), with a frequency between 20 and 30 cm^{-1}, presents the largest contribution to $\beta_L^v(0)$. It is of 47% in the case of $HOchchNO_2$ and of 68% for the molecule $HOnchNO_2$. However, the first normal mode, apart HO-Ph-Ph-NO_2 where its contribution is of 13% to $\beta_L^v(0)$, does not contribute in the other cases. Furthermore, the vibrational motions with frequencies 3000 cm^{-1}, or more, present negligible contributions to this property.

4. Conclusion

In the present investigation, electronic and vibrational contributions to the polarizability and first hyperpolarizability of a set of push-pull π-conjugated chromophores were determined, at the HF/6-31G, and discussed.

It was shown that the D/A-type pair changes the geometrical structure, the charge distribution along the molecular backbone and influences the electrical properties, dipole moment and polarizabilities. It, also, turns out that, for a given inter-ring linker, apart CH=CH, the longitudinal static β^e and β^v vary in the same order with the nature of the acceptor group (COH, NO_2 and CF_3). In all cases, the pair HO/NO_2 gives the largest values of these properties.

Upon the inter-ring torsion angle ϕ, for the pair HO/NO_2, the electronic $\beta_L^e(0)$ increases when this angle decreases. However, no correlation was found between ϕ and the vibrational static $\beta_L^v(0)$.

A study upon the SOM expression, due to Bishop and Kirtman, was made. It was showed that, only a few normal modes contribute to the vibrational first hyperpolarizability. Thus, more than 50% of the total vibrational contribution to β is due to three, two or even to only one normal mode. The vibrational motion at frequencies of, or more than, 3000 cm^{-1} do not contribute in any way to $\beta_L^v(0)$.

The results presented in this investigation are only qualitative, because they were carried out at the double harmonic oscillator approximation. However, they may be improved by taking into account the anharmonic contributions. Also, considering the solvent and the electronic correlation effects, as well as, extending the basis set can play an important role for agreement with experiment.

References

[1] D.S. Chemla and J. Zyss, eds, *Nonlinear Optical Properties of Organic Molecules and Crystals*, Academic Press, New Yark, 1987.

[2] D.M. Bishop and P. Norman, in: *Handbook of advanced Electronic and photonic Materials and devices*, (Vol. 9), H.S. Nalwa, ed., Nonlinear optical, 2001, Chap. I; B. Champagne and B. Kirtman, in: *Handbook of advanced Electronic and photonic Materials and devices*, (Vol. 9), H.S. Nalwa, ed., Nonlinear optical, 2001, Chap. II.

[3] D.M. Bishop, in: *Molecular vibration and nonlinear optics, Adv. Chem. Phys.*, (Vol. 104), d. I. Prigogine and Stuart A. Rice., eds, © John Wiley & Sons, Inc., 1998.

[4] J. Marti, J.L. Andrés, J. Bertran and M. Duran, *Mol. Phys.* **80** (1993), 625.

[5] J. Marti and D.M. Bishop, *J. Chem. Phys.* **99** (1993), 3860.

[6] B.J. Orr and J.F. Ward, *Mol. Phys.* **20** (1971), 513.

[7] D.M. Bishop, B. Kirtman and B. Champagne, *J. Chem. Phys.* **107** (1997), 5780.

[8] D.M. Bishop and B. Kirtman, *J. Chem. Phys.* **95** (1991), 2646.

[9] D.M. Bishop and B. Kirtman, *J. Chem. Phys.* **97** (1992), 5255.

[10] D.M. Bishop, J.M. Luis and B. Kirtman, *J. Chem. Phys.* **108** (1998), 10013.

[11] D.M. Bishop and S.A. Solunac, *Phys. Rev. Lett.* **55** (1985), 1986.

[12] L. Adamowicz, *Bartlett, J. Chem. Phys.* **84** (1986), 4988; **86** (1987), 7250.

[13] D.M. Bishop, M. Hasan and B. Kirtman, *J. Chem. Phys.* **103** (1995), 4157.

[14] B. Kirtman, J.M. Luis and D.M. Bishop, *J. Chem. Phys.* **108** (1998), 10008.

[15] D.S. Elliott and J.F. Ward, *Mol. Phys.* **51** (1984), 45.

[16] B. Champagne, *Chem. Phys. Lett.* **261** (1996), 57.

[17] J.M. Luis, M. Duran, B. Champagne and B. Kirtman, *J. Chem. Phys.* **113** (2000), 5203.

[18] D.M. Bishop, B. Champagne and B. Kirtman, *J. Chem. Phys.* **109** (1998), 9987.

[19] B. Champagne, *In. J. Quantum Chem.* **65** (1997), 689.

[20] B. Kirtman and B. Champagne, *In. Rev. Phys. Chem.* **16** (1997), 389.

[21] D.M. Bishop and B. Kirtman, *Phys. Rev. B* **56** (1997), 2273.

[22] A. Saal and O. Ouamerali, *Mol. Struc.* **14** (2003), 479.

[23] P. Zuliani, M. Del Zoppo, C. Castiglioni, G. Zerbi, C. Andraud, T. Brotin and A. Collet, *J. Phys. Chem.* **99** (1995), 16242.

[24] P. Zuliani, M. Del Zoppo, C. Castiglioni, G. Zerbi, S.R. Marder and J.W. Perry, *J. Chem. Phys.* **103** (1995), 9935.

[25] C. Castiglioni, M. Del Zoppo and G. Zerbi, *Phys. Rev. B* **53** (1996), 13319.

[26] A. Painelli, *Chem. Phys. Lett.* **285** (1998), 352.

[27] M.S. Wong, C. Bosshard and P. Günter, *Adv. Mater.* **9** (1997), 837.

[28] M.S. Wong, C. Bosshard, F. Pan and P. Günter, *Adv. Mater.* **8** (1996), 677.

[29] B. Champagne, J.M. Luis, M. Duran, J.L. Andrés and B. Kirtman, *J. Chem. Phys.* **112** (2000), 1011.

[30] E.A. Perpète, B. Champagne and D. Jacquemin, *J. Mol. Struct.* (*Theochem*) **529** (2000), 65.

[31] B. Champagne and B. Kirtman, *Chem. Phys.* **254** (1999), 213.

[32] D. Hamoutène, *Thèse de magister-USTHB*, Algiers 1992.
[33] D.M. Bishop, *Rev. Mod. Phys.* **62**(2) (1990), 343.
[34] J. Zyss, *J. Chem. Phys.* **70**(7) (1979), 3333.
[35] Y. Daoudi and P.J. Bonifassi, *J. Mol. Struc.* (*Theochem*) **451** (1998), 277.
[36] Y. Daoudi and P.J. Bonifassi, *J. Mol. Struc.* (*Theochem*) **425** (1998), 147.
[37] H.S. Kim, M. Cho and S.J. Jeon, *J. Chem. Phys.* **107** (1997), 1936.
[38] D.M. Bishop and B. Kirtman, *Phys. Rev. B* **56** (1997), 2273.
[39] R. Cammi. B. Mennucci and J. Tomasi, *J. Am. Chem. Soc.* **120** (1998), 8824.
[40] W. Bartkowiak and T. Misiaszek, *Chem. Phys.* **261** (2000), 353.
[41] J. Lipinski and W. Bartkowiak, *Chem. Phys.* **245** (1999), 263.
[42] M. Cho, *J. Phys. Chem. A* **102** (1998), 703.
[43] K. Clays, E. Hendrickx, M. Triest, T. Verbiest, C. Dehu and J.L. Brédas, *Science* **265** (1994), 632.
[44] G. Bourhill, J.L. Brédas, L.-T. Cheng, S.R. Marder, F. Meyers, J.W. Perry and B.G. Teimann, *J. Am. Chem. Soc.* **116** (1994), 2619.
[45] B. Kirtman and D.M. Bishop, *Chem. Phys. Lett.* **175** (1990), 601.
[46] D.S. Elliott and J.F. Ward, *Mol. Phys.* **51** (1984), 45.
[47] D.M. Bishop, M. Hasan and B. Kirtman, *J. Chem. Phys.* **103** (1995), 4157.
[48] Gaussian 98, Revision A.9, M.J. Frisch, G.W. Trucks, H.B. Schlegel, G.E. Scuseria, M.A. Robb, J.R. Cheeseman, V.G. Zakrzewski, J.A. Montgomery, Jr., R.E. Stratmann, J.C. Burant, S. Dapprich, J.M. Millam, A.D. Daniels, K.N. Kudin, M.C. Strain, O. Farkas, J. Tomasi, V. Barone, M. Cossi, R. Cammi, B. Mennucci, C. Pomelli, C. Adamo, S. Clifford, J. Ochterski, G.A. Petersson, P.Y. Ayala, Q. Cui, K. Morokuma, D.K. Malick, A.D. Rabuck, K. Raghavachari, J.B. Foresman, J. Cioslowski, J.V. Ortiz, A.G. Baboul, B.B. Stefanov, G. Liu, A. Liashenko, P. Piskorz, I. Komaromi, R. Gomperts, R.L. Martin, D.J. Fox, T. Keith, M.A. Al-Laham, C.Y. Peng, A. Nanayakkara, M. Challacombe, P.M.W. Gill, B. Johnson, W. Chen, M.W. Wong, J.L. Andres, C. Gonzalez, M. Head-Gordon, E.S. Replogle and J.A. Pople, Gaussian, Inc., Pittsburgh PA, 1998.
[49] B. Kirtman and M. Hasan, *J. Chem. Phys.* **96** (1992), 470.
[50] G.J.B. Hurst, M. Dupuis and E. Clementi, *J. Chem. Phys.* **89** (1988), 385.
[51] A. Zerroukhi, *Synthèse et caratŕisation de copolymr̀es pour des applications en optique nonlinéaire*, thèse, 93Ly010105, Saint Etienne University, France, 1993.
[52] J.P. Monthéard, B. Boinon and A. Zerroukhi, *J. Applied Polymer Science* **44** (1992), 1307.
[53] J.P. Monthéard, B. Boinon, A. Zerroukhi and A. Cachard, *Polymer* **33** (1992), 3756.
[54] A. Zerroukhi, A. Trouillet, D. Blanc, B. Boinon, A. Cachard and J.P. Monthéard, *J. Applied Polymer Science* **51** (1994), 1165.
[55] M. Barzoukas, A. Fort, G. Klein, C. Serbutoviez, L. Oswald and J.F. Nicoud, *Chem. Phys.* **164** (1992), 395.

Journal of Computational Methods in Sciences and Engineering 4 (2004) 345–355
IOS Press

Electric dipole polarizability and hyperpolarizability of NCCN, NCCP, and PCCP

Tadeusz Pluta* and Piotr Zerzucha
Institute of Chemistry, University of Silesia, Szkolna 9, PL-40 006 Katowice, Poland

Abstract. Components of the static electric dipole polarizabilty (α), and second hyperpolarizability (γ) tensors have been determined by accurate *ab initio* calculations for three molecules containing two conjugated carbon – nitrogen and/or carbon – phosphorus triple bonds. For the NCCP molecule the dipole moment (μ) and non-vanishing components of the first hyperpolarizability (β) tensor have also been calculated. Electron correlation effects have been taken into account by the second-order Many Body Perturbation Theory (MBPT(2)), and coupled cluster (CCSD and CCSD(T)) calculations. The basis set developed by Sadlej (Pol) and designed to be used in electric properties calculations together with its extension (HyPol) have been applied and compared to the results obtained with the standard aug-cc-PVTZ sets of Dunning.

Keywords: Electric dipole polarizability, hyperpolarizability, electron correlation, π- conjugated molecules

Mathematics Subject Classification: 81V55

1. Introduction

Electric response properties of atoms and molecules are known to account for a variety of their physico-chemical features and are underlying the molecular interpretation of different phenomena like refraction, absorption and scattering of electromagnetic radiation, intermolecular interactions [1,2], to mention just a few best known examples. To define the so – called dipole properties one usually expands the field dependent (induced) dipole moment $\mu_i(F_j)$ or the perturbed total energy $E(F_j)$ into a Taylor series with respect to the electric field strength. The electric dipole polarizability tensor (α_{ij}), characterizes linear dependence of the induced dipole moment (μ_i) on the applied external electric field F_j, $(i, j = x, y, z)$. The leading higher-order terms are referred to as the first (β_{ijk}) and second hyperpolarizabilities (γ_{ijkl}), respectively [2]. Because of their relevance for nonlinear optical (NLO) phenomena [3,4] these higher-order terms are objectives of particularly intense experimental and theoretical interest.

In the NLO-related applications one looks for materials with a very strong nonlinear response to the external electromagnetic field. At the molecular level this leads to the search for molecules with large values of the hyperpolarizabilities β and γ [4–6]. Among molecules of potential use in NLO-related applications, the organic molecules with spatially extended, conjugated π-bonds are of particular

*Corresponding author. E-mail: tp@us.edu.pl.

interest [8]. Accurate calculations of β and γ can provide useful data which are often difficult to obtain experimentally.

Recently, Miadoková, Kellö and Diercksen (MKD) [7] have estimated components of the α and γ tensors for a group of linear dicyanopolyacetylenes, $\mathrm{NC}-(\mathrm{CC})_{\mathrm{n}}-\mathrm{CN}$, $n=0,1,2,3$. For the whole set of the investigated molecules they used moderately sized basis sets derived from the original Pol sets developed by Sadlej [9,10] and the second-order many-body perturbation theory (MBPT(2) often referred to as MP2, Møller-Plesset (MP2). The first member of the series was additionally investigated by using the CCSD and CCSD(T) levels of approximations. The calculated value of the parallel component of the second hyperpolarizability γ_{zzzz} along the chain with three conjugated triple carbon – carbon bond, i.e. $n=3$, has been found to be impressively large, exceeding the value of $1.5 \cdot 10^{6}$ a.u., or in the SI units $9.3 \cdot 10^{-59}$ ($\mathrm{C}^4\mathrm{m}^4\mathrm{J}^{-3}$).

Systematic computational study of (hyper)polarizabilities of well – selected groups of molecules with conjugated π-bonds and different heteroatoms undoubtedly contributes to our knowledge and understanding of potentially useful NLO-relevant systems at the molecular level. The choice of systems for the present study (NCCN, NCCP, and PCCP) is motivated by some unusual features of their electronic structure. The first of them has already been investigated by Miadoková et al. [7] and is included here because of some differences in computational details and the comparative character of the present study.

In molecules with triple bonds the electron correlation contribution to the dipole polarizability α has been shown to be very small and negative, i.e., correlated values for both components are slightly smaller than the respective SCF values [11,12]. The same pattern can be observed for conjugated dienes [13]. The presence of the CN group modifies this pattern. However, the overall correlation contribution to the dipole polarizability remains very small. For the series of XCN (X = halogen atom) molecules the electron correlation contribution to the rotational average of the dipole polarizability, α_{av}, as estimated at the MP2 level of approximation, amounts to less than 3 per cent [14]. Similar conclusions can be drawn from the results for small cyano-substituted organic molecules [11] and linear dicyanopolyacetylenes [7]. This observation could be of great practical significance for it means that uncorrelated, low-cost SCF calculations could yield reasonably accurate results for the polarizability tensor α. The main factor affecting the accuracy of the calculated data is then the basis set choice. This finding and attractive consequences which may follow are much in contrast with conclusions obtained for other systems. Usually a significant amount of the electron correlation must be taken into account in order to obtain reasonable values of the (hyper)polarizability tensor [6]. Therefore, it would be of important practical consequences to find out whether the chemically related CP group displays the same pattern of the negligible electron correlation contribution to the dipole polarizability α. One should mention, however, that the correlation corrections for the second hyperpolarizability tensor γ do not seem to follow this pattern, and their contribution to γ of the NCCN molecule has been found to be significant [7].

The two main problems which arise in calculations of (hyper)polarizabilities are the choice of the basis set to describe properly the molecular wavefunction in the presence of an external electric field and taking into account electron correlation effects. Let us briefly address the latter problem first. We have employed advanced but standard and well-documented quantum chemical methods, namely the coupled cluster (CC) method including single and double excitations (CCSD) [22] and the improved CC scheme in which triple excitations are treated approximately (CCSD(T)) [23]. In the coupled cluster method the dynamical electron correlation is introduced in a very efficient way through infinite partial summation of important terms originating in the many-body perturbation theory. The CCSD – based methods have been proved to give excellent results for molecular properties [24,25]. The MBPT(2) results, which are obtained as a by – product in the process of the iterative solution of the CC equations can be regarded as a valuable alternative to those of the CC methods, especially for larger systems. Usually the MBPT(2) results are acceptably accurate while the corresponding caluclations are much less demanding [6,22].

The choice of the basis set appears to be the most important factor in achieving high accuracy in electric properties calculations. In this respect the approach pursued by Maroulis and his coworkers [12–14] is oriented towards obatining the most accurate data for small systems and consists of a careful, step-by-step extension and optimization of some thoughtfully selected parent sets of functions. The resulting basis sets are large, designed for particular systems and when employed with advanced *ab initio* methods may yield state-of-the art results for many small molecules. This approach for obvious reasons cannot be applied for larger systems. We have adopted a different approach here by using two much smaller sets: Pol [9,10] and their extension HyPol [26] designed for hyperpolarizability calculations. Sadlej [9] used the so-called basis set polarization method to develop small-sized basis sets designed especially for electric properties calculations. Contraction coefficients of the Pol set are determined by modelling the behavior of a single Gaussian-type orbital in the external electric field [27–29]. The Pol basis sets have been shown to yield accurate results not only for the dipole polarizability α but also for properties like Raman intensites (geometric derivatives of the polarizability tensor) [30], and recently polarizabilities calculated by the B3LYP density functional method, e.g. [31]. The recent review of Pol basis sets and their generation can be found in Ref. [32].

For the purpose of calculations of molecular hyperpolarizabilities the original Pol sets have been augmented with the second-polarization shell (*f* orbitals for C,N) and made additionally more diffuse. This extension is known as the HyPol sets [26] and has been successfully used in several applications [33–36]. The size of the HyPol basis [5s3p3d2f] for the N,C atoms and [7s5p3d2f] for phosphorus is not particularly large and makes the HyPol sets applicable to larger systems.

To study the above-mentioned aspects of the interplay between the electronic structure and electric properties, and the possible NLO-oriented applications, we have chosen three molecules with a pair of conjugated triple bonds: NCCN cyanogen (1,4-diazabutadiyne), and its phosphorus analogues NCCP (1,4-azaphosphabutadiyne or C-cyanophosphaethyne), and PCCP (formally called 1,4-diphosphabutadyine). Cyanogen has been a subject of many detailed theoretical studies [11,15,16]. Its optimized geometry obtained at the CCSD(T) level with an extended basis set has been determined and it agrees very well with experimental data [15]. Ding et al. [17] have recently analyzed the potential energy surfaces of many possible isomers of C_2NP, while Pham-Tran et al. [18] analyzed potential energy surfaces of the C_2NP isomers at the CCSD(T) and B3LYP levels of approximation. Both groups have found that the linear NCCP structure is by far the global energy minimum. The PCCP geometry has been determined by Jones et al. [19] using the quadratic Configuration Interaction method with single- and double- excitations and the augmented 6-311G basis set (QCISD/6-311G(2d)), and by the Bickelhaupts [20] in their detailed density functional analysis of the stability of the C_2P_2 isomers. Both investigations point to the symmetric, linear structure of PCCP as the global energy minimum.

The electric properties of the three molecules, except for NCCN, are less known. The dipole polarizability of NCCN has been calculated by Kobayashi et al. [16] using the coupled cluster approach with a variety of basis sets. Fowler and Diercksen have carried out the fourth-order MBPT calculations [11]. The most recent results are those obtained by Miadoková et al. in the CCSD(T) approximation with ammended Pol – type basis sets. Bizzocchi et al. [21] calculated the dipole moment of NCCP at the CCSD(T)/cc-pVQZ level of approximations. To the best of our knowledge no data for polarizability and hyperpolarizabilities of NCCP and PCCP seem to be available.

2. Computational methods

The total energy of a molecule perturbed by the external static homogenous electric field F_i can be expressed as:

$$E = E^0 - \mu_i F_i - \frac{1}{2}\alpha_{ij}F_iF_j - \frac{1}{6}\beta_{ijk}F_iF_jF_k - \frac{1}{24}\gamma_{ijkl}F_iF_jF_kF_l - \dots \quad (1)$$

where F_i denotes the external field strength along the i-th direction, μ, α, β, and γ, with appropriate cartesian coordinate labels as subscripts, denote the dipole moment, dipole polarizability, and first and second hyperpolarizability tensors, respectively. Eq. (1) is simpler than the general expression given by Buckingham [2] owing to the neglect of higher multipole moments and the assumed homogeneity of the perturbing electric field. The summation over repeated indices is implied.

Only the static, i.e., frequency independent electric (hyper)polarizabilities are considered in this paper. Their calculations are carried out at the SCF, MBPT(2), CCSD and CCSD(T) levels of approximation [41, 22]. The electron correlation contribution to (hyper)polarizabilities is calculated for valence electrons only.

The results presented in this paper have been obtained by using both analytical and numerical methods. At the SCF level α and β tensors have been calculated analytically employing *Gaussian 98* [37] program. The second hyperpolarizability γ i.e., the fourth derivative of the total perturbed energy has been obtained in a mixed numerical-analytical procedure. In the first step the field dependent polarizability tensor $\alpha_{ii}(F_j)$ has been obtained analytically for selected values of F ranging from 0.0010 to 0.0025 a.u. Then, $\alpha_{ii}(F_j)$ has been differentiated twice to yield γ_{iijj}. The first derivative represents the off – diagonal component iij of the first hyperpolarizability tensor β and has been used to obtain its xxz component for the NCCP molecule.

The same two – step procedure has been applied to calculate all components of the γ (and β for NCCP) tensor at the MP2 level where the analytical routine to determine second-order derivatives is available in the *Gaussian 98* package. All properties at the CCSD and CCSD(T) levels of theory have been calculated numerically using the values of the external field strengths listed above. The numerical accuracy of both numerical and mixed analytical-numerical schemes is sufficiently high to make meaningful all decimals of the data reported in this paper. The spherical Gaussian functions are used in all cases.

We used the *Gaussian 98* suite of programs [37] for all SCF and MP2 results reported here. The CC data have been obtained using the *AcesII* package [38]. In the course of the CC calculations MBPT(2) and SCF results are obtained and the numerical finite – field procedure is used to yield electric response properties. The differences between numerical values obtained by *AcesII* and analytical (or mixed analytical-numerical) results of *Gaussian 98* are within the precision of the reported data.

We have employed the newest generation of the Pol basis sets [26,32] which differs from the early Pol basis sets [9,10] by the use of general contractions in the core region. One should mention, however, that the electric properties calculated with either of the two sets are only marginally different. The HyPol sets are those described earlier [26]. The details of the Pol and HyPol basis can be obtained directly from the author or from the web page [39].

At the SCF and MP2 levels we have also performed comparative calculations using one of the correlation-consistent basis sets of Dunning [40], the aug-cc-pVTZ set. For the C and N atoms this set consists of [5s4p3d2f] contracted functions and is slightly larger than our HyPol set ([5s3p3d2f]). The good performance of the aug-cc-pVXZ ($X = T, Q, \dots$) basis sets in the dipole polarizability calculations has been reported [40] and it is of interest to study their efficiency in calculations of hyperpolarizabilities.

Table 1
Dipole properties of the NCCN molecule[a)]. Calculations with Pol and HyPol basis sets and their comparison with the aug-cc-PVTZ results. All values in a.u

		α		$\gamma^{b)}$		
		zz	xx	$zzzz$	$xxxx$	$xxzz$
SCF	Pol	53.66	21.39	6.5	2.4	1.2
	HyPol	53.68	21.79	6.0	2.2	1.0
	aug-cc-PVTZ	53.80	21.74	5.9	1.7	0.9
MBPT(2)	Pol	52.14	21.49	9.7	2.5	1.4
	HyPol	52.10	22.02	9.2	2.4	1.3
	aug–cc–PVTZ	51.94	21.79	8.8	1.7	1.1
	MKD [c)]	54.11	21.72	10.36	2.63	1.45
CCSD	Pol	52.18	21.46	8.8	2.5	1.3
	HyPol	52.16	21.96	8.2	2.3	1.2
	MKD [c)]	54.25	21.73	9.38	2.58	1.37
CCSD(T)	Pol	53.49	21.73	9.7	2.6	1.4
	HyPol	53.43	22.25	9.0	2.5	1.3
	MKD [c)]	55.68	22.00	10.26	2.76	1.47

[a)]The molecule lies on the z axis of the coordinate system. Experimental geometry data taken from [15].
[b)]The listed results should be multiplied by 10^3.
[c)]The work of Miadoková, Kellö, and Diercksen (MKD) [7].The basis used is the standard, segmented Pol basis set.

3. Results and discussion

3.1. Electric properties of NCCN

The polarizability α and second hyperpolarizability γ tensors for the NCCN molecule are presented in Table 1. The experimental geometry used in our calculations was taken from the work of Botschwina and coworkers [21]. The values of the dipole polarizability can be compared to the results of Kobayashi et al. [16], Fowler and Diercksen [11], and the most recent and complete study by MKD [7].

The SCF value of the parallel component of the dipole polarizability α_{zz} equals 53.66 with the Pol basis [9,10] and is essentially the same as the result (53.68 a.u.) obtained with the larger HyPol set [26]. The aug-cc-PVTZ set [40] of the size [5s4p3d2f] gives the value of 53.80 a.u., confirming the earlier findings with respect to the performance of the Dunning's basis sets in calculations of electric dipole polarizabilities. However, one should note that at least at the SCF level of approximation not too much is gained in comparison with the result calculated in much smaller Pol set ([5s3p2d]). The α_{zz} is well described by the bases sets used in the present work. The extensive compilation of different results in the paper by MKD [7] shows only small differences between results obtained with various bases.

For the perpendicular component α_{xx} the overall pattern is similar, although the difference between the Pol and HyPol results becomes slightly larger: 21.39 and 21.79 a.u., respectively. The small discrepancies between our Pol results and the Pol results presented by MKD are due to small differences in geometries used in these two studies. One can see almost perfect agreement between the HyPol and aug-cc-PVTZ results, the difference is less than 0.05 a.u. for the both components of the dipole polarizability tensor.

The electron correlation contribution to α_{zz} is very small and negative. At the most advanced level of our computations, CCSD(T), the result is smaller than the SCF value by 0.25 a.u. for the HyPol and 0.17 a.u. for the Pol basis. The respective contributions obtained by MKD are slightly larger but for all the bases employed are smaller than 0.7 a.u. The electron correlation effect on the perpendicular component

α_{xx} of the polarizability is small and positive. At the CCSD(T) level of approximation the effect is 0.34 a.u., and 0.46 a.u. for the Pol and HyPol set, respectively. The value reported by MKD reads 0.23 a.u. for the Pol basis.

The results of the MBPT(2) (MP2) calculations are very close to those of the CCSD method. The electron correlation lowers the value of the *zz* component, while increasing slightly the parallel component of the dipole polarizability. The same pattern can be observed in data obtained by MKD. Kobayashi et al. [16] performed CCSD calculations on all isomers of $(NC)_2$. With the modest cc-PVDZ basis set they obtained considerably underestimated results, in particular for the α_{xx} component. The results of their MP2 computations with the truncated aug-cc-PVTZ basis set are in good agreement with our data. Fowler and Diercksen [11] used many-body perturbation theory to calculate α for NCCN using the standard, i.e., segmented Pol basis. The electron correlation contributions obtained in the various orders of perturbation theory were shown to be always small. Also at the full MBPT(4) level the corrections are small and positive.

On the basis of our results and comparisons with other data one can safely conclude that the dipole polarizability α components are correctly determined and the SCF results for the parallel component obtained with the HyPol basis are less than 1 per cent above the correlated value. For α_{xx} the correlated value is larger than SCF result by less than 2 per cent of the final correlated value of the dipole polarizability.

For the second hyperpolarizability the only data available for comparison seem to be the recent results of MKD [7]. On passing from Pol to Hypol basis sets one can see significant changes already at the level of the SCF approximation. The HyPol result for γ_{zzzz} is by 500 a.u. (8 per cent) lower than the value obtained with the Pol set. For the *xxxx*, and *xxzz* components this lowering amounts to only 200 a.u. However, this makes as much as 20 per cent of the HyPol value for γ_{xxzz}. Let us recall that the Pol basis sets have been shown to generally overestimate the calculated values of the second hyperpolarizability tensor [26].

Also the aug-cc-PVTZ values are systematically lower than the corresponding Pol or HyPol sets. The largest discrepancy occurs for the *xxzz* component. The HyPol basis set leads to 2200 a.u. which is to be compared with 1700 a.u. for the aug-cc-PVTZ. The aug-cc-PVTZ basis set is definitely less flexible than the HyPol set, in particular for the *xxxx* component. Electron correlation contributions turn out to be quite important for the second hyperpolarizability tensor. The correlation contribution to γ_{zzzz} from the CCSD(T) calculations with the HyPol set amounts to 3200 a.u. (33 per cent of the total value). The corresponding result for the Pol basis set is equal to 3000 a.u. The correlation effects are significantly less important for the *xxxx* component: 300 (200) a.u. or 12 per cent (8 per cent) with the HyPol (Pol) set. Similar effect is seen for the off – diagonal component, although in this case the percentage of the correlation contribution is higher: 14 per cent for Pol and 23 per cent for HyPol sets. Since the correlated level calculations with the aug-cc-PVTZ basis become quite demanding the corresponding comparison of the electron correlation contributions is limited only to the MP2 level of approximation. At the MP2 level the Dunning set shows the pattern similar to that observed in calculations with the HyPol or Pol basis sets. There is a large electron correlation effect on the parallel component of the γ tensor, 2900 a.u. (32 per cent), much smaller for the off – diagonal part, 200 a.u. (18 per cent), and none for the *xxxx* component. The last result strongly suggests insufficient flexibility of the aug-cc-PVTZ basis set.

The results of the MBPT(2), CCSD and CCSD(T) calculations by MKD [7] are displayed in Table 1. These results have been obtained at the slightly different geometry and using the segmented version of the Pol basis sets. In spite of these differences they agree well with our Pol basis set results. However, they are systematically higher than the present values, i.e. by 560 a.u., 160 a.u., and 70 a.u. for *zzzz*, *xxxx* and

Table 2
Dipole properties of the NCCP molecule[a]. Calculations with Pol and HyPol basis sets and their comparison with the aug-cc-PVTZ results. All values in a.u

		μ	α		β		$\gamma^{b)}$		
	Basis set	z	zz	xx	zzz	zxx	$zzzz$	$xxxx$	$xxzz$
SCF	Pol	1.5524	93.72	36.13	198	57.5	20.5	7.3	3.6
	HyPol	1.5581	93.71	36.50	212	58.7	20.0	7.8	3.8
	aug-cc-PVTZ	1.5732	94.06	36.42	212	60.8	19.3	5.8	3.4
	aug-cc-PVQZ [c]	1.5820							
MBPT(2)	Pol	1.2627	90.47	35.96	85.0	53.4	29.5	8.1	4.5
	HyPol	1.2748	90.55	36.64	103	55.8	28.9	8.7	4.5
	aug-cc-PVTZ	1.2955	90.50	36.25	97.9	56.6	27.4	6.0	3.8
	aug-cc-PVQZ [c]	1.3060							
CCSD	Pol	1.3721	89.40	35.56	100	50.8	26.4	7.6	4.1
	HyPol	1.3818	89.52	36.18	117	51.0	25.7	8.0	4.1
	aug-cc-PVQZ [c]	1.4170							
CCSD(T)	Pol	1.3199	91.30	35.81	125	54.0	28.8	8.0	4.4
	HyPol	1.3290	91.39	36.49	143	54.3	28.0	8.4	4.5
	aug–cc–PVQZ [c]	1.3570							

[a] The molecule lies on the z axis of the coordinate system. Geometry taken from Ref. [21].
[b] The listed results should be multiplied by 10^3.
[c] The work of Botchswina and coworkers [21].

$xxzz$ components of γ, respectively. This shows the known fact [26] that the Pol basis sets overestimate the hyperpolarizability values. One should also mention that the electronic contributions to the γ tensor can be further affected by the possible large values of the vibrational hyperpolarizability [42].

3.2. *Electric properties of NCCP*

The results of the present work for the NCCP molecule are presented in Table 2. The knowledge of electric properties of this molecule is very limited. Bizzocchi et al. carried out a careful experimental and computational study on NCCP including the millimeter – wave spectroscopy and CCSD(T) calculations using aug-cc-PVXZ bases [21]. Their values of the dipole moment at the equilibrium bond distance, μ_e, listed in Table 2, can be compared to our results. At the SCF level both Pol and HyPol agree well with the SCF aug-cc-PVQZ data of Bizzochi et al.. Our HyPol value (1.5732 a.u.) is smaller by 0.015 a.u. (0.061 D) than the aug-cc-PVQZ result by Bizzocchi et al. [21]. Electron correlation is seen to make significant contributions to the final values of the dipole moment. MBPT(2) calculations lower the SCF value by 0.283 a.u. (HyPol), and 0.278 a.u. for the aug-cc-PVTZ set. As shown by our CCSD(T) calculations, these MBPT(2) results are underestimated. The CCSD(T) HyPol value of 1.329 a.u. is in very good agreement with the accurate aug-cc-PVQZ data and the recommended value for the μ_e (1.357 a.u.). The electron correlation contribution amounts to approximately 17 per cent for the HyPol set and can be compared with the value of 16.5 per cent obtained by Bizzocchi et al. with the much larger aug-cc-pVQZ set.

No comparative data on the dipole polarizability and higher – order polarizabilities of NCCP seem to be available. The present dipole polarizability results reveal highly anisotropic character of NCCP with the parallel component zz being *ca.* 2.5 times larger than the corresponding α_{xx} value, independently of the level of theory or the basis set used. The electron correlation contribution behaves similarly as in the case of NCCN. Its overall effect is small and negative for the zz component (−2.32 a.u. or −2.5 per cent in CCSD(T) HyPol calculations) and almost negligible for α_{xx} (-0.01 a.u., or -0.03 per cent for the CCSD(T) method with the HyPol basis set).

The first hyperpolarizability tensor β is very difficult to estimate theoretically. Our Pol/HyPol calculations should be treated with caution. The SCF HyPol and aug-cc-PVTZ results for β_{zzz} are identical, while the Pol value is lower by 14 a.u., i.e., is less than 7 per cent of the HyPol value. The electron correlation is significant and amounts to about 50 per cent of the final value of β_{zzz}, leading to the CCSD(T) values of 143 a.u. (HyPol) and 125 a.u. (Pol). The pattern observed for the off – diagonal component β_{xxz} is very different. The correlation contribution is relatively small and amounts to 8 per cent in CCSD(T) HyPol calculations. Unlike in the case of the parallel component, the difference between CCSD(T) and MBPT(2) results for β_{xxz} is negligible.

Similarly to the case of NCCN the Pol and HyPol values for γ_{zzzz} are in close agreement, whereas the values obtained with the aug-cc-PVTZ basis set are systematically lower. The difference between the SCF HyPol result and the aug-cc-PVTZ for the perpendicular component is 2000 a.u., i.e., more than one fourth of the HyPol value. The usefulness of the aug-cc-PVTZ basis set for calculations of the second hyperpolarizability is therefore quite limited. For the usually less demanding *zzzz* component the discrepancy between Pol/HyPol and aug-cc-PVTZ results is smaller, though again the Dunning set yields lower values of this component.

The electron correlation contribution turns out to be of particular importance for γ_{zzzz}. The change due to the correlation contribution is 8000 a.u. for the HyPol basis and 8300 for the more compact Pol set. In both cases this makes about 29 per cent of the final value. The MBPT(2) values seem to overestimate the γ_{zzzz} component, while CCSD results grossly underestimate it. For the perpendicular *xxxx* component the electron correlation as computed at the CCSD(T) level of theory increases the SCF values by 700 a.u. (9%) for the Pol and 600 a.u. (8%) for the HyPol basis. Again, the aug-cc-PVTZ set leads to much lower values. With this basis set the percentage of the correlation contribution calculated at the MBPT(2) level amounts to modest 3 per cent compared to 10 per cent for the Pol and HyPol bases at the same level of approximation.

3.3. Electric properties of PCCP

Table 3 displays the calculated electric dipole properties for the PCCP molecule. The geometry of PCCP was optimized at the MP2/6-311G(d,p) level of approximation employing the *Gaussian 98* package. The $D_{\infty h}$ structure is a true minimum with the carbon – carbon and carbon – phosphorus bonds of the length of 1.360 Å, and 1.583 Å, respectively. These values are in good agreement with the QCISD/6-311g(d,p) results of Jones et al. [19]: 1.37 and 1.583 Å, respectively. It should be noted that the carbon–carbon bond in PCCP is shorter than in NCCN by 0.024 Å.

The dipole polarizability tensor α is highly anisotropic with the parallel *zz* component 3.2 times larger than the perpendicular *xx* one at the SCF level, and 3.0 times according to the CCSD(T) HyPol calculations. The electron correlation contribution is negative for α_{zz}, but unlike in the case of NCCN, is of considerable magnitude: -12.6 a.u. or 8 per cent in CCSD(T) calculations with the HyPol basis set. On the other hand, the perpendicular *xx* component is only slightly affected by taking into account the dynamic electron correlation. The effect computed at the most advanced CCSD(T) level with the HyPol basis is less than 1 a.u. (2 per cent) and is negative. The same negligible effect is found at all levels of correlated calculations and for Pol, HyPol and aug-cc-PVTZ basis sets.

The second hyperpolarizability tensor γ is even more anisotropic than the dipole polarizability. For all basis sets used in this study the SCF values of γ_{zzzz} exceed 66,000 a.u. and are about 5 times larger

Table 3
Dipole properties of the PCCP molecule[a)]. Calculations with Pol and HyPol basis sets and their comparison with the aug-cc-PVTZ results. All values in a.u

		α		$\gamma^{b)}$		
	Basis set	*zz*	*xx*	*zzzz*	*xxxx*	*xxzz*
SCF	Pol	171.3	53.04	67.0	15.1	14.8
	HyPol	171.3	53.36	66.5	17.4	15.0
	aug-cc-PVTZ	172.1	53.25	66.2	12.3	12.8
MBPT(2)	Pol	156.5	52.03	99.4	16.4	14.9
	HyPol	157.0	52.87	98.6	18.3	14.7
	aug-cc-PVTZ	157.0	52.22	94.9	12.6	12.4
CCSD	Pol	155.4	51.37	92.4	15.1	13.8
	HyPol	156.0	52.09	90.9	16.7	13.5
CCSD(T)	Pol	158.2	51.55	94.6	15.8	14.6
	HyPol	158.7	52.38	93.2	17.4	14.3

a)The molecule lies on the z axis of the coordinate system. Geometry optimized at the MP2/ 6-311G(d,p) level.
b)The listed values are to be multiplied by 10^3.

than γ_{xxxx}. Since the aug-cc-PVTZ is not sufficiently flexible to describe properly the perpendicular components of the second hyperpolarizability the anisotropy becomes higher than for Pol and HyPol sets. The electron correlation increases the parallel component up to 93,200 a.u. (29 per cent of the total correlated value). The MBPT(2) results seem to overestimate γ_{zzzz} which is rather common in the case of very large electron correlation contributions. Compared to the correlation effect on the $zzzz$ component, the electron correlation contribution for γ_{xxxx} is quite modest.

It is worthwhile to mention certain features of the calculated values of the *xxxx* and *xxzz* components of γ. The CCSD(T) calculations with the Pol basis yield γ_{xxxx} larger than SCF result 700 a.u., while our presumably best calculations at the CCSD(T) level with the HyPol set show no electron correlation contribution at all. This becomes reversed at CCSD level. The MBPT(2) calculations give a reasonable magnitude of the effect, i.e., the increase by 1300 a.u. with Pol basis sets and by 900 a.u. for the HyPol set. The electron correlation contribution is small and negative for the off – diagonal *xxzz* component. At the CCSD(T) level the second hyperpolarizability values decrease by 700 a.u. (5 per cent) for HyPol and by 200 a.u. (1 per cent) for Pol sets. Because of the already mentioned insufficient flexibility of the aug-cc-PVTZ set, the corresponding results for the *xxxx* and *xxzz* components of γ are systematically lower then the HyPol values. There seems to be no simple explanation of the observed electron correlation effect on perpendicular components of γ in PCCP. The basis set dependence of these data may only indicate that even the largest HyPol basis sets need to be further augmented.

4. Final conclusions

Electric dipole response properties: dipole polarizability (α), first (β), and second hyperpolarizability (γ) have been systematically studied for three molecules: NCCN, NCCP and PCCP. The parallel *zz* component of the dipole polarizability increases significantly across this series; from 53.5 a.u. for NCCN to over 158 a.u. for PCCP. The perpendicular component increases less: from 22.2 a.u.(NCCN) to 52.4 a.u. for PCCP. Considering the accuracy of the computational methods used in this work and the performance of the HyPol basis set we are confident that the dipole polarizability α tensor components for all three molecules are accurately determined at the CCSD(T) calculations with the HyPol set. All systems show quasi – 1D character of the dipole polarizability with the parallel component significantly larger then the *xx* component. Electron correlation contribution to the α tensor for all three systems is very small, this fact can be practically exploited in future calculations of electric properties of longer chains of the conjugated π-systems with CN/CP groups. The usage of the HyPol set seems to be the minimum requirement for obtaining accurate and reliable data for the second hyperpolarizability. In this respect the often used aug-cc-PVTZ basis set turns out to be inadequate. Its further systematic extension in the spirit of the Dunning's recipe [40] would obviously help. However, this would considerably increase its size.

Similarly to the dipole polarizability of molecules investigated in this study, the second hyperpolarizability is also a highly anisotropic, quasi – 1D property. The value of the *zzzz* component increases approximately ten times; from 9000 a.u. for NCCN to 93,000 a.u. for the PCCP molecule. The increase for the *xxxx* component is far less dramatic: from 2500 a.u. (NCCN) to 15800 a.u. for PCCP. The dynamic electron correlation effects are important for this property, generally their contribution is significantly large and positive for the *zzzz* component and much smaller for the other two components. It also follows from the present study that realistic calculations of the second hyperpolarizability require very careful incorporation of the correlation effects. Moreover, to improve upon the reliability of the calculated values for perpendicular components of γ one should further explore the basis set effects.

The data in Table 1 to 3 are given in atomic units, the most convenient from the computational point of view. For the dipole polarizability α 1 a.u. equals $1.6488 \cdot 10^{-41}$ $\mathrm{C^2m^2J^{-1}}$, for the β hyperpolarizability 1 a.u. equals $3.2064 \cdot 10^{-53}$ $\mathrm{C^3m^3J^{-2}}$, while for the second-hyperpolarizability tensor γ the conversion factor reads $6.2354 \cdot 10^{-65}$ $\mathrm{C^4m^4J^{-3}}$.

Acknowledgements

Part of the computations were performed at the Computational Center of the Silesian University of Technology, Gliwice in the framework of the supported project *'Theoretical determination of molecular electric and optical properties'*. The authors wish to thank Professor Andrzej J. Sadlej of the Nicolaus Copernicus University, Toruń, Poland, for helpful discussions and interest in this project.

References

[1] K.D. Bonin and V.V. Kresin, *Electric-Dipole Polarizabilities of Atoms*, Molecules and Clusters, World Scientific, Singapore, 1997.
[2] A.D. Buckingham, *Adv. Chem. Phys.* **12** (1967), 107.
[3] P.N. Prasad and D.J. Williams, *Introduction to Nonlinear Optical Effects in Molecules and Polymers*, Wiley, New York, 1991.
[4] D.S. Chemla and J. Zyss, eds, *Nonlinear optical properties of organic molecules and crystals*, Vols 1, 2, Academic Press, New York, 1987.
[5] Y. Luo, H. Ågren, P. Jørgensen and K.V. Mikkelsen, *Adv. Quantum Chem.* **26** (1995), 165.
[6] D.P. Shelton and J.E. Rice, *Chem. Rev.* **94** (1994), 3.
[7] I. Miadoková, V. Kellö and G.H.F. Diercksen, *Chem. Phys. Lett.* **287** (1998), 509.
[8] D.R. Kanis, M.A. Ratner and T.J. Marks, *Chem. Rev.* **94** (1994), 195.
[9] A.J. Sadlej, *Coll. Czech. Chem. Commun.* **53** (1988), 1995.
[10] A.J. Sadlej, *Theor. Chim. Acta* **79** (1991), 123.
[11] P.W. Fowler and G.H.F. Diercksen, *Chem. Phys. Lett.* **167** (1990), 105.
[12] G. Maroulis and A. Thakkar, *J. Chem. Phys.* **95** (1991), 9060.
[13] G. Maroulis, C. Makris, U. Hohm and U. Wachsmuth, *J. Phys. Chem.* **A103** (1999), 4359.
[14] G. Maroulis and C. Pouchan, *Chem. Phys. Lett.* **215** (1997), 67.

[15] P. Botschwina, *J. Chem. Phys.* **99** (1993), 6217.
[16] R. Kobayashi, H. Koch and P. Jørgensen, *J. Chem. Phys.* **101** (1994), 4956.
[17] Y. Ding, Z. Li, X. Huang and C. Sun, *Theor. Chem. Acc.* **107** (2001), 1.
[18] N.-N. Pham-Tran, B. Hajgató, T. Veszpremi and M.T. Nguyen, *Phys. Chem. Chem. Phys.* **3** (2001), 1588.
[19] L.A. Jones, E.P.F. Lee, P. Soldán and T.G. Wrigth, *Phys. Chem. Chem. Phys.* **1** (1999), 391.
[20] F.M. Bickelhaupt and F. Bickelhaupt, *Chem. Eur. J.* **5** (1999), 162.
[21] L. Bizzocchi, C. Degli Esposti and P. Botschwina, *J. Chem. Phys.* **113** (2000), 1465.
[22] R.J. Bartlett, in: *Advanced Series in Physical Chemistry, Methods in Computational Chemistry*, (Vol. 2), D.R. Yarkony, ed., World Scientific, Singapore, 1995, p. 1047; T.D. Crawford and H.F. Schaefer III., *Rev. Comput. Chem.* **14** (2000), 33.
[23] K. Raghavachari, G.W. Trucks, J.A. Pople and M. Head-Gordon, *Chem. Phys. Lett.* **157** (1989), 479.
[24] N. Vaval, A.B. Kumar and S. Pal, *Int. J. Mol. Sci.* **2** (2001), 89.
[25] N. Oliphant and R.J. Bartlett, *J. Chem. Phys.* **100** (1994), 6550.
[26] T. Pluta and A.J. Sadlej, *Chem. Phys. Lett.* **297** (1998), 391.
[27] A.J. Sadlej, *Chem. Phys. Lett.* **47** (1977), 50.
[28] A.J. Sadlej, *Acta Phys. Polon.* **A53** (1978), 297.
[29] B.O. Roos and A.J. Sadlej, *Chem. Phys.* **94** (1985), 43.
[30] M.D. Halls and H.B. Schlegel, *J. Chem. Phys.* **111** (1999), 8819.
[31] M. Le Guennec, K. Evain and B. Illien, *J. Mol. Struct.* (*Theochem*) **542** (2001), 167.
[32] I. Černušák, V. Kellö and A.J. Sadlej, *Coll. Czech. Chem. Commun.* **68** (2003), 211.
[33] U. Eckart, M. Fülscher, L. Serrano – Andrés and A.J. Sadlej, *J. Chem. Phys.* **113** (2000), 6235.
[34] T. Pluta and A.J. Sadlej, *J. Chem. Phys.* **114** (2001), 136.
[35] T. Pluta, *Mol. Phys.* **99** (2001), 1535.
[36] P. Neogrády, M. Medveď, I. Černušák and M. Urban, *Mol. Phys.* **100** (2002), 541.
[37] Gaussian 98 (Rev.A1), M.J. Frisch, G.W. Trucks, H.B. Schlegel, G.E. Scuseria, M.A. Robb, J.R. Cheeseman, V.G. Zakrzewski, J.A. Montgomery, R.E. Stratmann, J.C. Burant, S. Dapprich, J.M. Millam, A.D. Daniels, K.N. Kudin, M.C. Strain, O. Farkas, J. Tomasi, V. Barone, M. Cossi, R. Cammi, B. Menucci, C. Pomelli, C. Adamo, S. Clifford, J. Ochterski, G.A. Petersson, P.Y. Ayala, Q. Cui, K. Morokuma, D.K. Malick, A.D. Rabuck, K. Raghavachari, J.B. Foresman, J. Cioslowski, J.V. Ortiz, B.B. Stefanov, G. Liu, A. Liashenko, P. Piskorz, J. Komaromi, R. Gomperts, R.L. Martin, D.J. Fox, T. Keith, M.A. Al – Laham, C.Y. Peng, A. Nanayyankara, C. Gonzalez, M. Challacombe, P.M.W. Gill, B.G. Johnson, W. Chen, M.W. Wong, J.L. Andres, M. Head – Gordon, E.S. Relplogle and J.A. Pople, *Gaussian*, Inc. Pittsburgh PA, 1998.
[38] AcesII (Release 3.0) is a program product of the Quantum Theory Project, University of Florida. Authors: J.F. Stanton, J. Gauss, J.D. Watts, M. Nooijen, N. Oliphant, S.A. Perrera, P.G. Szalay, W.J. Lauderdale, S.R. Gwaltney, S. Beck, A. Balková, D.E. Bernholdt, K.-K. Baeck, P. Rozyczko, H. Sekino, C. Hober and R.J. Bartlett. Integral packages included are VMOL (J. Almlöf and P.R. Taylor); VPROPS (P.R. Taylor); ABACUS, T. Helgaker, H.J.Aa. Jensen, P. Jørgensen, J. Olsen and P.R. Taylor, 1998.
[39] T. Pluta, e-mail address, tp@us.edu.pl, from the web page http://molpir.fns.uniba.sk/HyPol.txt.
[40] D.E. Woon and T.H. Dunning, Jr., *J. Chem. Phys.* **100** (1994), 2975.
[41] M. Urban, I. Černušák, V. Kellö and J. Noga, in: *Methods in Computational Chemistry*, (Vol. 1), S. Wilson, ed., Plenum, New York, 1987, p. 117.
[42] D.A. Bishop, *Adv. Chem. Phys.* **104** (1998), 1.

Journal of Computational Methods in Sciences and Engineering 4 (2004) 357–364
IOS Press

Ab initio calculation of the nonlinear susceptibility $\chi^{(2)}$ of a crystal surface

Isabelle Baraille, Clovis Darrigan and Michel Rérat*
Laboratoire de Chimie Théorique et Physico-Chimie Moléculaire, UMR 5624 – FR2606, Université de Pau, 64000 Pau, France

Abstract. The polarizability and first hyperpolarizability of a crystal surface are calculated from a sum over states method using crystalline orbitals of the CRYSTAL program. The linear and nonlinear optical properties of the (001) LiF surfaces are evaluated layer by layer, showing that the response of inner layers to an electric field tends effectively to the bulk response value while the $\chi^{(2)}$ susceptibility components responsible of second harmonic generation (SHG) are no longer equal to zero at the surface of the crystal.

Keywords: Quantum chemistry, surface, nonlinear optics, lithium fluoride

Mathematics Subject Classification: Primary 81V80, Secondary 78A60

1. Introduction

On the surface of a material, nonlinear optical interactions differ distinctly from those in the bulk material because of the lower symmetry of the atomic sites at the surface. For example, in any material with an inversion center as tridimensional systems, the nonlinear electric susceptibility $\chi^{(2)}$ responsible for second harmonic generation (SHG) vanishes and no second-harmonic is generated. When considering atoms or molecules at the surface of this material for which the inversion center is now absent, a nonzero contribution of $\chi^{(2)}$ appears. It follows that second-harmonics are generated by theese atoms or molecules, and the signal is then a powerful probe of the vicinity of a surface, or atoms or molecules adsorbed on it. This is the reason why the experimental study of SHG from surfaces has been persued intensely in recent years [1–3] (see also Ref. [4] and references therein).

Moreover, the last progress in *ab initio* calculation of periodic wavefunctions allows to determine response properties of crystalline systems perturbed by an electric field. The linear susceptibility or dielectric constant can be accurately obtained at different levels of the density functional theory (DFT), using time-dependent coupled method with plane wavefunctions as basis sets generally. More recently, the method implemented in the ADF-BAND program [5] permits the use of linear combination of atomic orbitals (LCAO). A finite field (FF) perturbation method has been also implemented in the CRYSTAL98 code [6] using the same kind of LCAO basis set [7].

Dynamic nonlinear susceptibilities can also be obtained from a sum over states (SOS) method using crystalline orbitals computed by the CRYSTAL98 code as it has been already done in our team [8].

*Corresponding author. E-mail: michel.rerat@univ-pau.fr.

1472-7978/04/$17.00

This method is less accurate than the previous coupled ones but it allows to determine easily the relative behaviour of each $\chi^{(2)}$ (and $\chi^{(3)}$) component versus the electric field frequency.

In this work, our SOS code has been modified in order to evaluate the contribution of each layer of a two dimensional system to linear and nonlinear electric properties. This partition per layer permits to refine the analysis by evaluating the contribution of the top layers to SHG. It is also interesting to mention that a recent time-dependent Hartree-Fock (TDHF) study (Ref. [9]) of these optical properties has been done for clusters of molecules using a similar decomposition procedure over the atomic orbitals of each molecule. In Section 2, the SOS method and computational details are presented. In Section 3, in order to test the advantages of the partition per layer, we calculate the polarizability and first hyperpolarizability of the (001) LiF surface. This example was chosen from previously published study to allow technical analysis and verification rather than for any intrinsic interest. Indeed, the studied properties are nearly additive for ionic systems as LiF.

2. Methodological details

The surface properties described here are obtained by simulating the semi-infinite crystal by the slab model. In order to evaluate the influence of the thickness of the slab, several systems (S_N) with different numbers of atomic layers N ($1 \rightarrow 8$) parallel to the (001) face have been considered and their electric properties compared to the bulk ones. All the studied systems correspond to unrelaxed slabs in which the experimental parameter of the bulk ($a = 7.5936$ a.u.) has been maintained. The geometry optimization has just been carried out for a slab of 4 atomic layers. In this procedure, the vertical rumpling of the different ions has been taken into account as an essential part of the relaxation phenomena occuring at the surface and more specifically, as a feature that could exert a non negligible influence on the polarizability and hyperpolarizability of the system. The corresponding atomic coordinates of the 4 non equivalent atoms in the S_4 unit cell have been simultaneously computed using an optimization code developed in our laboratory (PENTE [10]). This program which is linked to the CRYSTAL98 code carries out the optimization of any parameter (geometrical parameters or basis set exponents) relative to the total energy of the system, by a Newton-Raphson method.

The crystalline orbitals expanded in terms of localized atomic gausssian basis set, in a way close to the L.C.A.O. method, have been obtained from the *ab initio* periodic code CRYSTAL [6]; The eigenvalues equations can be solved using different hamiltonians (Hartree-Fock, LDA, GGA and hybrids such as B3LYP). All the calculations to be discussed in the following have been performed at the LDA level; convergence checks to be shown in the following refer to LiF for which one of the moderately large basis sets (labelled b_1 in the following) used in a previous paper [7] has been adopted: a 6-1G contraction for the Li atom (corresponding to s and sp shells) and a 7-311G(d) contraction (one s, three sp and one d shells) for the F atom; the exponents of the most diffuse sp functions are 0.525 (Li) and 0.137 (F) bohr^{-2}; the d shell exponent is 0.6 bohr^{-2}. An additional sp shell (with exponent 0.25 bohr^{-2}) has been added to describe the Li atom, in order to measure the basis set effect (basis labelled b_2). Standard conditions have been adopted for the SCF part of the calculation; the SCF convergence threshold has however been set to 10^{-8}, and a shrinking factor 8 has been used for the irreducible Brillouin zone sampling. In the second step, polarizability and first hyperpolarizabilities per unit cell of the system have been calculated from a SOS program, as in Ref. [8], using the occupied and unoccupied crystalline orbitals obtained from the CRYSTAL98 code. The expressions of the component α_{uv} and β_{uvw} of the static polarizability and first hyperpolarizability of any periodic systems are:

$$\alpha_{uv} = \sum_{P_{u,v}} \sum_{K} \Omega_K \sum_{i_{oc},j_{un}} \frac{< i_{oc}|u|j_{un} >_K < j_{un}|v|i_{oc} >_K}{(\varepsilon_{j_{un}} - \varepsilon_{i_{oc}})_K} \tag{1}$$

$$\beta_{uvw} = \sum_{P_{u,v,w}} \sum_{K} \Omega_K \left[\sum_{i_{oc},j_{un},k_{un}} \frac{< i_{oc}|u|j_{un} >_K < j_{un}|v - < i_{oc}|v|i_{oc} > \delta_{j,k}|k_{un} >_K < k_{un}|w|i_{oc} >_K}{(\varepsilon_{j_{un}} - \varepsilon_{i_{oc}})_K (\varepsilon_{k_{un}} - \varepsilon_{i_{oc}})_K} - \sum_{i_{oc} \neq k_{oc},j_{un}} \frac{< i_{oc}|u|j_{un} >_K < i_{oc}|v|k_{oc} >_K < j_{un}|w|k_{oc} >_K}{(\varepsilon_{j_{un}} - \varepsilon_{i_{oc}})_K (\varepsilon_{j_{un}} - \varepsilon_{k_{oc}})_K} \right] \tag{2}$$

where $\sum_{P_{u,v(,w)}}$ means sum of permutation over u and v (and w). K is a vector of the irreducible Brillouin zone (IBZ), Ω_K is its geometric weight. $\varepsilon_{i(k)_{oc}}$ and $\varepsilon_{j(k)_{un}}$ are the eigenvalues of the corresponding occupied $i(k)_{oc}$ and unoccupied $j(k)_{un}$ eigenvectors for this K-point. The index u, v and w correspond to the projections of the length operator when the dipole hamiltonian gauge is chosen. The dynamic electric susceptibility formulas of $\chi^{(1)}$ and $\chi^{(2)}$ in the same approximations can be found in Ref. [8]. The expressions of $\chi^{(3)}$ for different nonlinear optic processes are reported in Darrigan's thesis [11].

In this work, we have also calculated these properties for slabs by layer, in annihilating transition moments when atomic orbitals (or Bloch functions) of the occupied LCAO crystalline orbitals i_{oc} do not belong to the studied layer. It means also that the only part of the occupied i_{oc} crystalline orbitals developped on the atomic orbitals belonging to the layer is considered. However, the resulting occupied i'_{oc} orbitals for which unchanged $\varepsilon_{i_{oc}}$ eigenvalues are attributed, are no longer orthogonal to the unoccupied j_{un} orbitals. Then, a translation of the frame origin to the Z-layer position is necessary (the applied electric field being in the z-direction orthogonal to the slab), corresponding to the nuclear part of the electric dipole perturbation which cannot be dropped out any more. In Eqs (1) and (2), operators $u(v,w)$ become $u(v,w) - Z\delta_{u(v,w),z}$ where $\delta_{u(v,w),z} = 1$ if $u(v,w) = z$ and 0 if not. For example, the static polarizability α^Z_{uv} for Z-layer becomes:

$$\alpha^Z_{uv} = 2 \sum_{K} \Omega_K \sum_{i'_{oc},j_{un}} \frac{< i'_{oc}|u - Z\delta_{u,z}|j_{un} >_K < j_{un}|v - Z\delta_{v,z}|i'_{oc} >_K}{(\varepsilon_{j_{un}} - \varepsilon_{i_{oc}})_K} \tag{3}$$

Atomic polarizability and hyperpolarizability can be also obtained by limiting the linear combinations of atomic orbitals (for occupied crystalline orbitals) to those of the considered atom, and by a translation of the frame origin on its coordinates. This way to evaluate atomic (hyper)polarizability in a crystal and electric property by layer lets suppose that these properties are additive, which is not exact but nearly true for ionic systems like the cubic LiF crystal. However, additivity would be completely confirmed if we let unchanged $< i_{oc}|$ (instead of $< i'_{oc}|$) in the "emission" term of the scattering α^Z_{uv} expression, and if we only changed the "absorption" term with $|i'_{oc} >$. But we think that it would cause small variations of the layer-property values for LiF, just like intermolecular (off-diagonal or mixed) contributions to molecular (hyper)polarizability are small for the clusters studied by Botek and Champagne (Ref. [9]).

3. Results

3.1. Polarizability

The uncoupled SOS method applied to the LiF bulk leads, via the formula of dilute solutions (ε = 1 + 4πNα), to a dielectric constant value equal to 2.37 (α = 12.54 a.u per LiF formula). This value is larger than that obtained by the finite field coupled method [7] with the same b_1 middle size basis set (2.02), and than the experimental one (1.96) [12]. Note that in the finite field approach of Ref. [7], the dielectric constant is computed directly from the macroscopic polarization of the electron density using Poisson's equation, avoiding like this the problem of transition from the microscopic property (α) toward the macroscopic one (ε) in condensed matter. However, when a scissor operator is used (0.12 hartree) in the SOS expression (Eq. (1)) for increasing the gap as in Ref. [13], the ε value becomes equal to 2.19 and corresponds to α = 10.39 a.u. for one LiF formula. It is also interesting to note that the dielectric constant value is equal to 1.98 if the length operator is replaced by the velocity one in Eq. (1) (without any gap correction). But in this operator gauge choice, the sum over oscillator strengths is much less than the number of electrons by cell and the Thomas-Reiche-Kuhn relation is far to be checked.

Then, in the following, we use both the length operator and the scissor operator allowing also to calculate partial polarizability and hyperpolarizability, even if their absolute values are not so accurate. In this work, our aim is to study the behaviour of these properties from internal layer (or bulk) to the top layer of the (001) LiF surface.

For all the LiF slabs (S_N with N going from 1 to 8), we have noticed that the polarizability property was practically additive: for instance, the sum over the layer polarizabilities for S_4 is equal to 39.68 a.u. for the parallel component (α_{xx} or α_{yy}) and 35.34 a.u. for the perpendicular one (α_{zz}), while the values obtained for the whole slab of 4 layers are 39.79 and 36.02 a.u., respectively. The same additivity can be observed for the Li and F atoms: the polarizability components of a layer is practically equal to the sum of the Li and F atomic polarizabilites of this layer, according to the ionic character of this system.

In Fig. 1, we have reported the polarizability components, perpendicular (α_{zz}) and parallel (α_{xx}) to the layer, for the inner and top layers of the unrelaxed S_N slabs when N goes from 1 to 8. For N equal to 1 and 2, both layers are identical. Convergence is reached for the S_8 system. For the inner layers, the parallel component (α_{xx}) tends rapidly to the bulk value (9.5 a.u. with the b_1 basis set) obtained in the same conditions while the perpendicular component (α_{zz}) converges toward 8.9 a.u. instead of 9.5 a.u. as can be expected. Even if the environment of the inner layer is nearly the same as in the bulk, the 2D and 3D calculations lead to different results because the z direction perpendicular to the surface is treated differently according to the translational properties of each system. In the 2D approach, the translational symmetry concerns only the x and y in plane directions. The z direction does not satisfy the periodic boundary conditions (Born Von Karman conditions). The crystalline orbitals are developped on Bloch sums relative to k vectors in the irreducible Brillouin zone (IBZ) with two components while the IBZ is tridimentional for the bulk. To remedy to this default, the converged zz-component has been forced to be equal to the bulk value. For the b_1 basis set, a multiplicative factor of 1.07 is necessary to fit the α_{zz} value of the S_8 inner layer to the bulk one. In practice, it means that α_{zz} values must be increased by one atomic unit, approximatively. When all the other α_{zz} values are corrected by this factor, as it has been done in Fig. 1, we notice that the polarizability is practically isotropic at the crystal surface too (1% of anisotropy for N=8). In any case, with or without this α_{zz}-correction, the polarizability of the LiF top layer is smaller than in the bulk (8% smaller). The use of the most extended basis set b_2 of Ref. [7] (dash lines in Fig. 1) leads to the same conclusion even if the difference between inner and outer polarizability values is reduced and the slight anisotropy ($\alpha_{zz} > \alpha_{xx}$) at the crystal surface inverted.

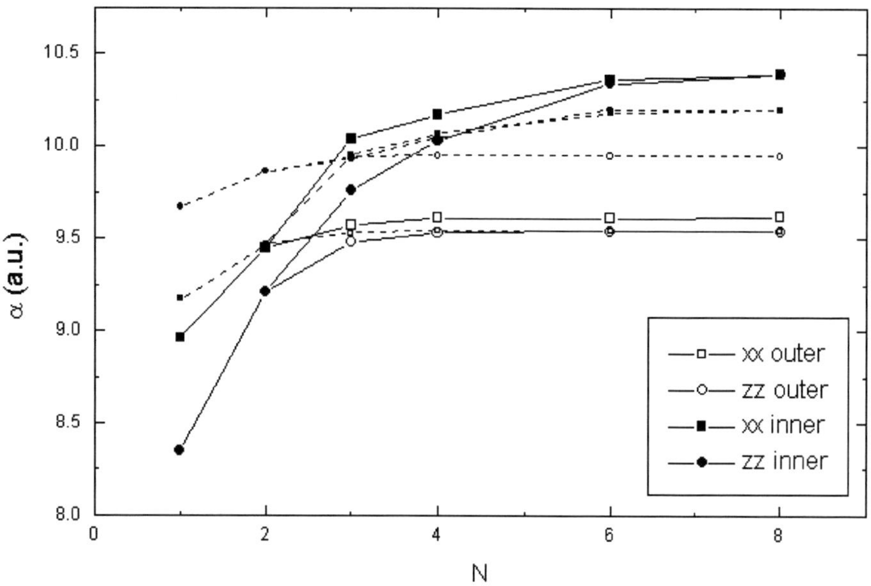

Fig. 1. Polarizability components of the inner and outer layers for a slab of N layers (full lines: middle size b_1 basis set, dash lines: extended b_2 basis set).

However this conclusion is drastically changed when the basis set is extended by adding a layer of ghost atoms to the S_N slab at 3.796 a.u. above the top atomic layer. The resulting system (labelled S_{NG}) is very similar to the $S_{(N+1)}$ slab, except that the outer layer has ghost atoms with a sp (0.1 a.u.) orbital, in place of the Li and F atoms. The component values of the polarizability per layer are reported in Table 1, for the S_4 system with two additional layers of ghost atoms at $Z = 9.489$ and -9.489 a.u. (S_{4G}). These results are compared to those obtained for the S_4 and S_6 unrelaxed surfaces. Note that no shift of α_{zz} values has been done. As expected, if we allow the electrons to be polarized toward the vacuum medium, the outer layer polarizabilities are increased with respect to the S_4 and S_6 slabs (11% and 20% for the xx and zz components respectively in the case of the S_4 slab and, surprisingly, they become larger than the inner ones. These results confirm the importance of the choice of the basis sets when trying to describe the electron density response at the crystal surface, according to the direction of the applied field. An additional d-orbital (0.6 a.u.) centered on ghost atoms increases a little bit the polarizability at the surface. The other point is the influence of the geometrical relaxation on the components of the polarizability. As shown in Table 1, the vertical rumpling of the different ions (+0.016 a.u and −0.044 a.u. for F^- and Li^+ respectively in the top layer ($Z = 5.5652$ a.u.) and −0.008 a.u and +0.006 a.u. for F^- and Li^+ respectively in the inner layer ($Z = 1.8544$ a.u.)) does not really change the values obtained for the unrelaxed slab. It seems that, in this case, the analysis of the polarizabilities can be reduced to the unrelaxed systems, the influence of the geometrical factor being very slight.

3.2. *First hyperpolarizability*

The first hyperpolarizability β is null for the bulk and slabs because of their inversion symmetry structure, while its contribution per layer is different from zero. Obviously, the algebric sum of β over

Table 1
Inner and outer polarizability components of slabs of 4 (S_4) and 6 (S_6) layers, and β components at the crystal surface. The middle size basis set is used (see text) for Li and F, and a sp(0.1) orbital is centered on each ghost atom. All the values are in atomic units

Layer position	S_4+ ghost atoms				S_4				S_6			
Z	α_{xx}	α_{zz}	β_{xxz}	β_{zzz}	α_{xx}	α_{zz}	β_{xxz}	β_{zzz}	α_{xx}	α_{zz}	β_{xxz}	β_{zzz}
1.898	10.32	9.25			10.20	9.06			10.39	9.34		
	10.34[a]	9.28[a]			10.25[b]	8.93[b]						
5.693	10.67	10.33	−1.3	+8.8	9.64	8.61	+1.3	−4.2	10.23	9.11		
	10.87[a]	10.43[a]	−1.5[a]	+9.1[a]	9.56[b]	8.57[b]	+2.2[b]	−3.6[b]				
9.489									9.64	8.62	+1.3	−4.2

[a]values obtained with an additional d(0.6) orbital on each ghost atom.
[b]values obtained when the geometry of the slab is optimized. In this case, the layer position ($Z = 5.5652$ and $Z = 1.8544$ a.u. for the outer and inner layer, respectively) passes through the middle of outer Li and F atoms (thickness: $Z\pm$ 0.1 a.u.).

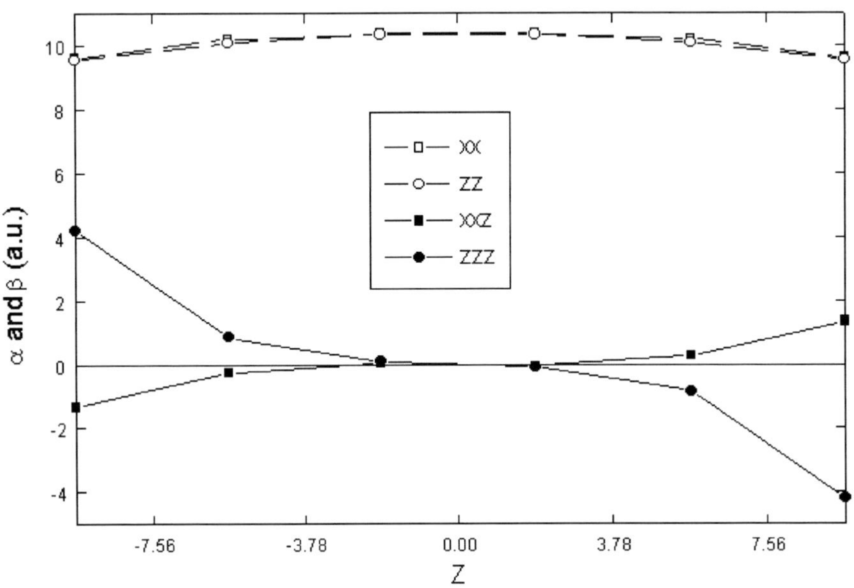

Fig. 2. Polarizability and hyperpolarizability components of each layer in a slab of 6 layers. The origin $Z = 0$ corresponds to the middle of the slab, and $|Z| = 9.489$ a.u. to the surface layer positions.

all the layers of the slab is null. The β_{zzz} and β_{xxz} components for each layer of the S_6 slab have been reported versus the Z layer position in Fig. 2, the smallest values of $|Z|$ corresponding to the inner layers and the largest ones to the top layers (the middle size b_1 basis set is used). Polarizability components have been also plotted in this figure, and as expected the polarizability function versus Z is symmetric while the first hyperpolarizability is antisymmetric. We can see also that both β components of the inner layers decrease rapidly to zero when entering into the bulk.

In Fig. 3, the β components of the outer layer have been plotted for the S_N slabs when no ghost atoms layer is added. They converge rapidly with N, and the same remark could be done for the inner layer β components, not represented in this figure, which decrease even more rapidly to zero when N increases since the symmetry of the inner layer tends to the bulk one. From the figure, the ratio β_{zzz}/β_{xxz} deduced

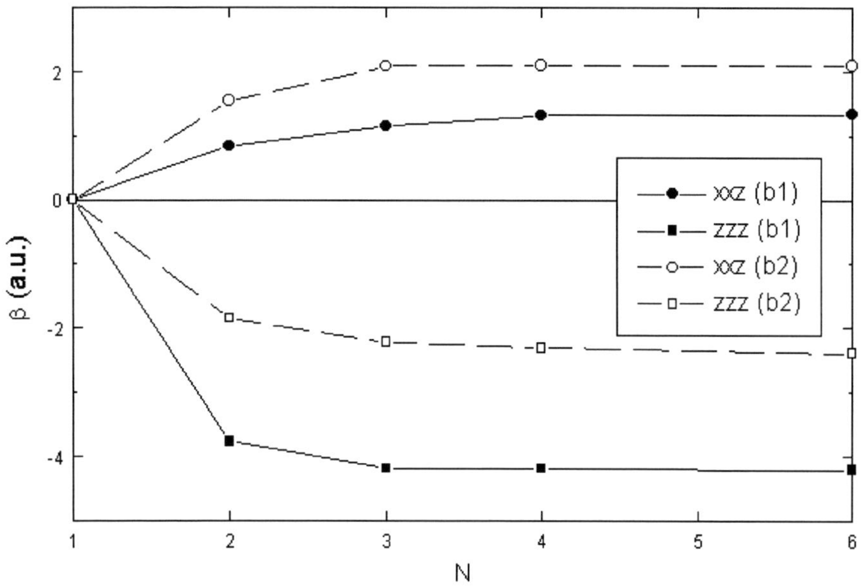

Fig. 3. Hyperpolarizability components of the outer layer in a slab of N layers (full lines: middle size b_1 basis set, dash lines: extended b_2 basis set).

at the convergence is equal to -3.2 when calculations are done with the middle size basis set of Ref. [7] (called b_1; full lines). However, this ratio becomes equal to -1.1 when the most extended basis set (called b_2; dash lines in the figure) is used. It underlines the difficulty to describe the response of the extern electronic cloud to the field in the vacuum direction as well as that of the electron density in the infinite two-dimensional space of the surface. Effectively, when adding now ghost atom layers with sp (0.1 a.u.) orbitals above the surface of the crystal, both signs of β-components change (see Table 1) and lead also to a negative ratio equal to -6.8 (and -6.2 if d(0.6 a.u.) orbitals are added on ghost atoms).

4. Conclusion

To our knowledge, this work is a first attempt to calculate the first hyperpolarizability at the surface of a cubic crystal, using the *ab initio* LCAO-SCF (DFT) method, implemented in the CRYSTAL98 program. Our sum over states method uses the crystalline orbitals and their eigenvalues obtained by the LCAO-DFT periodic approach to calculate the linear and non linear polarizability. It has been modified in order to determine each atomic layer contribution. However, the new analytical expressions obtained under the hypothesis of additivity let suppose that the system is ionic. Their application on the cubic LiF system shows that the principle of additivity of these properties by layer is effectively checked. It follows also that the isotropic character of the bulk polarizability of the cubic ionic LiF system is practically kept for polarizability by layer, even at the surface of the slab.

In order to know if the polarizability components increase from inner to outer layer, it is important to use extended basis sets, particularly for well describing electron density at the surface (and its response to an electric field) on the vacuum side. In this aim, a ghost atom layer has been added above each considered systems at the same distance of the first layer below the surface of the slab. In this case, the polarizability at the surface becomes a little bit larger than the bulk value of the inner layers for LiF. As for the first hyperpolarizability by layer, which is no more equal to zero near the surface as expected, its calculation is even more sensitive to the basis set used for Li anf F, and particularly to the presence or not of ghost atom layers in our calculation. For ionic systems, it follows that SHG experiment should only enhance the ions of the surface on a very weak thickness. Finally a geometry optimization has been also investigated, showing that the electric properties are not drastically changed for this ionic system.

In the future, we plan to study other ionic and more covalent real systems with a particular care on the important choice of basis sets for describing surface properties.

References

[1] T. Schneider, D. Wolfframm, R. Mitzner and J. Reif, *Applied Surface Sciences* **154–155** (2000), 565.
[2] Y.M. Chang, L. Xu and H.W.K. Tom, *Phys. Rev. Lett.* **78** (1997), 4649.
[3] S. Yamada and I.-Y.S. Lee, *Analytical Sciences* **14** (1998), 1045.
[4] D.L. Mills in *Nonlinear Optics; Basic Concepts*, (Ch. 8: Nonlinear Optical Interactions at Surfaces and Interfaces), Ed., Springer-Verlag, Berlin, 1998.
[5] F. Kootstra, P.L. de Boeij and J.G. Snijders, *Phys. Rev. B*, **62** (2000), 7071.
[6] V.R. Saunders, R. Dovesi, C. Roetti, M. Causà, N.M. Harrison, R. Orlando and C.M. Zicovich-Wilson, *CRYSTAL98 User's Manual*, University of Torino, Torino, 1998.
[7] C. Darrigan, M. Rérat, G. Mallia and R. Dovesi, *J. Comp. Chem.* **24** (2003), 1305.
[8] M. Rérat, W.D. Cheng and R. Pandey, *J. Phys. Condens. Matter* **13** (2001), 343.
[9] E. Botek and B. Champagne, *Chem. Phys. Lett.* **370** (2003), 197.
[10] PENTE, A. Dargelos Laboratoire de Chimie Théorique et Physico-Chimie Moléculaire, UMR 5624, 1999.
[11] C. Darrigan, *thesis, Université de Pau*, France, 2001.
[12] E.D. Palik, *Handbook of Optical Constants of Solids*, Academic, NY, 1985.
[13] W.Y. Ching, F. Gan and M.Z. Huang, *Phys. Rev. B* **52** (1995), 1596.

Journal of Computational Methods in Sciences and Engineering 4 (2004) 365–380
IOS Press

The molecular electric quadrupole moment and electric-field-gradient induced birefringence (Buckingham effect) of Cl_2

Chiara Cappelli[a,**], Ulf Ekström[a,***], Antonio Rizzo[a,*] and Sonia Coriani[b]
[a]*Istituto per i Processi Chimico-Fisici del CNR, Area della Ricerca, via G. Moruzzi 1, I-56126 Pisa, Italy*
[b]*Dipartimento di Scienze Chimiche, Università degli Studi di Trieste, Via L. Giorgieri 1, I-34127 Trieste, Italy*

Accepted 13 May 2003

Abstract. An *ab initio* investigation of the molecular properties rationalizing the electric-field-gradient induced birefringence (Buckingham effect) for Cl_2 is presented. The quadrupole moment is determined using hierarchies of basis sets and wave-function models. The electric dipole polarizability, the dipole – dipole – quadrupole and dipole – dipole – magnetic dipole hyperpolarizabilities are determined exploiting a Coupled Cluster Singles and Doubles (CCSD) response approach. The properties are zero-point vibrationally averaged, and the contribution of excited ro-vibrational states accounted for. To this end, the interatomic $^1\Sigma_g^+$ ground state potential has been computed at CCSD plus perturbative triples – CCSD(T) – level employing a large augmented correlation consistent basis set. The effect of relativity is estimated at the Dirac-Hartree-Fock level. Our best value for the quadrupole moment of Cl_2 is (2.327 ± 0.010) au and it is in excellent agreement with experiment which, after revision and dependent on the procedure employed for correcting the original estimate of (2.24 ± 0.04) au of Graham et al., [Mol. Phys., **93**, 49, (1998)], ranges from (2.31 ± 0.04) au to (2.36 ± 0.04) au.

Keywords: Molecular quadrupole moment, birefringence, electric field gradient, electric dipole (hyper)polarizabilities, coupled cluster theory, relativistic effects, molecular vibrations

PACS: 33.15.-e, 33.15.Kr, 33.55.Fi, 31.15.Ar, 31.25.Nj

1. Introduction

One of the most successful experimental methodologies to determine molecular quadrupole moments is based on the measurement of the linear birefringence – that is, the anisotropy of the refractive index for two components of plane linearly polarized light – observed when radiation propagates through a (low density) fluid sample with a component at right angles with respect to an external electric field gradient.

The methodology was proposed in 1959 by Buckingham [1], who also derived the first semiclassical molecular theory of the experiment, and it was applied for the first time in 1963 by Buckingham and Disch

*Corresponding author. Tel.: +39 050 315 2456; Fax: +39 050 315 2442; E-mail: rizzo@ipcf.cnr.it.
**Current affiliation: POLYLAB, INFN, Pisa.
***Current affiliation: University of Linköping, Sweden.

to the determination of the quadrupole moment of CO_2 [2]. For this reason we follow the suggestion coming from the experimentalists working on the field [3], and refer to the effect as to Buckingham effect, or Buckingham birefringence. Buckingham's original theory was reformulated by Buckingham and Longuet-Higgins in 1968 to account for the effect on dipolar molecules, where the quadrupole moment is origin dependent [4]. A different, despite a few formal analogies, semiclassical theory of the experiment was proposed in 1991 by Imrie and Raab [5].

Ab initio response theory can be used to obtain reliable and accurate results for all molecular properties that enter the molecular expression for the induced anisotropy of the refractive index according to two above-mentioned semiclassical theories of the effect, the one due to Buckingham and coworkers [4, 6] (BLH), and the one laid down by Imrie and Raab [5] (IR). For non-dipolar linear molecules these properties are the molecular quadrupole moment, the electric dipole polarizability and the hyperpolarizabilities contributing to the so-called temperature-independent term. Beside the determination of state-of-the-art reference values of the quadrupole moments of several linear molecules [7–11], the use of well-established *ab initio* methodologies – in particular at the coupled cluster level of theory [12] – for the calculation of molecular properties has permitted an analysis of the numerical differences occurring between the results obtained within the two semiclassical theories [4,5]. These should, in principle, describe the same observables and had previously been assumed to agree [9,10,13–15]. In the case of carbon monoxide the study of Ref. [15] in particular has shown, at the end of a sophisticated procedure involving very accurate calculations and the revision of experimental data, that excellent agreement could be obtained between the prediction of BLH theory and experiment, whereas IR theory failed to produce values falling within the error bars associated with the measurement. The publication of several *ab initio* results by our group on Buckingham effect in the last few years has resulted in a renewed interest by experimentalists. Very recently Ritchie and his group, motivated also by the results of Refs. [16] and [7], repeated the measurement of the Buckingham birefringence of N_2 [3] and revised the estimate of the quadrupole moment given in Ref. [17]. This led to an overall excellent agreement between *ab initio* calculations and experiment.

In this paper we concentrate on the study of the temperature dependence of Buckingham birefringence of Cl_2. This involves the accurate determination of its quadrupole moment, and of its electric and mixed electric-magnetic (hyper)polarizabilities. Buckingham birefringence in Cl_2 was studied experimentally by Graham, Imrie and Raab in 1998 [17]. The authors have determined an electric quadrupole moment of $(10.07 \pm 0.16) \times 10^{-40}$ C m^2, corresponding to $(2.24 \pm 0.04)\ ea_0^2$ in atomic units. This estimate was obtained by performing a measurement at a single temperature of 27 °C, and it relies on assuming an anisotropy of the electric dipole polarizability of Cl_2 taken from experiment [18] and a contribution due to the temperature independent term, b, as estimated *ab initio* via an approximate Coupled Hartree Fock approach by Amos [19]. Here we obtain a far more accurate value of b, and use it, together with an equally accurate estimate of the anisotropy of the electric dipole polarizability, to revise the estimate for the quadrupole moment obtained by Graham, Imrie and Raab. The revised value is then compared with the one obtained *ab initio*. To this end we employ throughout the paper a Coupled Cluster approach extending to singles and doubles plus perturbative triples [20,21] – CCSD(T) – for the determination of the quadrupole moment, and to CCSD frequency dependent response, up to quadratic, for the calculation of the electric dipole polarizability, dipole – dipole – quadrupole and dipole – dipole – magnetic dipole hyperpolarizabilities. Estimates are obtained for the effect of relativity and for that of molecular vibrations on the properties.

2. Definitions

According to the semiclassical theories of the birefringence induced by an electric field gradient (EFGB) [4–6], the anisotropy observed for a linear non dipolar molecule, Δn, for light propagating along the Z direction can be written as

$$n_X - n_Y = \Delta n = \frac{N\nabla E}{2\epsilon_0}\left\{b + \frac{2}{15kT}\Theta\Delta\alpha\right\} \tag{1}$$

where $\Theta \equiv \Theta_{zz}$ is the symmetry unique component of the permanent (traceless) quadrupole moment and $\Delta\alpha = \alpha_{zz} - \alpha_{xx}$ is the anisotropy of the electric dipole polarizability. $N = \frac{N_A}{V_m}$ is the number density, N_A is Avogadro's number, V_m is the molar volume, $\nabla E = \nabla_X E_X$ the laboratory field gradient, k the Boltzmann constant, ϵ_0 the vacuum permittivity, and T the temperature. The temperature-independent term b is a combination of various hyperpolarizabilities which depends on the formulation [4,5,9,13] and so we write

$$\begin{aligned} b_{\mathrm{BLH}} &= \frac{1}{15}\left(3b_{\alpha\beta,\alpha\beta} - b_{\alpha\alpha,\beta\beta}\right) - \frac{1}{15}\left(3ß_{\alpha,\beta\alpha,\beta} - ß_{\alpha,\beta\beta,\alpha}\right) - \frac{2}{3\omega}\epsilon_{\alpha\beta\gamma}J'_{\alpha,\beta,\gamma} \\ &= B - \mathcal{B}_{\mathrm{BLH}} - \frac{2}{3\omega}J' \end{aligned} \tag{2}$$

$$\begin{aligned} b_{\mathrm{IR}} &= \frac{1}{15}\left(3b_{\alpha\beta,\alpha\beta} - b_{\alpha\alpha,\beta\beta}\right) - \frac{1}{30}\left(ß_{\alpha,\alpha\beta,\beta} + 3ß_{\alpha,\beta\beta,\alpha}\right) - \frac{1}{3\omega}\epsilon_{\alpha\beta\gamma}J'_{\alpha,\beta,\gamma} \\ &= B - \mathcal{B}_{\mathrm{IR}} - \frac{1}{3\omega}J' \end{aligned} \tag{3}$$

where [9,10,13,16]

$$b_{\alpha\beta,\gamma\delta} = \langle\langle\hat{\mu}_\alpha;\hat{\mu}_\beta,\hat{q}_{\gamma\delta}\rangle\rangle_{\omega,0} \tag{4}$$

$$ß_{\alpha,\beta\gamma,\delta} = \langle\langle\hat{\mu}_\alpha;\hat{q}_{\beta\gamma},\hat{\mu}_\delta\rangle\rangle_{\omega,0} \tag{5}$$

$$J'_{\alpha,\beta,\gamma} = i\langle\langle\hat{\mu}_\alpha;\hat{m}_\beta,\hat{\mu}_\gamma\rangle\rangle_{\omega,0}\ , \tag{6}$$

in the usual notation employed to represent frequency dependent quadratic response functions. Above $\hat{\mu}$, $\hat{q}$ and $\hat{m}$ indicate the electric dipole, traced electric quadrupole and magnetic dipole operators, respectively. Subscripts BLH and IR were used to distinguish between Buckingham's and Imrie and Raab's formulation, respectively. Sum – over – state expressions for the responses in Eqs (4) to (6) are given in Ref. [4].

From a single temperature measurement of the birefringence (Δn), according to Eq. (1), the quadrupole moment of the non dipolar linear molecule can be obtained as

$$\Theta = \frac{15kT}{2\Delta\alpha}\left\{\frac{2\epsilon_0\Delta n}{N\nabla E} - b\right\} \tag{7}$$

once $\Delta\alpha$ and b are known from independent sources.

3. Computational details

The wavefunction methods employed in the non-relativistic calculations of the quadrupole moment are Hartree-Fock (SCF), second-order Møller-Plesset (MP2) [22], Coupled Cluster Singles and

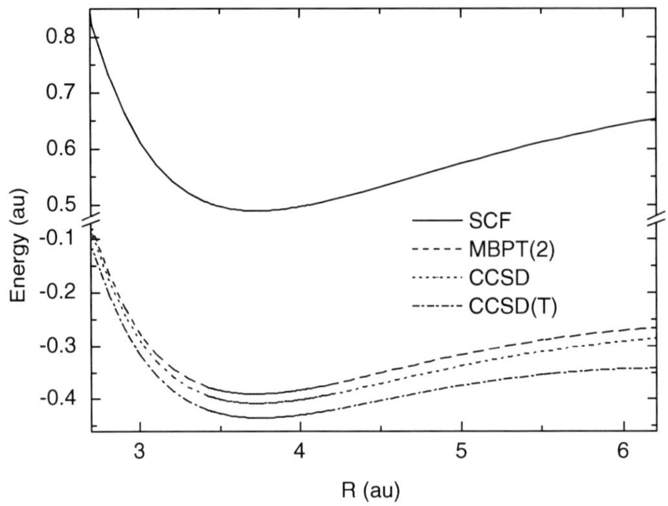

Fig. 1.

Doubles (CCSD) [23] and Coupled Cluster Singles and Doubles plus perturbative triples corrections (CCSD(T)) [20,21]. The (non relativistic) frequency dependent properties were determined at the CCSD level.

All non-relativistic calculations were performed with a development version of the DALTON program [24] which contains the CCSD(T) first-order property module described in Ref. [25]. For some of the calculations of the $^1\Sigma_g^+$ ground state potential curve of Cl_2 ACESII was also employed [26].

The chosen basis sets are the augmented versions of Dunning and co-workers correlation consistent polarized valence sets aug-cc-pVXZ, x-aug-cc-pVXZ [27,28], polarized core-valence sets aug-cc-pCVXZ [29,30] and revised polarized valence sets aug-cc-pV(X+d)Z and d-aug-cc-pV(X+d)Z [31] – with X = D,T,Q,5 and x = d,t.

The $1s$ electrons were kept frozen in the non relativistic calculations, except for some of cases where the valence correlation-consistent sets were employed, and in which the frozen-core approximation was also used. The acronyms "f$1s$" and "fc" were used to indicate the calculations with frozen $1s$ and with the whole core kept frozen, respectively.

The experimental equilibrium geometry R=1.987060 Å [32] (3.7550 au) was used.

The frequency-dependent properties were determined for $\omega = 0.07200$ au which corresponds to a wavelength of 632.8 nm.

All relativistic corrections were calculated using DIRAC [33], a fully relativistic four component program. A development version of this program was used for the quadratic response calculations [34]. The effects of relativity have been estimated by a comparison between standard Hartree – Fock and four component Dirac-Hartree-Fock (DHF) calculations. The correlation consistent basis sets used throughout this study are optimized for nonrelativistic calculations, and therefore they had to be used uncontracted in this comparison. Basis sets for the small components in the DHF case were generated from the kinetic balance condition.

In order to perform the rotational and vibrational averaging of the electronic properties, the potential energy curve for the $^1\Sigma_g^+$ ground electronic state of Cl_2 needed to be known, in particular around the equilibrium geometry. Since we are not aware of other *ab initio* accurate studies, we determined a non relativistic curve for the $X^1\Sigma_g^+$ state of Cl_2 in a wide range of internuclear distances. To this end we

Table 1
Cl_2. Non relativistic CCSD(T)/f1s electronic energy, $V(R)$, for the $X^1\Sigma_g^+$ state in the aug-cc-pV(5+d)Z basis set, Atomic units. The absolute energy is obtained by subtracting 919.50 E_h

R	CCSD(T)	R	CCSD(T)	R	CCSD(T)
2.611	−0.005802	3.731	−0.435071	4.201	−0.420025
2.711	−0.116986	3.751	−0.435096	4.301	−0.414403
2.811	−0.202866	3.755	−0.435091	4.401	−0.408488
2.911	−0.268803	3.771	−0.435037	4.501	−0.402445
3.011	−0.319011	3.791	−0.434897	4.601	−0.396409
3.111	−0.356787	3.811	−0.434681	4.701	−0.390489
3.211	−0.384717	3.831	−0.434393	4.801	−0.384771
3.311	−0.404833	3.851	−0.434036	4.901	−0.379322
3.411	−0.418747	3.871	−0.433615	5.001	−0.374192
3.421	−0.419849	3.891	−0.433133	5.101	−0.369417
3.431	−0.420904	3.911	−0.432593	5.201	−0.365021
3.441	−0.421911	3.931	−0.431998	5.301	−0.361019
3.451	−0.422873	3.951	−0.431352	5.401	−0.357417
3.461	−0.423791	3.971	−0.430658	5.501	−0.354213
3.471	−0.424665	3.981	−0.430293	5.551	−0.352759
3.481	−0.425497	3.991	−0.429917	5.601	−0.351400
3.491	−0.426287	4.001	−0.429531	5.701	−0.348962
3.501	−0.427037	4.011	−0.429135	5.801	−0.346882
3.511	−0.427747	4.021	−0.428728	5.901	−0.345139
3.531	−0.429052	4.031	−0.428311	6.001	−0.343710
3.551	−0.430211	4.041	−0.427885	6.101	−0.342569
3.571	−0.431229	4.051	−0.427451	6.201	−0.341691
3.591	−0.432113	4.061	−0.427007	6.301	−0.341048
3.611	−0.432871	4.081	−0.426094	7.000	−0.341128
3.631	−0.433507	4.101	−0.425149	8.000	−0.346524
3.651	−0.434028	4.121	−0.424175	10.000	−0.354874
3.671	−0.434440	4.141	−0.423174	20.000	−0.365113
3.691	−0.434747	4.161	−0.422147		
3.711	−0.434956	4.181	−0.421097		

employed the aug-cc-pV(5+d)Z basis set and computed the energy at SCF, MP2, CCSD and CCSD(T) levels for some eighty internuclear distances between 2.611 au and 20 au. The 1s orbital was again kept frozen in the post-Hartree Fock calculations. The results are shown in Fig. 1 and those obtained employing the CCSD(T) wavefunction model and actually used to determine the vibrational corrections to the electronic properties are given in Table 1. In Table 2 the spectroscopic constants determined from the curves are collected.

The relevant properties, i.e. the quadrupole moment Θ, the tensor elements of the electric dipole polarizability α at a wavelength of 632.8 nm and the relevant tensor elements of Eqs (4), (5) and (6) were then computed at CCSD/f1s level at eleven internuclear distances, equally spaced between $R = 3.255$ au and $R = 4.255$ au, around the equilibrium distance which, for our CCSD(T) potential curve, lies at $R_{eq} = 3.7468$ au, see Table 2. The basis set chosen for these calculations was the d-aug-cc-pV(T+d)Z set. Zero point vibrational averages and the matrix elements of the electronic properties between the lowest eight vibrational levels supported by the CCSD(T) potential curve, and, for each vibrational state, between the lowest seven rotational sublevels, were computed using VIBROT, which is part of the MOLCAS program suite [35].

Unless specified otherwise, atomic units will be used all throughout the next Section. Conversion factors between atomic units, SI units and esu are based on the compilation of Ref. [36], see also Ref. [37].

Table 2

Cl_2. Spectroscopic constants and term values $G(v)$ for the first eight vibrational levels of the ground electronic non relativistic interatomic potential of the $X^1\Sigma_g^+$ state, as obtained *ab initio* and compared to experiment [32]. Notations for the errors on the experimental values are those of Ref. [32], i.e. a subscript last digit means an uncertainty which may considerably exceed ± 10 units of the last decimal place

	SCF	MP2	CCSD	CCSD(T)	Exp
R_e (au)[a]	3.7275	3.7146	3.7315	3.7468	3.756_6
D_e (au)[b]	0.283	0.202	0.153	0.069	0.09239
D_0 (au)	0.282	0.201	0.152	0.068	
ω_e (cm^{-1})[c]	617.37	594.82	586.36	568.81	559.7_2
$\omega_e x_e$ (cm^{-1})[d]	1.964	2.251	2.289	2.521	2.67_5
$\omega_e y_e$ (10^{-2} cm^{-1})[e]	−1.14	−1.16	−1.19	−1.53	-0.6_7
B_e (10^{-1} cm^{-1})	2.4779	2.4952	2.4727	2.4525	2.439_9
α_e (10^{-3} cm^{-1})	1.17	1.34	1.35	1.44	1.4_9
γ_e (10^{-5} cm^{-1})	0.13	0.06	0.02	−0.24	-0.1_7
δ_e (10^{-7} cm^{-1})	1.60	1.75	1.76	1.82	
β_e (10^{-10} cm^{-1})	−1.08	1.79	2.65	7.46	
v			$G(v)$		
0	308.2	296.9	292.6	283.8	279.2
1	921.6	887.1	874.4	847.5	833.5
2	1531.0	1472.8	1451.4	1406.0	1382.5
3	2136.3	2053.8	2023.7	1959.3	1926.0
4	2737.4	2630.1	2591.2	2507.2	2464.0
5	3334.3	3201.5	3153.8	3049.7	2996.4
6	3926.8	3768.1	3711.4	3586.6	3523.3
7	4515.0	4329.6	4264.0	4117.8	4044.6

[a]Cf. 3.772 au (CASSCF); 3.750 au (Møller-Plesset Epstein-Nesbet, MP-EN), Ref. [45];
[b]Cf. 0.08243 au (CASSCF); 0.1074 au (MP-EN), Ref. [45];
[c]Cf. 558.1 au (CASSCF); 567.0 au (MP-EN), Ref. [45];
[d]Cf. 3.10 au (CASSCF); 3.11 au (MP-EN), Ref. [45];
[e]Cf. 1.6 au (CASSCF); 3.4 au (MP-EN), Ref. [45].

4. Discussion

4.1. The molecular electric quadrupole moment

The equilibrium values for the quadrupole moment $\Theta_e \equiv q_e$ obtained using the hierarchies of wave-function models SCF, MP2 [22], CCSD [23] and CCSD(T) [20,21] and of basis sets specified in Section 3 are collected in Table 3.

For all series of bases and methods, a large decrease in the quadrupole moment is observed moving from the double zeta sets to the triple zeta sets, this showing the relative inadequacy of the double zeta quality basis sets. The quadrupole moment in the doubly augmented sets is overall smaller than the corresponding one obtained with the singly augmented sets. For all methods the results in the x-aug-cc-pVXZ sets appear to converge less systematically with respect to the cardinal number X than those obtained in the revised x-aug-cc-pV(X+d)Z sets.

Focusing on the basis set convergence for the CCSD(T) model, we see that for both the aug-cc-pVXZ and the d-aug-cc-pVXZ series the change in Θ_e when increasing X by one ($\Theta_{e,X+1} - \Theta_{e,X}$) decreases in absolute value as X increases. However, whereas in the aug-cc-pVXZ series the effect decreases at each increase in X, for the d-aug-cc-pVXZ series the difference $(Q - T)$ is about of the same order as the difference $(5 - Q)$. A look at the SCF values reveals that this behavior is mainly related to the

Table 3
Cl_2. Systematic investigation of the equilibrium molecular electric quadrupole moment Θ_e, # c.b.f. indicates the number of contracted basis functions. Atomic units

Basis/Frozen orbital space	# c.b.f.	SCF	MP2	CCSD	CCSD(T)
aug-cc-pVDZ/f1s	54	2.71344	2.68923	2.61353	2.62439
aug-cc-pVTZ/f1s	100	2.30215	2.39459	2.30796	2.32390
aug-cc-pVQZ/f1s	168	2.28562	2.38139	2.29689	2.31574
aug-cc-pV5Z/f1s	262	2.28562	2.38236	2.29918	2.31850
d-aug-cc-pVDZ/f1s	72	2.57757	2.57353	2.49797	2.51223
d-aug-cc-pVTZ/f1s	132	2.26008	2.34512	2.25800	2.27525
d-aug-cc-pVQZ/f1s	218	2.26829	2.36404	2.27901	2.29790
d-aug-cc-pV5Z/f1s	334	2.28109	2.37836	2.29519	2.31465
t-aug-cc-pVDZ/f1s	90	2.56234	2.56068	2.48463	2.49996
t-aug-cc-pVTZ/f1s	164	2.25396	2.33991	2.25243	2.27000
t-aug-cc-pVQZ/f1s	268	2.26768	2.36356	2.27852	2.29744
aug-cc-pV(D+d)Z/f1s	64	2.77566	2.71533	2.63595	2.64716
aug-cc-pV(T+d)Z/f1s	110	2.33264	2.41666	2.32957	2.34443
aug-cc-pV(Q+d)Z/f1s	178	2.30247	2.39435	2.30896	2.32703
aug-cc-pV(5+d)Z/f1s	272	2.29035	2.38172	2.29787	2.31744
d-aug-cc-pV(D+d)Z/f1s	82	2.63625	2.59900	2.51889	2.53379
d-aug-cc-pV(T+d)Z/f1s	142	2.28807	2.36463	2.27665	2.29274
d-aug-cc-pV(Q+d)Z/f1s	228	2.28472	2.37658	2.29081	2.30890
d-aug-cc-pV(5+d)Z/f1s	344	2.28410	2.37555	2.29176	2.31138
aug-cc-pCVDZ/f1s	72	2.71659	2.67569	2.60213	2.61422
aug-cc-pCVTZ/f1s	150	2.32379	2.40276	2.31470	2.33104
aug-cc-pCVQZ/f1s	268	2.30539	2.38909	2.30323	2.32217
d-aug-cc-pCVDZ/f1s	90	2.58350	2.56375	2.48976	2.50543
d-aug-cc-pCVTZ/f1s	182	2.28560	2.35617	2.26801	2.28550
d-aug-cc-pCVQZ/f1s	318	2.28679	2.37063	2.28442	2.30342
aug-cc-pVDZ/fc	54	2.71345	2.68655	2.61088	2.62175
aug-cc-pVTZ/fc	100	2.30215	2.39160	2.30323	2.31883
aug-cc-pVQZ/fc	168	2.28562	2.38007	2.29186	2.31028
d-aug-cc-pVDZ/fc	72	2.57757	2.57126	2.49569	2.50990
d-aug-cc-pVTZ/fc	132	2.26008	2.34181	2.25294	2.26989
d-aug-cc-pVQZ/fc	218	2.26829	2.36273	2.27400	2.29241

SCF contribution: the $(Q - T)$ difference is smaller that the $(5 - Q)$ one, whereas for the correlation contribution – defined as the difference between CCSD(T) and SCF – a decrease is observed at each increment of X. The convergence of the SCF quadrupole moment improves remarkably within the revised valence sets d-aug-cc-pV(X+d)Z, and this is also true at CCSD(T) level, where $(\Theta_{e,X+1} - \Theta_{e,X})$ reduces by one order of magnitude at each increase in the cardinal number.

Similar to what already observed in our previous studies of quadrupole moments [7–11], the difference between the results in the triply and doubly augmented basis sets decreases steadily as we go from the double-zeta level to the quadruple-zeta level, e.g. from 0.012 au to 0.0004 au. Clearly, as X increases, the effect of diffuse basis functions beyond the double augmented level becomes negligible.

To investigate the importance of core-correlation we looked at the differences between the results obtained in the x-aug-cc-pCVXZ with only the $1s$ electrons frozen – as recommended by the authors of Ref. [29] – and those obtained in the x-aug-cc-pVXZ, both with the $1s$ electrons and with all core electrons frozen. As seen from Table 3, for the CCSD(T) model the difference between core-valence results and valence results with only the $1s$ electron frozen is rather contained and, for $X = Q$, equal to 0.006 au. When comparing the core-valence/f1s results and the valence/fc ones the difference is about twice as large, at least for $X = Q$. This supports the use of the valence sets in connection with the f1s approximation for the calculation of the quadrupole moment of Cl_2. We do not consider the difference between the core-valence sets and the modified valence sets since the former have been obtained as extension of the valence sets, and prefer to only on the difference between valence and core-valence sets of the same type.

For the discussion of the convergence within the hierarchy of wavefunction models we turn to the d-aug-cc-pV(5+d)Z basis set results and note that the change in Θ_e from SCF to CCSD is only 0.0077 au (less than 0.5%), whereas that observed going from CCSD to CCSD(T) amounts to 0.01962 au (0.9% increase with respect to the CCSD value). The effect of triple excitations as approximately described by CCSD(T) is thus remarkably larger (ca. 250%) than of the effect of single and double excitations. This makes it very difficult to give an estimate of the correlation effects beyond CCSD(T). However, since the overall correlation effect is so small – the CCSD(T) values are about 1.2% larger than the SCF values (with the larger basis sets) – we estimate it (rather conservatively) to be less than 0.01 au, i.e. half the difference between CCSD and CCSD(T) results in the d-aug-cc-pV(5+d)Z basis. Since we cannot determine *a priori* its sign, we shall treat it as an uncertainty of 0.01 au, see below. We finally note that, the correlation effects are significantly overestimated by the MP2 model for all basis sets with the exception of the inadequate X = D and X = (D + d) ones.

To conclude, our best estimate for the non relativistic equilibrium quadrupole moment of Cl_2 is (2.317 $\pm$ 0.010) au. This was obtained taking the CCSD(T)/d-aug-cc-pV(5+d)Z/f1s value (2.31138) and adding the difference between the CCSD(T)/d-aug-cc-pCVQZ/f1s value (2.30342 au) and the CCSD(T)/d-aug-cc-pVQZ/f1s value (2.29790 au) to account for the addition of core-correlation functions. The error bar was obtained taking the difference between d-aug-cc-pV(5+d)Z and d-aug-cc-pV(Q+d)Z as basis set error at the CCSD(T) level (ca 0.00248), and half the difference between CCSD and CCSD(T) results in the d-aug-cc-pV(5+d)Z/f1s basis as correlation error (0.00981), and treating them as standard deviations. This estimate of the error should be considered rather conservative.

4.2. The electric dipole polarizability and the b contribution to the birefringence

4.2.1. The electric dipole polarizability

The values of the anisotropy of the frequency dependent electric dipole polarizability, $\Delta\alpha$ obtained at the CCSD level for a wavelength of 632.8 nm using the hierarchies of basis sets specified in Section 3 are collected in Table 4.

A general increase in $\Delta\alpha$ is observed moving from the double zeta sets to the triple zeta and quadruple zeta sets. Also, moving from the singly to the doubly augmented sets $\Delta\alpha$ decreases in the case of the double zeta sets whereas the opposite happens for the triple and quadruple zeta sets. In particular, differences up to 0.1 au arise going from the singly augmented double zeta sets to the corresponding triple zeta sets, whereas the differences become three times larger (up to 0.3 au) going from the doubly augmented double zeta to the corresponding triple zeta sets. On the contrary going from the singly augmented triple zeta to the corresponding quadruple zeta sets and from the doubly augmented triple zeta to the corresponding quadruple zeta sets, the differences remain in the range 0.10–0.20 au. This behavior shows once again the inadequacy of the double zeta quality basis sets.

The effect of augmentation decreases as we go from the triple zeta level to the quadruple zeta level (from 0.086 au to 0.077 au for the "standard" sets, and from 0.124 au to 0.079 au for the revised sets). This behavior parallels that seen for the quadrupole moment and as before it can be ascribed to the diminishing influence of the diffuse basis functions.

The convergence trend of $\Delta\alpha$ is the same for the "standard" x-aug-cc-pVXZ and the revised x-aug-cc-pV(X+d)Z valence sets, as well as for the "f1s" and "fc" approximations. Smaller $\Delta\alpha$ values are obtained in the frozen core calculations.

Table 4
Cl_2. CCSD results for the frequency dependent properties entering the molecular expression for Buckingham birefringence. λ = 632.8 nm. Atomic units

Basis/Frozen orbital space	$\Delta\alpha$[a]	b_{IR}	b_{BLH}
aug-cc-pVDZ/f1s	17.5831	−126.4414	−134.3529
aug-cc-pVTZ/f1s	17.6540	−206.8147	−168.0796
aug-cc-pVQZ/f1s	17.8088	−253.4045	−187.4126
d-aug-cc-pVDZ/f1s	17.5157	−220.1274	−199.6747
d-aug-cc-pVTZ/f1s	17.7402	−293.5700	−202.0177
d-aug-cc-pVQZ/f1s	17.8863	−298.0110	−198.4185
aug-cc-pV(D+d)Z/f1s	17.5736	−126.5692	−135.6330
aug-cc-pV(T+d)Z/f1s	17.6605	−210.8285	−172.3264
aug-cc-pV(Q+d)Z/f1s	17.8264	−255.1371	−190.4288
d-aug-cc-pV(D+d)Z/f1s	17.5031	−220.6589	−200.8384
d-aug-cc-pV(T+d)Z/f1s	17.7849	−294.7426	−203.8504
d-aug-cc-pV(Q+d)Z/f1s	17.9051	−297.6785	−199.2765
aug-cc-pCVDZ/f1s	17.6145	−125.7410	−133.5559
aug-cc-pCVTZ/f1s	17.7125	−205.7948	−168.0630
aug-cc-pCVQZ/f1s	17.8547	−251.1041	−186.8063
aug-cc-pVDZ/fc	17.5722	−126.1295	−133.8929
aug-cc-pVTZ/fc	17.6211	−205.8312	−166.7967
aug-cc-pVQZ/fc	17.7642	−252.4419	−186.0801
d-aug-cc-pVDZ/fc	17.5039	−219.5536	−199.0577
d-aug-cc-pVTZ/fc	17.7081	−292.1507	−200.5406
d-aug-cc-pVQZ/fc	17.8415	−297.0078	−197.0683

[a]Cf. $\Delta\alpha = 25.63$ au (TDHF), 25.30 (SOPPA), from Ref. [38]; $\Delta\alpha = 17.56$ au (Exp) from Ref. [18]).

To investigate the importance of core-correlation we analyzed, as for the quadrupole, the differences between the results obtained with the aug-cc-pCVXZ/f1s and those obtained with the aug-cc-pVXZ bases keeping both the 1s and all core electrons frozen. Parallel to what was observed above for the quadrupole moment, the differences between valence/fc and core-valence/f1s are in some cases almost twice as large as those seen in the comparison involving only f1s calculations.

From the discussion so far, our best estimate for $\Delta\alpha$ is 17.9510 au, as obtained by taking the d-aug-cc-pV(Q+d)Z/f1s value, 17.9051 au, and adding the difference between the aug-cc-pCVQZ/f1s value (17.8547 au) and the aug-cc-pVQZ/f1s value (17.8088 au) to account for core-correlation. This value is far closer to experiment than the time dependent Hartree-Fock (TDHF) or second order polarization propagator approximation (SOPPA) results obtained by Oddershede and Svendsen [38] for the same wavelength, see Table 4. Although the static limit of the electric dipole polarizability of Cl_2 has been studied often [39–41], to our knowledge the frequency dependence has seldom been analyzed. It has been very recently the indirect subject of studies of Raman scattering involving also two of the present authors, see Refs. [42–44]. Basis set and electron correlation effects [42], including those of triple excitations [43], and relativistic effects [44], have been analyzed and discussed for wavelengths other than 632.8 nm, and of interest in the domain of Raman scattering spectroscopy. The current estimate (non relativistic and not yet including the effect of molecular vibrations) is approx. 0.4 au higher (2.2% off) than experiment, which records a $\Delta\alpha = 17.56$ au [18]. See below for a discussion of relativistic and molecular vibration effects.

4.2.2. The b contribution to the birefringence

The values of the temperature-independent term b, see Eqs (1) to (3), obtained at the CCSD level using the hierarchies of basis sets specified in Section 3 are collected in Table 4. Both the results obtained in BLH and IR formulations of the theory are given. The wavelength is again $\lambda = 632.8$ nm.

b decreases moving from the double zeta to the triple and quadruple zeta sets. Looking in detail at Table 4, we see a strong decrease in the value (more than 60% for IR, around 25% for the BLH) going from the double zeta to the triple zeta sets in the case of the singly augmented sets. The rate of decrease lowers to about 30% for IR and 1% for BLH in the case of the doubly augmented sets. Moving from the triple to the quadruple zeta sets, a further decrease in observed (about 20% for IR and 10% for BLH) for the singly augmented sets, whereas in the case of the doubly augmented sets the results appear to be roughly at convergence (only 1–2% of difference going from the doubly augmented triple zeta to the doubly augmented quadruple zeta sets). In this case, IR predicts a decrease of b, whereas BLH yields an increase.

Once again, the trends described above are similar for "standard" and revised valence sets, as well as for the core-valence sets. Larger values of b are obtained in the frozen core calculations.

Comparing the differences between the aug-cc-pCVXZ/f1s results and those yielded by the aug-cc-pVXZ/f1s or the aug-cc-pVXZ/fc we note that the behavior is here less systematic than in the case of the quadrupole moment and of the polarizability anisotropy. It is thus difficult to draw conclusions in this case on the superiority of either the "f1s" or the "fc" approximation when adopted in connection to the use of valence basis sets. Nonetheless we employ the same procedure used above for the other properties to account for core correlation on b, noting that the overall effect of the correction due to core electrons is anyway very small, not exceeding 1% in the worst case.

Our best estimates are $b_{\mathrm{IR}} = -295.378$ au and $b_{\mathrm{BLH}} = -198.670$ au. They were obtained by taking the d-aug-cc-pV(Q+d)Z/f1s value, -297.679 au (IR) and -199.277 au (BLH), respectively, and adding the difference between the aug-cc-pCVQZ/f1s value (-251.104 au for IR and -186.807 au for BLH) and the aug-cc-pVQZ/f1s value (-253.405 au for IR and -187.413 au for BLH) to account for core-correlation. The difference between the two formulations is huge, ≈ 100 au. Also, it is more than evident that the value of b is far larger, in absolute, than that estimated in 1982 by Amos [19] using an approximate Coupled Hartree-Fock method (-46.56 au) and employed by Graham and co-workers in Ref. [17] to determine the quadrupole moment of Cl_2 from their single-temperature measurement of Buckingham birefringence. The consequences of this will be clear below.

4.3. Relativistic corrections on properties

To assess the importance of relativistic effects on our observables we carried out full four-component Dirac-Hartree-Fock calculations, using the Dirac-Coulomb Hamiltonian, on the quadrupole moment, the electric dipole polarizability and the hyperpolarizability terms entering the expression of b. The results are shown in Table 5. We employed the fully uncontracted versions of the aug-cc-pVXZ (for Θ_e) and d-aug-cc-pVXZ (for all properties) basis sets (X = D,T,Q), and we are comparing here the fully relativistic results with those obtained with the same basis sets at Hartree-Fock level. Relativistic corrections computed at the SCF level cannot in principle be directly applied to the quantities computed with correlated methods. They provide though an estimation of the errors due to relativity in the correlated results. On the quadrupole moment of Cl_2 we carried out also "semi"-relativistic calculations, where a numerical differentiation of the energy containing the one-electron Darwin and Mass-Velocity first-order correction terms was performed in presence of an external electric field gradient. This approach

Table 5
Cl_2. Investigation of relativistic effect on the equilibrium molecular properties. Atomic units. The properties were calculated at Hartree – Fock and Dirac-Hartree-Fock level, using fully uncontracted basis sets in both cases

Basis set	HF	DHF	increase (%)
	Θ_e		
aug-cc-pCVDZ	2.69592	2.72492	1.08
aug-cc-pCVTZ	2.31792	2.34679	1.25
aug-cc-pCVQZ	2.30449	2.33321	1.25
d-aug-cc-pVDZ	2.55608	2.58480	1.12
d-aug-cc-pVTZ	2.25845	2.28766	1.29
d-aug-cc-pVQZ	2.27400	2.30303	1.28
	$\Delta\alpha$		
d-aug-cc-pVDZ	18.9887	19.0659	0.41
d-aug-cc-pVTZ	18.9290	19.0010	0.38
d-aug-cc-pVQZ	18.9145	18.9861	0.38
	b_{BLH}		
d-aug-cc-pVDZ	−163.674	−168.821	3.14
d-aug-cc-pVTZ	−167.117	−171.778	2.79
d-aug-cc-pVQZ	−181.891	−187.094	2.86
	b_{IR}		
d-aug-cc-pVDZ	−201.905	−206.012	2.03
d-aug-cc-pVTZ	−266.376	−270.282	1.47
d-aug-cc-pVQZ	−268.917	−272.847	1.46

represents an approximation which was already employed in previous cases [7–9,11], and which is applied here directly to the electron correlated wavefunction model.

The effects of relativity on the quadrupole moment of Cl_2, estimated as Θ_e(DHF)− Θ_e(HF), increase Θ_e by approximately 1.2–1.3%, quite independent of the quality of the basis set. The "semi"-relativistic calculations at the CCSD/d-aug-cc-pCVTZ/f1s level yielded a quadrupole moment of 2.2994 au, which should be compared with the corresponding non relativistic value of 2.26801 au. This corresponds to an increase by $\approx +$ 1.4%, not too different thus from the full four component DHF estimate. On the other hand, at the CCSD(T)/d-aug-cc-pCVTZ/f1s level the result was $\Theta_e = 2.31711$ au, versus 2.30342 au (+ 0.6%), an after all remarkable reduction with respect to CCSD. We will take this percentage as the estimate of relativistic correction on the best value for the equilibrium quadrupole moment of Cl_2 determined in Section 4.1.

The relativistic effect on $\Delta\alpha$ is, from Table 5, quite small (0.3–0.4%), as already discussed in full detail in Ref. [44]. As far as the hyperpolarizability terms are concerned, the largest effect is on the J' term, which is present in the expression for both b_{BLH} and b_{IR}. This causes in turn a larger (2–3%) effect on b_{BLH}, than on b_{IR} (1–2%), since in BLH the two dipole-dipole-quadrupole contributions B and $\mathcal{B}_{\mathrm{BLH}}$, see Eq. (3), tend to cancel each other as the frequency goes to zero. In Section 4.5 we will use these estimates, in terms of percentage variation, to correct the equilibrium values.

4.4. *Vibrational corrections on properties*

As mentioned in Section 3, the study of the effect of molecular vibrations on the observables implied the determination of an accurate electronic ground state potential curve for Cl_2. The curves obtained at SCF, MP2, CCSD and CCSD(T) levels are shown in the Figure. The CCSD(T) potential curve is also given explicitly in Table 1. The spectroscopic constants and term values are listed in Table 2, where they are compared with experiment and with those characterizing the complete active space SCF (CASSCF) and multi-reference perturbation-theory (MR-PT) curves obtained quite recently for the $X\,^1\Sigma_g^+$ of Cl_2 by Angeli and co-workers [45] employing a $< 17s12p5d4f\ |7s7p5d4f >$ ANO basis set of 162 contracted functions. The effect of triple excitations is quite substantial. The ground state potential computed at CCSD(T)/f1s level yields term values for the lowest eight vibrational levels on the average between 1.6 and 1.8% off experiment. SCF overestimates these values by about 10%, whereas both MP2 and CCSD are off, on the average, by 6 and 5%, respectively. The equilibrium distance obtained at CCSD(T) level is about one hundredth of an atomic unit too short. Also, the effect of triples is especially evident for D_e. CCSD(T) is in excellent agreement with Angeli and co-workers mixed Møller-Plesset Epstein-Nesbet multi-reference perturbation CI results [45].

The effect of molecular vibration and of quantized rotation on the anisotropy of the refractive index induced by an electric field gradient was discussed by Buckingham and Pariseau in Ref. [46], see also Refs. [9] and [13]. The quantization of molecular rotation reduces the alignment of the quadrupolar molecule in the field gradient, thus influencing the inverse-temperature term, and it introduces an extra contribution arising from centrifugal distortion. Moreover, both the temperature independent term b and the properties entering the inverse-temperature term should be zero-point vibrationally averaged. If only the ground vibrational state $|v_0\rangle$ of the electronic ground state is populated, Eq. (1) can be rewritten as

$$\Delta n = \frac{N\nabla E}{2\epsilon_0}\left\{\langle v_0|b|v_0\rangle + \frac{2}{15kT}\langle v_0|\Theta|v_0\rangle\langle v_0|\Delta\alpha|v_0\rangle F(T) + C(T)\right\}, \tag{8}$$

Above $\langle v_0|P|v_0\rangle$ indicates the average over $|v_0\rangle$ of the electronic property P (for the electronic ground state), a function of the nuclear displacement. The function $F(T)$ takes the expression [46]

$$F(T) = 1 - \sigma + \frac{8}{15}\sigma^2 + \cdots \tag{9}$$

where $\sigma = \frac{hc_0B_0}{kT}$ involves the rotational term B_0 (c_0 indicates here the speed of light in vacuum). The centrifugal term $C(T)$ is essentially independent of T, and its magnitude is roughly proportional to $\frac{B_0}{\omega_0}$, the ratio of the rotational term and the vibrational frequency ω_0 [46].

By employing the results obtained at CCSD(T)/aug-cc-pV(5+d)Z/f1s level given in Table 2 we determine a value of $B_0 = 1.114 \times 10^{-6}$ au, which leads to an estimate of $\sigma = 1.288 \times 10^{-3}$ and, through Eq. (9), to $F(T) = 0.9987$. Again from Table 2 we obtain for $\frac{B_0}{\omega_0}$ a value of 4.30×10^{-4}. Thus the reduction of the temperature dependent contribution in Eq. (1) due to quantized rotation is of the order of a part per thousand, one order of magnitude smaller than computed for carbon monoxide in Ref. [9], and far smaller than the 25% reduction obtained by Buckingham and Pariseau for molecular hydrogen [46]. Centrifugal distortion is even less important.

The effect of zero-point vibrational average (ZPVA) on the electronic properties entering Eq. (1) can be discussed with reference to Table 6. The equilibrium values are those obtained using the d-aug-cc-pV(T+d)Z set at CCSD/f1s level and at the wavelength of 632.8 nm. The averages are taken at 0K and at 273.15 K, the latter introducing the effect of the rotational fine structure of the ground electronic state lowest vibrational level. Differences between ZPV averages taken at 0K and at 273.15 K are negligible, and will not be discussed any further. ZPVA decreases the equilibrium value of Θ by less than 0.2%, whereas the anisotropy of the electric dipole polarizability increases by $\approx$ 0.3%. The dipole-dipole-quadrupole hyperpolarizability contributions, B, $\mathcal{B}_{\rm BLH}$ and $\mathcal{B}_{\rm IR}$ all increase slightly, less than 0.2%. The largest effect is felt by the dipole-dipole-magnetic dipole response contribution J', which increases by

Table 6
Cl_2. Effect of molecular vibrations. Atomic units

	Eq. value	ZPVA(0 K)	ZPVA(273.15 K)	%	Pure Vib.	%
Θ	2.2858	2.2818	2.2819	−0.18		
α_{zz}	43.4606	43.5074	43.5085	0.11		
α_{xx}	25.6757	25.6761	25.6763	0.00		
α_{ave}^{a}	31.6040	31.6199	31.6204	0.05		
$\Delta\alpha$	17.7849	17.8313	17.8322	0.26		
B	−694.98	−696.11	−696.12	0.16	−8.09	1.33
$\mathcal{B}_{\rm IR}$	−497.98	−498.81	−498.81	0.17	−1.10	0.39
$\mathcal{B}_{\rm BLH}$	−686.62	−687.76	−687.77	0.17	−4.46	0.82
J'	−7.04	−7.14	−7.14	1.49		
$b_{\rm IR}$	−294.74	−296.52	−296.53	0.60	−6.98	2.97
$b_{\rm BLH}$	−203.85	−206.77	−206.79	1.43	−3.62	3.21

[a] $\alpha_{ave} = \frac{1}{3}(2\alpha_{xx} + \alpha_{zz})$.

$\approx$ 1.5%. Due to near cancellation between the B and $\mathcal{B}_{\rm BLH}$ contributions in Eq. (1), this non negligible effect on J' transfers almost unaltered to $b_{\rm BLH}$. Combining the ZPV averages on the three contributions to the T-independent part leads instead to a 0.60% increase on $b_{\rm IR}$.

Turning to the analysis of the contribution due to the excited vibrational states, generally identified as pure-vibrational (PV) contribution [9,13,47,48]. General expressions for the vibrational dipole-dipole-quadrupole and dipole-dipole-magnetic dipole hyperpolarizabilities entering the definition of b were briefly introduced in Ref. [9] and given in detail in Ref. [13]. For non dipolar molecules with no permanent magnetic dipole moment, non vanishing PV contributions exist only for the tensors b and ß. Using the notations of Ref. [48] we can write

$$b^{v}_{\alpha,\beta,\gamma\delta}(\omega_\sigma;\omega_1,\omega_2) = ß^{v}_{\alpha,\gamma\delta,\beta}(\omega_\sigma;\omega_2,\omega_1) = [\text{“}\alpha q\text{”}]_{\alpha,\beta,\gamma\delta} \tag{10}$$

where $\omega_\sigma = -\omega_1 - \omega_2$ and in the usual sum – over – state notation,

$$[\text{“}\alpha q\text{”}]_{\alpha,\beta,\gamma\delta} = \frac{1}{2\hbar^2}\sum_{k\neq 0}\left(\frac{\alpha^{0k}_{\alpha\beta}\,q^{k0}_{\gamma\delta}}{\omega_k - \omega_1} + 2\frac{q^{0k}_{\gamma\delta}\,\alpha^{k0}_{\alpha\beta}}{\omega_k + \omega_2} + \frac{\alpha^{0k}_{\alpha\beta}\,q^{k0}_{\gamma\delta}}{\omega_k + \omega_1 + \omega_2}\right) \tag{11}$$

Above $P^{ij}_{\alpha\beta}$ represents the matrix element, taken between vibrational states $|v_i\rangle$ and $|v_j$, of the $\alpha\beta$ tensor component of the electronic property P ($P = \alpha, q$), whereas ω_k indicates the vibrational excitation energy of state $|v_k\rangle$. The PV contributions computed here are such as to increase the absolute value of B and $\mathcal{B}$, with a percentage ranging from $\approx$ 0.4% for $\mathcal{B}_{\rm IR}$ to $\approx$ 1.4% for B. The overall effect on b is contained to roughly 3% for both semiclassical theories, see Table 6.

4.5. *Comparison with an experimentally deduced quadrupole moment*

An experimental estimate of the quadrupole moment, Θ^{exp}, of Cl_2 has been reported by Graham et al. [17] based on EFGB measurements at 300 K. A mean value for $\{\Theta + 15bkT/2\Delta\alpha\}$ of $(9.99\pm0.16)\times 10^{-40}$ Cm^2 – (2.23 ± 0.04) au – was obtained. Using the hyperpolarizability correction term $b^{Amos} = -0.079\times10^{-60}$ $\text{C}^3\text{m}^4\text{J}^{-2}$ (– 46.56 au) computed by Amos [19], who employed an approximation to the Coupled Hartree-Fock method, and a value of $\Delta\alpha^{exp} = 2.89\times10^{-40}\text{C}^2\text{m}^2\,\text{J}^{-1}$ (17.5 au) at 632.8 nm as reported by Bridge and Buckingham [18], a value of $\Theta^{exp} = (10.07\pm0.16)\times10^{40}$ Cm^2 – (2.24 ± 0.04) au – was derived.

Such an estimate for the quadrupole moment can be revised on the basis of our calculated b and $\Delta\alpha$.

Let us first correct our best estimates for the electric dipole polarizability anisotropy and the b contribution, obtained in Sections 4.2.1 and 4.2.2, respectively, for relativistic and molecular vibration effects.

We add to the best estimate for $\Delta\alpha$ in Section 4.1 (17.9510 au), a 0.38% to account for relativistic effects (see Table 5, d-aug-cc-pVQZ basis) and a 0.26% to account for ZPVA (see Table 6). We obtain a value of $\Delta\alpha^{BE} = 18.0659$ au, approximately 2.9% higher than experiment [18]. We note in passing that triple excitations are needed in order to improve agreement with experiment, see Ref. [43].

As far as b_{IR} is concerned, we add to the best estimates of Section 4.2.2 ($b_{\mathrm{IR}} = -295.378$ au) a 1.46% to account for relativistic effect (Table 5, d-aug-cc-pVQZ basis), 0.60% to account for ZPVA and 2.97% due to PV (see Table 6), to obtain a final value of $b_{\mathrm{IR}}^{BE} = -310.236$ au. In BLH formulation, starting from $b_{\mathrm{BLH}} = -198.670$ au, we add in sequence 2.86%, 1.43% and 3.21%, which yields a final estimate of $b_{\mathrm{BLH}}^{BE} = -213.571$ au.

The revised value of experiment, Θ_{rev}^{exp}, can now be obtained from the equation

$$\Theta_{rev}^{exp} = \Theta^{exp} + \frac{15kT}{2}\left[\frac{b^{Amos}}{\Delta\alpha^{exp}} - \frac{b^{BE}}{\Delta\alpha^{BE}}\right] = (2.23 \pm 0.04) - \frac{15kT}{2}\frac{b^{BE}}{\Delta\alpha^{BE}} \tag{12}$$

For $T = 300$ K, this yields $\Theta_{rev,\mathrm{IR}}^{exp} = (2.35 \pm 0.04)$ au, and $\Theta_{rev,\mathrm{BLH}}^{exp} = (2.31 \pm 0.04)$ au, which both correct substantially the original estimate by Graham and co-workers of (2.24 ± 0.04) au, a 5.5% and 3.8% increase, respectively.

If we employ, for our revision, the experimental datum for the anisotropy of the electric dipole polarizability, $\Delta\alpha^{exp}$, we obtain alternative estimates of $\Theta_{rev,\mathrm{IR}}^{exp} = (2.36 \pm 0.04)$ au and $\Theta_{rev,\mathrm{BLH}}^{exp} = (2.32 \pm 0.04)$ au, which do not change substantially the overall picture.

Our best estimate of the electronic quadrupole moment of Cl_2 prior to analysis of the relativistic and molecular vibration correction was given in Section 4.1 as $\Theta_e = (2.317 \pm 0.010)$ au. Adding to this value a 0.6% to account for relativity and subtracting a 0.18% due to ZPVA (Table 6) yields a best estimate Θ_0^{BE} of (2.327 ± 0.010) au, in excellent agreement with both $\Theta_{rev,\mathrm{IR}}^{exp}$ and $\Theta_{rev,\mathrm{BLH}}^{exp}$.

Acknowledgments

This work has been supported by the European Research and Training Network "Molecular Properties and Molecular Materials" (MOLPROP), contract No. HPRN-CT-2000-00013. Some of the calculations were carried out on the computers of the Department of Chemistry of the University of Pisa.

Notes in proofs

In the long period of time since this manuscript was accepted for publication a few important developments have been recorded in the literature of the subject. First of all the discrepancies between the two theories of Buckingham birefringence mentioned in this paper (BLH vs. IR) have been resolved in favour of the former. Two papers have appeared in the literature (R.E. Raab and O.L. de Lange, *Mol. Phys.* **101** (2003), 3467 and O.L. de Lange and R.E. Raab, *Mol. Phys.* **102** (2004), 125) reconciling the two formulations, and showing that the correct expression for the electric-field-gradient birefringence is that of Ref. (26), due to Buckingham and Longuet-Higgins. As a consequence the discussion of IR results in this paper becomes obsolete, and only the results obtained in BLH formulation should be taken into account55. Also, after the present manuscript was accepted, a paper, bearing several similarities with our work, on the calculation of the quadrupole moment of Cl_2 by Junquera and co-workers (J.M. Junquera-Hernández, J. Sánchez-Marín, V. Pérez-Mondéjar and A. Sanchez de Merás, *Chem. Phys. Lett.* **378** (2003), 211) has appeared.

References

[1] A.D. Buckingham, Direct method of measuring molecular quadrupole moments, *J. Chem. Phys.* **30** (1959), 1580.

[2] A.D. Buckingham and R.L. Disch, The quadrupole moment of the carbon dioxide molecule, *Proc. Roy. Soc. A* **273** (1963), 275.

[3] G.L.D. Ritchie, J.N. Watson and R.I. Keir, Temperature dependence of electric field-induced birefringence (Buckingham effect) and molecular quadrupole moment of N_2. Comparison of experiment and theory, *Chem. Phys. Lett.* **370** (2003), 376.

[4] A.D. Buckingham and H.C. Longuet-Higgins, The quadrupole moments of dipolar molecules, *Mol. Phys.* **14** (1968), 63.

[5] D.A. Imrie and R.E. Raab, A new molecular theory of field gradient induced birefringence used for measuring electric quadrupole moments, *Mol. Phys.* **74** (1991), 833.

[6] A.D. Buckingham and M.J. Jamieson, The birefringence induced in spherical molecules by an electric field gradient, *Mol. Phys.* **22** (1971), 117.

[7] A. Halkier, S. Coriani and P. Jørgensen, The molecular electric quadrupole moment of N_2, *Chem. Phys. Lett.* **294** (1998), 292.

[8] A. Halkier and S. Coriani, On the molecular electric quadrupole moment of C_2H_2, *Chem. Phys. Lett.* **303** (1999), 408.

[9] A. Rizzo, S. Coriani, A. Halkier and C. Hättig, *Ab initio* study of the electric-field-induced birefringence of a polar molecule: CO, *J. Chem. Phys.* **113** (2000), 3077.

[10] S. Coriani, A. Halkier, A. Rizzo and K. Ruud, On the molecular electric quadrupole moment and electric-field-gradient-induced birefringence of CO_2 and CS_2, *Chem. Phys. Lett.* **326** (2000), 269.

[11] A. Halkier and S. Coriani, State-of-the-art *ab initio* calculations of the molecular electric quadrupole moments of hydrogen fluoride, *Chem. Phys. Lett.* **346** (2001), 329.

[12] O. Christiansen, C. Hättig and P. Jørgensen, Response functions from Fourier component variational perturbation theory applied to a time-averaged quasienergy, *Int. J. Quantum Chem.* **68** (1998), 1.

[13] S. Coriani, A. Halkier and A. Rizzo, The electric-field-gradient-induced birefringence and the determination of molecular quadrupole moments, in: *Recent Res. Devel. Chem. Physics*, (Vol. 2), S. Pandalai, ed., Trivandrum, Kerala, India: Transworld Scientific, 2001, p. 1.

[14] S. Coriani, *Ab initio determination of molecular properties*, Ph.D. dissertation, Aarhus University, Århus, Denmark, July, 2000.

[15] S. Coriani, A. Halkier, D. Jonsson, J. Gauss, A. Rizzo and O. Christiansen, On the electric field gradient induced birefringence and electric quadrupole moment of CO, N_2O and OCS, *J. Chem. Phys.* **118** (2003), 7329.

[16] S. Coriani, C. Hättig, P. Jørgensen, A. Rizzo and K. Ruud, Coupled cluster investigation of the electric-field-induced birefringence of H_2, N_2, C_2H_2 and CH_4, *J. Chem. Phys.* **109** (1998), 7176.

[17] C. Graham, D.A. Imrie and R.E. Raab, Measurement of the electric quadrupole moments of CO_2, CO, N_2, Cl_2 and BF_3, *Mol. Phys.* **93** (1998), 49.

[18] N.J. Bridge and A.D. Buckingham, The polarization of laser light scattered by gases, *Proc. Roy. Soc.* (*London*) **A 295** (1966), 334.

[19] R.D. Amos, A theoretical study of hyperpolarizability effects in the measurement of molecular quadrupole moments, *Chem. Phys. Lett.* **85** (1982), 123.

[20] K. Raghavachari, G.W. Trucks, J.A. Pople and M. Head-Gordon, A fifth order comparison of electron correlation theories, *Chem. Phys. Lett.* **157** (1989), 479.

[21] R.J. Bartlett, J.D. Watts, S.A. Kucharski and J. Noga, Non-iterative fifth-order triple and quadruple excitation energy corrections in coupled-cluster theory, *Chem. Phys. Lett.* **165** (1990), 513.

[22] C. Møller and M.S. Plesset, Note on an approximation treatment for many-electron systems, *Phys. Rev.* **46** (1934), 618.

[23] G.D. Purvis and R.J. Bartlett, A full coupled cluster singles and doubles model: The inclusion of disconnected triples, *J. Chem. Phys.* **76** (1982), 1910.

[24] T. Helgaker, H.J.Aa. Jensen, P. Jørgensen, J. Olsen, K. Ruud, H. Ågren, A.A. Auer, K.L. Bak, V. Bakken, O. Christiansen, S. Coriani, P. Dahle, E.K. Dalskov, T. Enevoldsen, B. Fernandez, C. Hättig, K. Hald, A. Halkier, H. Heiberg, H. Hettema, D. Jonsson, S. Kirpekar, R. Kobayashi, H. Koch, K.V. Mikkelsen, P. Norman, M.J. Packer, T.B. Pedersen, T.A. Ruden, A. Sanchez de Merás, T. Saue, S.P.A. Sauer, B. Schimmelpfennig, K.O. Sylvester-

Hvid, P.R. Taylor and O. Vahtras, DALTON, an *ab initio* electronic structure program, Release 1.2, 2001, see http://www.kjemi.uio.no/software/dalton/dalton.html.

[25] K. Hald, A. Halkier, P. Jørgensen, S. Coriani, C. Hättig and T. Helgaker, A Lagrangian, integral-density direct formulation and implementation of the analytic CCSD and CCSD(T) gradients, *J. Chem. Phys.* **118** (2003), 2985.

[26] J.F. Stanton, J. Gauss, J.D. Watts, W.J. Lauderdale and R.J. Bartlett, The ACESII program system, *Int. J. Quantum Chem. Symp.* **26** (1992), 879.

[27] D.E. Woon and T.H. Dunning, Jr., Gaussian basis sets for use in correlated molecular calculations. III. The atoms aluminium through argon, *J. Chem. Phys.* **98** (1993), 1358.

[28] D.E. Woon and T.H. Dunning, Jr., Gaussian basis sets for use in correlated molecular calculations. IV. Calculation of static electrical response properties, *J. Chem. Phys.* **100** (1994), 2975.

[29] K.A. Peterson and T.H. Dunning, Jr., Accurate correlation consistent basis sets for molecular core-valence correlation effects: The second row atoms Al-Ar, and the first row atoms B-Ne revisited, *J. Chem. Phys.* **117** (2002), 10548.

[30] The basis sets were obtained from the Extensible Computational Chemistry Environment Basis Set Database, Version 1.0, as developed and distributed by the Molecular Science Computing Facility, Environmental and Molecular Sciences Laboratory which is part of the Pacific Northwest Laboratory, P.O. Box 999, Richland, Washington 99352, USA, and funded by the U.S. Department of Energy. PNL is a multiprogram laboratory operated by Batelle Memorial Institute for the U.S. Department of Energy under Contract No. DE-AC06-76RLO, 1830.

[31] T.H. Dunning, Jr., K.A. Peterson and A.K. Wilson, Gaussian basis sets for use in correlated molecular calculations. X. The atoms aluminium through argon revisited, *J. Chem. Phys.* **114** (2001), 9244.

[32] K.P. Huber and G.H. Herzberg, *Molecular Spectra and Molecular Structure. Constants of Diatomic Molecules*, New York: Van Nostrand-Reinhold, 1979.

[33] T. Saue, V. Bakken, T. Enevoldsen, T. Helgaker, H.J.A. Jensen, J.K. Laerdahl, K. Ruud, J. Thyssen and L. Visscher, DIRAC, a relativistic *ab initio* electronic structure program, Release 3.2, 2000, see http://dirac.chem.sdu.dk.

[34] P. Norman and H.J.Aa. Jensen, *J. Chem. Phys.* **121** (2004), 6145.

[35] K. Andersson, M.R.A. Blomberg, M.P. Fülscher, G. Karlström, R. Lindh, P.Å. Malmqvist, J. Olsen, B.O. Roos, A.J. Sadlej, M. Schütz, L. Seijo, L. Serrano-Andrés, P.E.M. Siegbahn and P.O. Widmark, *MOLCAS version 4*, Lund University, Sweden, 1997.

[36] I. Mills, T. Cvitaš, K. Homann, N. Kallay and K. Juchitsu, *IUPAC, Physical Chemistry Division: Quantities, units and symbols in Physical Chemistry*, Oxford: Blackwell Science, 1993.

[37] P.J. Mohr and B.N. Taylor, CODATA recommended values of the fundamental physical constants: 1998, *Rev. Mod. Phys.* **72** 2000, 351.

[38] J. Oddershede and E.N. Svendsen, Dynamic polarizabilities and Raman intensities of CO, N_2, HCl and Cl_2, *Chem. Phys.* **64** (1982), 359.

[39] G. Maroulis, Accurate dipole polarizability for $Cl_2(X^1\Sigma_g^+)$, *Mol. Phys.* **77** (1992), 1085.

[40] G. Maroulis, Electric properties of chlorine, *J. Mol. Struct.* (*Theochem*) **279** (1993), 79.

[41] J. Dougherty and M.A. Spackman, Accurate prediction of static dipole polarizabilities with moderately sized basis sets. Molecules containing p-block atoms from boron to iodine, *Mol. Phys.* **82** (1994), 193.

[42] M. Pecul and A. Rizzo, Linear response Coupled Cluster calculation of Raman scattering cross sections, *J. Chem. Phys.* **116** (2002), 1259.

[43] M. Pecul and S. Coriani, The effect of triple excitations in coupled cluster calculations of Raman scattering cross-sections, *Chem. Phys. Lett.* **355** (2002), 327.

[44] M. Pecul and A. Rizzo, Relativistic effects on the electric polarizabilities and their geometric derivatives for hydrogen halides and dihalogens – a Dirac-Hartree-Fock study, *Chem. Phys. Lett.* **370** (2003), 578.

[45] C. Angeli, R. Cimiraglia and J.-P. Malrieu, On a mixed Møller-Plesser Epstein-Nesbet partition of the Hamiltonian to be used in multireference perturbation configuration interaction, *Chem. Phys. Lett.* **317** (2000), 472.

[46] A.D. Buckingham and M. Pariseau, Molecular quadrupole moments. Quantum corrections to the classical formulæ, *Trans. Faraday Soc.* **62** (1966), 1.

[47] D.M. Bishop, Molecular vibrational and rotational motion in static and dynamic electric fields, *Rev. Mod. Phys.* **62** (1990), 343.

[48] D.M. Bishop and B. Kirtman, A perturbation method for calculating vibrational dynamic dipole polarizabilities and hyperpolarizabilities, *J. Chem. Phys.* **95** (1991), 2646.

Journal of Computational Methods in Sciences and Engineering 4 (2004) 381–397
IOS Press

The calculation of excited-state polarizabilities of solvated molecules

Kenneth Ruud[a,*], Benedetta Mennucci[b], Roberto Cammi[c] and Luca Frediani[c]
[a]*Department of Chemistry, University of Tromsø, N-9037 Tromsø, Norway*
[b]*Dipartimento di Chimica e Chimica Industriale. Universitá di Pisa, Via Risorgimento 35, 56126 Pisa, Italy*
[c]*Dipartimento di Chimica Generale ed Inorganica. Universitá di Parma, viale delle Scienze, 43100 Parma, Italy*

Abstract. We briefly review the different *ab initio* methods that have been introduced for the calculation of excited-state polarizabilities of molecules in solution. Emphasis is put on the conceptual differences between the methods, emphasizing the strengths and weaknesses of the different approaches. A general discussion of the use of dielectric continuum methods in the modeling of linear and non-linear electric properties of ground- and excited state polarizabilities is given. Particular attention is given to the notion of *equilibrium* and *non-equilibrium* solvation models. We discuss the results of the few theoretical calculations that have been presented in the literature at the time of this review, and also give a few new results.

Keywords: Excited-state polarizability, solvated molecules, *Ab initio* calculation

1. Introduction

The excited states of molecules are important in molecular reactions and for understanding the behavior of molecules in the presence of electromagnetic fields. As such, there is a need for developing experimental and theoretical tools that allow the properties and characteristics of these molecular states to be probed.

One characteristic feature of an excited state is the excited state polarizability. Whereas the excited state dipole moment has received much attention in the literature (see Refs. [1–3] and references therein), little is currently known about the excited state polarizabilities, experimentally as well as theoretically. There are many reasons for this lack of information. One important reason is the fact that for polar molecules, the molecular dipole moment often dominates the interaction of a molecule in its excited state with the surroundings or an external electromagnetic field. Furthermore, differences in the dipole moment of the ground and lowest-lying excited states have been shown to be an important measure for the effectiveness of molecules to display large hyperpolarizabilities, which has prompted a number of studies of this facet of the properties of excited states [2]. Secondly, few experimental ways of directly determining the excited state polarizabilities exist, and experimentally derived excited-state polarizabilities are often extracted from experimental data using models with a large number of parameters, leaving the quality

*Corresponding author. E-mail: kenneth.ruud@chem.uit.no.

1472-7978/04/$17.00

of the estimated excited-state polarizability uncertain. We note, however, that recent advances in Stark spectroscopy have made it possible to directly obtain information about the excited-state polarizability for non-polar molecules in the gas phase [4–8].

Challenges also face the theoretical determination of excited-state polarizabilities. Although accurate theoretical static and dynamic ground-state polarizabilities now can be calculated for small and medium-sized molecules [3,9] (as also discussed elsewhere in this issue), the situation is less favorable for excited state polarizabilities. There are several reasons for this: (1) Many electronic structure methods are designed to calculate ground electronic state wave functions and cannot easily be extended to excited states, examples of such methods are Hartree–Fock (HF), Density Functional Theory (DFT), and coupled-cluster (CC) theory methods. (2) Excited electronic states often display large static *and* dynamic correlation effects and in some cases a non-degenerate perturbation theory analysis of the excited state polarizability may not be adequate. (3) Although HF, DFT, and CC methods can be extended to the study of second-order excited-state properties through the use of high-order response theory [10–13], this comes with a rather high additional computational cost. (4) Semi-empirical methods are in general not adequate for the study of excited-state second-order properties, making it difficult even to obtain qualitative insight into the nature of excited-state polarizabilities.

However, the challenges facing both theory and experiment must be met since there is a need for an increased knowledge about the properties of electronic excited states, and in particular the polarizabilities. This can easily be illustrated by considering a non-polar molecule in a non-polar solvent. The dominating interactions will then be dispersive. Understanding the excitation spectrum of the solute will involve an understanding of the dispersive interactions present both in the ground and the excited electronic states of the molecule, which are related to the polarizabilities of the two states.

Another example of the importance of excited-state polarizabilities is in the understanding of solvatochromic shifts of non-polar molecular in different non-polar innocuous solvents or glasses [14,15]. The solvatochromic shifts are correlated with the difference in the polarizability of the lowest-lying excited state and the ground electronic state of the solute molecule.

Although recent developments in Stark spectroscopy allows excited-state polarizabilities to be obtained directly from an analysis of the observed spectra of the molecules in the gas phase, most excited-state polarizabilities are determined for molecules in solvents or in glasses. This adds to the complications facing theory in the modeling of the experimental situation, and this will be the main focus of this review. To the best of our knowledge, all theoretical studies of excited-state polarizabilities of solvated molecules [16,17] have focussed on the use of dielectric continuum methods [18], and we will in the following limit ourselves to giving a brief presentation of these models for calculating excited-state polarizabilities.

The rest of this paper is divided as follows: In Section 2 we give the basic outline of response theory, and how this can be applied to calculate excited-state polarizabilities. In Section 3 we outline the basic ansätze for the dielectric continuum model, and illustrate how this approach gives rise to a new variational problem that can be cast into a well-known framework and combined with response theory. In Section 4.2 we briefly present and discuss the findings of the only two *ab initio* studies of excited-state polarizabilities of solvated molecules we are aware of, before we in Sections 4.3 and 4.4 present a few new results. We give some concluding remarks and outlook in Section 5.

2. The use of response theory to calculate excited-state polarizabilities

In this section we will briefly outline an approach for analytical calculations of static and frequency-dependent polarizabilities in the electronic ground and excited states. We will not go into the details of the theory with respect to the exact parameterization of the wave function, focusing here instead on the general features of the theory. We will in particular contrast the determination of the excited state polarizability using the linear response function – which will require the reference state to be that of the excited state of interest – or by using the cubic response function, in which case we only need to determine the ground-state wave function explicitely.

In the case of an exact wave function, we may express the linear response function as a sum over all excited states of our reference system [19]

$$-\langle\langle A;B\rangle\rangle_{\omega} = \mathcal{P}\sum_{n\neq 0}\frac{\langle 0|A|n\rangle\langle n|B|0\rangle}{\omega_{n0}-\omega}. \quad (1)$$

$\mathcal{P}$ here indicates a summation over all cyclic permutations of the operators entering the response function and their associated frequencies. ω_{n0} is the energy difference between the state $|n\rangle$ and the reference state $|0\rangle$. This form of the linear response function is often referred to as the *spectral representation* of the response function, and can be seen to be identical to the result obtained from time-dependent perturbation theory [20].

We note that by selecting both A and B to be the electric dipole operators, we can calculate the dynamic polarizability as

$$\alpha_{\alpha\beta} = -\langle\langle \mu_\alpha;\mu_\beta\rangle\rangle_{\omega}. \quad (2)$$

Since the reference state in Eq. (1) can be any non-degenerate electronic state, we note that we can obtain the excited-state polarizability by calculating the linear response function in Eq. (2) for a wave function optimized for an excited electronic state.

In a similar manner, we find the quadratic response function to be [19]

$$-\langle\langle A;B,C\rangle\rangle_{\omega_1,\omega_2} = \mathcal{P}\sum_{k,l\neq 0}\frac{\langle 0|A|k\rangle\langle k|\overline{B}|l\rangle\langle l|C|0\rangle}{(\omega_{k0}+\omega_0)(\omega_{l0}-\omega_2)}, \quad (3)$$

and the cubic response function [19]

$$\begin{aligned}-\langle\langle A;B,C,D\rangle\rangle_{\omega_1,\omega_2,\omega_3} = {} & \mathcal{P}\sum_{k,l,m\neq 0}\frac{\langle 0|A|k\rangle\langle k|\overline{B}|l\rangle\langle l|\overline{C}|m\rangle\langle m|D|0\rangle}{(\omega_{k0}+\omega_0)(\omega_{l0}-\omega_2-\omega_3)(\omega_{m0}-\omega_3)}\\ & -\mathcal{P}\sum_{k,l\neq 0}\frac{\langle 0|A|k\rangle\langle k|B|0\rangle\langle 0|C|l\rangle\langle l|D|0\rangle}{(\omega_{k0}+\omega_0)(\omega_{k0}-\omega_1)(\omega_{l0}+\omega_2)},\end{aligned} \quad (4)$$

where $-\omega_0 = \omega_1+\omega_2+\ldots$ and $\langle k|\overline{B}|l\rangle = \langle k|B-\langle 0|B|0\rangle|l\rangle$.

Using the electric dipole operator for all operators in the above response functions, we can from the quadratic response function obtain the first hyperpolarizability $\beta_{\alpha\beta\gamma}$, and from the cubic response function we can obtain the second hyperpolarizability $\gamma_{\alpha\beta\gamma\delta}$. However, we will not discuss these properties any further here.

However, by taking a double residue of the cubic response function – that is, by considering the spectral resolution of the cubic response function as two of the optical frequencies approach an energy that corresponds to an electronic excitation from the reference state to a state $|e\rangle$, we observe that we can recover the difference between the excited-state and reference-state polarizability

$$\begin{aligned}&\lim_{\omega_2\to-\omega_e}(\omega_2+\omega_e)\lim_{\omega_3\to\omega_e}(\omega_3-\omega_e)\langle\langle A;B,C,D\rangle\rangle_{\omega_1,\omega_2,\omega_3}\\&=\langle 0|C|e\rangle\left(\langle\langle A;B\rangle\rangle^0_{\omega_1}-\langle\langle A;B\rangle\rangle^e_{\omega_1}\right)\langle e|D|0\rangle,\end{aligned} \tag{5}$$

where the superscripts on the linear response functions designate the nature of the reference state for these polarizabilities. By selecting the reference state to be the ground electronic state, we can determine the excited-state polarizability relative to the ground-state polarizability *without explicitly optimizing the excited state.* In this way we can determine the excited state polarizability even for a ground-state based method such as Hartree – Fock [13].

The spectral resolution of the response functions given above involves the summation over all excited states in the molecule. Such an approach would not be practicable, in part because it would require us to be able to calculate all excited states of the molecule, but also because such a sum-over-states approach has been shown to converge slowly with respect to the number of states included in the summation [21].

To avoid this problem, we can derive the response equations for an approximate reference space [19]. The key to succeed in such an approach is the solution of the response equations in the subspace spanned by the operators describing the time-development of the approximate state. The approximate wave function must then be described in the same manner as the exact state – that is, we must be able to determine the reference state as an eigenstate of the molecular Hamiltonian, and the time-development of the reference state must be described as a unitary transformation of this reference state.

These requirements are fulfilled for any variationally determined wave function, and implementations of the higher-order response functions have been presented for Hartree – Fock [22–24], MCSCF [19, 22,25] and DFT [26,27]. By a suitable choice of the parameterization of the reference state and the time-evolution of this state, the explicit summation over all excited states can be replaced by a set of linear transformations involving the unitary operators describing the time-evolution of the reference state. For instance can we symbolically represent the calculation of the linear response function through the set of equations

$$\langle\langle A;B\rangle\rangle_\omega=\mathbf{A}^{[1]}\mathbf{N}^{[B]}, \tag{6}$$

$$\left(\mathbf{E}^{[2]}-\omega\mathbf{S}^{[2]}\right)\mathbf{N}^{[B]}=\mathbf{B}^{[1]}. \tag{7}$$

In these equations, $\mathbf{A}^{[1]}$ represent a *property gradient* – that is, the first derivative of a molecular property with respect to the unitary parameters describing the time-evolution of the reference state. In a similar manner, $\mathbf{E}^{[2]}$ is an electronic Hessian, the second-derivative of the molecular energy with respect to the unitary parameters describing the time-evolution of the state, and $\mathbf{S}^{[2]}$ the second-derivative of the metric matrix. The explicit expressions for these matrices and vectors need not concern us here, but can be found for instance in Ref. [19] for the case of SCF and MCSCF wave functions. However, we note that the determination of the linear response function now is reduced to the solution of a set of linear equations, and similar expressions can be derived for the higher-order response functions.

It is worth noticing that the solution of the linear transformations in Eq. (7) is an iterative procedure, since $\mathbf{E}^{[2]}$, $\mathbf{S}^{[2]}$ and $\mathbf{A}^{[1]}$ are the known quantities in this equation. $\mathbf{E}^{[2]}-\omega\mathbf{S}^{[2]}$ is multiplied onto a suitably chosen trial vector $\boldsymbol{\sigma}$ and compared to the right-hand side $\mathbf{B}^{[1]}$. When the difference between the two vectors is small enough, the trial vector is considered to be equal to the *response vector* $\mathbf{N}^{[B]}$. This iterative procedure is chosen in order to avoid the explicit construction and inversion of the $\mathbf{E}^{[2]}$

and $\mathbf{S}^{[2]}$ matrices, since the dimensions of these matrices may become prohibitively large for instance for wave functions such as the multiconfigurational self-consistent field (MCSCF) approach. Indeed, modern implementations of the response formalism never construct $\mathbf{E}^{[2]}$ explicitly, but rather construct the product $\mathbf{E}^{[2]}\mathbf{N}^{[B]}$ directly [28,29], in a manner similar to that used in the direct CI approach [30,31].

Let us at this stage note that non-variational wave functions such as coupled-cluster (CC), configuration interaction (CI) or Møller – Plesset second-order perturbation theory may be cast in a variational form through the use of Lagrangian multipliers [32]. In this way response functions can also be defined for instance for coupled-cluster wave functions [33,34]. We also note that in the case of non-variational wave functions, the theory may be more conveniently recast in the form of derivatives of quasi-energies as described in detail in Ref. [35].

Before leaving the theoretical determinations of polarizabilities of molecules in the gas phase, let us just mention that one can obtain the static polarizabilities from any kind of molecular wave function by finite difference. This is in many respects the simplest approach, as all that is needed is to add the dipolar interaction between the electronic structure of a molecule and an external electric field to the molecular Hamiltonian, and then optimizing the molecular state in the presence of an applied electric field. By numerical differentiation of the energies obtained with suitable electric fields, the static polarizabilities can be obtained. Although easy to implement, the approach suffers from being limited to static polarizabilities, and care must furthermore be taken to ensure numerical stability of the calculated results. On a more technical note regarding the determination of excited-state polarizabilities, we recall that the application of electric field perturbations will reduce the symmetry of the molecular system, and in an MCSCF optimization it may be difficult to ensure that the proper excited state is obtained for all applied external field perturbations calculated in the lower molecular point group symmetry, in particular for highly excited molecular states.

3. Dielectric continuum models

As in the previous section, we will not go into the details of the dielectric continuum models for describing a solute molecule in a solvent, referring the interested reader instead to recent reviews [3,18, 36]. Our emphasis will be on the general approach, and in particular how the model can be seen to lead to modifications of the gas-phase Hamiltonian used in the response theory approach presented in the previous section.

The starting point for dielectric continuum models is the consideration of the solvent as a homogeneous dielectric medium with a permittivity ϵ. This dielectric medium is polarized by a solute molecule placed in a cavity C in the dielectric medium. This system is described by the Poisson and Laplace equations

$$-\nabla V = 4\pi\rho_M \qquad \text{in C,} \tag{8}$$

$$-\epsilon\nabla V = 0 \qquad \text{outside C,} \tag{9}$$

together with the proper boundary conditions

$$V_o - V_i = 0 \qquad \text{on } \Sigma, \tag{10}$$

$$\left(\frac{\partial V}{\partial n}\right)_i - \epsilon\left(\frac{\partial V}{\partial n}\right)_o = 0 \qquad \text{on } \Sigma, \tag{11}$$

where V is the electrostatic potential due to the presence of the charge distribution ρ_M of the solute molecule located inside the cavity. The potential is required to be continuous across the cavity boundary

Σ through Eq. (10), where the subscripts o and i indicate regions outside and inside of the cavity, respectively. Eq. (11) indicates a corresponding boundary condition for the dielectric displacement (i.e. the derivative of the potential along the direction n perpendicular to the cavity surface).

There are different approaches to solve this problem. We will here focus on one such method, known as the Polarizable Continuum Model (PCM) [37–41]. This method belongs to the class of apparent surface charge (ASC) continuum methods as it exploits an apparent surface charge to represent the solvent dielectric response.

In the PCM approach, the first step is the transformation of Eqs (8)–(9) into integral equations on the surface Σ that can be solved with standard numerical methods. In this framework, the solution of Eqs. (8)–(9) is given by the sum of two electrostatic potentials, one produced by ρ_M and the other due to an "apparent" surface charge distribution σ, placed on the interface, arising from the polarization of the dielectric medium

$$V(x) = V_M(x) + V_\sigma(x) = \int_{R^3} \frac{\rho_M(y)}{|x-y|} dy + \int_\Sigma \frac{\sigma(s)}{|x-s|} ds, \tag{12}$$

where the integral in the first term is taken over the entire three-dimensional space.

The numerical approach used to treat the ASC density exploits a partitioning of the surface into K small portions, called tesserae, of known area a_k, on which the charge σ is assumed constant. In this framework – which can be linked to the analogous techniques used in the fields of physics and engineering and known as the Boundary Element Method (BEM) – the integral form of $V_\sigma(x)$ in Eq. (12) is reduced to a finite sum running over the point charges representing the surface charge

$$V_\sigma(x) = \sum_k^K \frac{q_k(s_k)}{|x-s_k|} \Longleftrightarrow q_k(s_k) = \sigma(s_k) a_k, \tag{13}$$

where the vector s_k indicates the position on each tessera k where the constant value of σ is evaluated (usually, it identifies the center of the tessera and is called the representative point). In the most recent version of the PCM method, known as the Integral Equation Formalism (IEF), the apparent surface charges q_k are determined by the electrostatic potential produced by the solute charge ρ_M on the cavity surface and depend on the cavity geometry and on the solvent permittivity ϵ; in a contracted form we can thus write

$$\mathbf{q}(\rho_M) = \mathbf{Q}(\epsilon, \Sigma; \rho_M),$$

where $\mathbf{Q}$ is a square matrix of dimension $K \times K$ representing the solvent dielectric response.

The generalization of continuum models to quantum mechanical calculations implies the need to define an effective Hamiltonian, i.e. a Hamiltonian in which the solute-solvent interactions are added through the introduction of a solvent reaction potential

$$\hat{H}_{\text{eff}} |\Psi\rangle = \left[\hat{H}^0 + \hat{V^R}(\Psi)\right] |\Psi\rangle = E |\Psi\rangle, \tag{14}$$

where $\hat{H}^0$ is the Hamiltonian describing the isolated molecule, and V^R describes the *reaction potential* associated with the solvent. The treatment of this reaction potential operator $\hat{V^R}(\Psi)$ is delicate, as this term introduces a nonlinear character into the solute Schrödinger equation.

We recall that the nonlinear equation Eq. (14) is a direct consequence of the variational principle applied to the (free) energy functional $\mathcal{G}$ to be introduced in the presence of the solvent

$$\mathcal{G} = \langle \Psi | \hat{H}^0 | \Psi \rangle + \frac{1}{2} \langle \Psi | \hat{V^R} | \Psi \rangle , \tag{15}$$

where Ψ is the electronic wavefunction obtained as a solution of the Schrödinger equation Eq. (14).

The (free) energy functional $\mathcal{G}$ has a privileged role in this theory, as the solution of the Schrödinger equation is a minimum for this functional even though it is not the eigenvalue of the nonlinear Hamiltonian, here indicated as E. The difference between E and $\mathcal{G}$ has, however, a clear physical meaning; it represents the polarization work that the solute does in order to create the charge density inside the solvent.

In the case of responses of a molecular solute to time-dependent perturbations like those we are considering in this paper, a dynamic reformulation of the continuum solvation methods is required. The main point to consider is that what we call solvent dielectric response involves different physical processes, each evolving in time with a proper relaxation time. The most widely diffused approximation is to decompose such a response into two terms only, collecting all the "dynamic" and all the "inertial" processes, respectively. In particular, we indicate as dynamic response that which describes the response due to the bound electrons of the solvent molecules which can instantaneously adjust themselves to any change in the solute charge distribution. The inertial response can on the other hand roughly be associated with all motions involving the nuclei of the solvent molecules (vibrational relaxations, rotational and translational diffusion etc.). This inertial response is in general characterized by a much longer time scale compared to those involved in the changes of the solute description and/or of the electronic polarization.

The different time-scales characterizing the dynamic and the inertial responses give rise to two possible solute-solvent schemes: (a) both the dynamic and the inertial terms are equilibrated to the solute charge distribution, or (b) only the fast dynamic (or electronic) term instantaneously adapts to the electronic structure of the solute whereas the inertial term can be assumed to be frozen in the reference state configuration. These two alternative schemes are usually indicated as equilibrium and non-equilibrium solvation schemes, respectively.

In the PCM framework, the partitioning of the solvent response into inertial and dynamic components is shifted to the apparent charges. Generalizing Eq. (3) to a quantum mechanical description, these charges can be expressed as an expectation value of an apparent charge operator $\hat{\mathbf{Q}}(\epsilon, \Sigma)$. In terms of this operator, we can define "inertial" and "dynamic" apparent charges as: [42]

$$\mathbf{q}^{in}(\Psi_0) = < \Psi_0 | (\hat{\mathbf{Q}}(\epsilon_0, \Sigma) - \hat{\mathbf{Q}}(\epsilon_\infty), \Sigma) | \Psi_0 >, \tag{16}$$

$$\mathbf{q}^{d}(\Psi_1) = < \Psi_1 | \hat{\mathbf{Q}}(\epsilon_\infty, \Sigma) | \Psi_1 >, \tag{17}$$

where ϵ_0 and ϵ_∞ are the static and optical dielectric constants, respectively, and where Ψ_0 is the electronic wave function of the molecular solute immediately before the time-dependent perturbation is turned on and Ψ_1 is the actual wave function.

In this scheme, the solvent reaction operator of Eq. (14) may be expressed in terms of the two sets of apparent charges

$$\hat{V}^R(\Psi_1) = \hat{J} + \hat{X}^{in}(\Psi_0) + \hat{X}^d(\Psi_1), \tag{18}$$

$$\hat{J} = \hat{\mathbf{V}} \cdot \mathbf{q}^N, \tag{19}$$

$$\hat{X}^{in}(\Psi_0) = \hat{\mathbf{V}} \cdot \mathbf{q}^{in}(\Psi_0), \tag{20}$$

$$\hat{X}^d(\Psi_1) = \hat{\mathbf{V}} \cdot \mathbf{q}^d(\Psi_1) = \hat{\mathbf{V}} \cdot < \Psi_1 | \hat{\mathbf{Q}}(\epsilon_\infty) | \Psi_1 >, \tag{21}$$

where $\hat{\mathbf{V}}$ is a vector containing the electrostatic potential operator at the position of the charges. In Eq. (19), $\mathbf{q}^N$ is the vector containing the apparent charges induced by the solute nuclei.

The definition of the solvent operators given in Eqs (18)–(21) constitutes the basis for the introduction of solvent effects in the calculation of response properties of molecular solutes – both in their ground and excited states – by applying the formalism briefly sketched in the previous section. Here we do not report the details, which can be found in Ref. [17], but we recall that the partitioning into inertial and dynamic terms allows us to describe either the complete (i.e. static) or the incomplete (i.e. dynamic) response of the solvent. In this way we can get information about the properties of both relaxed excited states and Franck – Condon excited states, and we can compare directly with different spectroscopic data obtained through complementary methods.

Let us finally mention that if the cavity in which the solute is placed is chosen to have a simple geometric shape, such as a sphere or an ellipsoid, the use of the ASC method can be avoided, the interaction of the solute's charge distribution and the dielectric medium instead being evaluated from a multipolar expansion of the interaction operators [43–46]. In particular, truncation of the multipolar expansion of the interaction operators at the dipole moment term gives for a spherical cavity rise to the simple Onsager model [47]. The multipolar expansion approach can also be extended to include equilibrium and non-equilibrium solvation schemes, and an implementation of this approach for spherical cavities have been implemented for HF and MCSCF wave functions all the way to the cubic response function for an arbitrary order of the multipolar expansion [48–50]. In these methods, the problem of determining the interaction between the solute and the dielectric medium is shifted from determining apparent surface charges to the evaluation of one-electron multipolar integrals to (in principle) arbitrary order. Although the use of spherical cavities in many cases may be a suitable cavity model, it is much more crude than the cavity used in PCM and can in many case not be expected to provide a reasonable description of the solute/solvent interactions for molecules whose shape deviates significantly from that of a sphere. Furthermore, the problem of charge penetration – that is, the penetration of the solute's charge distribution through the boundaries of the cavity and into the dielectric continuum, which violates the basic ansatz in Eqs (8) and (9) is much more severe than in the PCM-IEF approach, and special procedures for penalizing the solute's wave function for such charge escape must be introduced [51,52].

4. Calculations of excited-state dipole moments and polarizabilities of solvated molecules

To the the best of our knowledge, there has only been two *ab initio* investigations of the excited state polarizability of a molecule in a solvent [16,17]. Both these investigations used a dielectric continuum model to describe the solvent, but whereas Ref. [16] calculated the excited-state polarizability from the cubic response function and a spherical cavity, Ref. [17] employed a linear response function from the explicitly optimized excited states and a molecule-shaped cavity.

Here we shall present a summary of the results reported in Ref. [17] for pNA together with new data computed for two other system, one being the closely related molecule nitrobenzene (NB), the other the non-polar molecule benzene.

Para-nitroaniline is a prototypical push – pull system and has therefore served as an important model for theoretical and experimental studies of linear and nonlinear optical properties. Push – pull organic molecules are characterized by a π-conjugated linkage of an electron-donating and an electron-accepting group. The optical properties of push – pull chromophores exhibit a strong dependence on solvation.

The ground and lowest electronically excited states of push – pull molecules are often described in terms of two molecular resonance structures, the neutral and the zwitterionic structures. According to this two-state model, the solvent polarity should strongly affect the relative weight of the two resonance structures in the different electronic states and thus significantly modify their properties.

Nitrobenzene (NB) is a prototypical aromatic nitro compound. The electronic properties of the molecule, including the nature of the excited states and their dynamics, have attracted much interest as have the effects of polar and apolar solvents on the properties of this molecule [53–55]. Benzene, on the other hand, is the model compound for a non-polar aromatic compound. It is also the smallest member of the polyacene family, and the excited state polarizability of the polyacenes have recieved much attention lately, both in the gas phase [56] and in glasses [57]. Since benzene does not possess a permanent dipole moment, it is a convenient test case for understanding the interactions between excited-state polarizabilities and an external medium.

The MCSCF electronic wave function of the three solutes was described by a complete active space (CAS) wave function. Following previous investigations, an active orbital subset where the 6 (for benzene), 10 (for NB) or 12 (for pNA) active π electrons are distributed in all possible ways among the 6 (for benzene), 10 (for NB) or 12 (for pNA) active orbitals, has been used. These active spaces correspond to the π-orbital valence space for the different molecules. The calculations of the excited-state polarizabilities of pNA and NB have been performed with Dunning's double-zeta basis set (DZV) plus semi-diffuse polarization functions (basis set "b" of Ref. [58]), with d orbital functions for C, N and O with exponent 0.2 and p orbital functions for H with exponent 0.1) This basis set contains 160 basis functions for NB and 180 for pNA and gives Hartree – Fock static and dynamic polarizabilities close to the Hartree-Fock limit [58]. This basis set was chosen in order to get comparable results for NB as those obtained in Ref. [17] for pNA. A standard DZV basis set has been used for the geometry optimizations in gas phase and in solution. For benzene, we have used the 6-31G basis set for the geometry optimizations and the augmented correlation-consistent polarized valence double-zeta (aug-cc-pVDZ) basis set for the calculation of the polarizabilities. All geometry optimizations have been done enforcing the C_{2v} planar symmetry constraint for pNA and NB, and D_{2h} symmetry for benzene.

1,4-Dioxane with permittivity $\epsilon(0) = 2.209$ and $\epsilon(\infty) = 2.022$, and acetonitrile with permittivity $\epsilon(0) = 36.64$ and $\epsilon(\infty) = 1.8060$, have been used as solvents.

The geometry of the three solutes in solution has been optimized using the analytical PCM-IEF-MCSCF gradients implemented in DALTON [59]. The molecular cavity containing the solute was obtained using interlocking spheres centered on the six carbon atoms of the aromatic ring and on all the atoms of the nitrile, the nitro and the amino groups. The radii of the spheres, $R_{CH} = 2.28$ Å, $R_C = 2.04$ Å, $R_N = 1.86$ Å, $R_O = 1.824$ Å , $R_{H(NH)} = 1.44$ Å , were obtained from the pertinent Bondi van der Waals radii, all multiplied by a factor of 1.2.

The excited state wave functions of all solutes both in gas phase and in solution have been optimized at the geometry of the ground state, and thus describe Franck-Condon excited states. Moreover, in the case of the polar solvent we have compared equilibrium and non-equilibrium solvation schemes in order to study the effects of the solvent reorganization on the response properties.

4.1. The polar molecules para-nitroaniline and nitrobenzene

In Table 1 we report the energies and the dipole moments of the ground and the lowest excited states of symmetry B_1 and A_1 for the two polar molecules in gas-phase and in the two solvents.

From the results in Table 1, we note that the 2^1A_1 excited state represents the charge transfer state for both *para*-nitroaniline and nitrobenzene, since the dipole moments of these states are more than twice

Table 1

Energies and dipole moments of the three lowest states of *para*-nitroaniline (from Ref. [17]) and nitrobenzene. Energies in Hartree and dipole moments in Debye. For a description of the active space and basis sets used, see the text

	1^1A_1		2^1A_1		1^1B_1	
	Energy	dip.mom.	Energy	dip.mom.	Energy	dip.mom.
			NB			
Gas	−434.236778	4.388	−434.012178	11.893	−434.061993	4.890
Dioxane	−434.240690	4.919	−434.029576	13.714	−434.066625	5.658
Acetonitrile/NonEq	–	–	−434.041038	14.252	−434.072206	6.530
Acetonitrile/Eq	−434.245458	5.540	−434.052089	15.630	−434.072619	6.935
			pNA			
Gas	−489.274392	6.067	−489.095943	17.237	−489.102532	6.185
Dioxane	−489.287961	6.890	−489.124209	19.360	−489.108753	6.998
Acetonitrile/NonEq	–	–	−489.140301	19.670	−489.116247	7.956
Acetonitrile/Eq	−489.288026	7.893	−489.154355	20.509	−489.116375	7.968

that of the ground state, whereas the changes in the dipole moments are much smaller for the $1B_1$ states, leading for both molecules to only a slight increase in the respective dipole moments relative to the ground state result.

Since the results for *para*-nitroaniline have been discussed in detail and compared to the results of Jonsson et al. [16] in another paper [17], we will in the next section only briefly review the most important findings for this molecule as obtained using the PCM-IEF-MCSCF linear response approach. In the subsequent sections we will discuss in somewhat more detail our results for nitrobenzene and benzene.

4.2. *The excited-state polarizabilities of para-nitroaniline*

In Table 2 we report the excited-state polarizabilities of the $2A_1$ state of pNA. We recall that the molecule is placed on the yz plane with its dipole axis along the y direction.

One interesting observation from Table 2 is that solvation actually leads to a rather substantial decrease in the isotropic polarizability. This reduction is caused by the strong decrease in the polarizability along the dipole axis (α_{yy}). The other two components of the polarizability behaves more 'normal' in the sense that they increase slightly upon solvation. The consequence of this decrease in the polarizability upon solvation is that the change in the ground-to-excited state polarizabilty becomes negative for the solvated molecules, whereas it is slightly positive for the molecule in the gas phase.

Negative $\Delta\alpha$ values are unusual and are most often seen in molecules that are zwitterionic in the excited state but not in the ground state [60]. This can easily be realized by considering the sum-over-states expression for the polarizability

$$\alpha_{yy} = 2\sum_{j\neq i} \frac{\omega_{ji}}{\omega_{ji}^2 - \omega^2} \langle i\,|y|\,j\rangle \langle j\,|y|\,i\rangle\,, \tag{22}$$

where y is a component of the electric dipole moment, i is the electronic state of interest and j are all other electronic (ground and excited) states, and ω is the frequency of the applied field (for a static field $\omega = 0$). Since we have a strong dipole transition from the excited $2A_1$ state back to the ground state, this term may dominate the overall change in the polarizability from the ground to the excited state, thus leading to a decrease in the polarizability upon excitation from the ground to the excited state.

To better analyze the results we recall that the large ground-to-excited state change observed in the static polarizability of the acetonitrile-solvated molecule is due to the fact that only for the static case the ground state polarizability is computed assuming an equilibrium response; for all the dynamic values as

Table 2
CASSCF values of the components of the dynamic polarizability tensor $\alpha(-\omega\ \omega)$/a.u. of the singlet excited state 2^1A_1 in gas and in solution, and ground-to-excited state change of the isotropic polarizabilities ($\Delta\alpha_{iso} = \alpha_{iso}^{2^1A_1} - \alpha_{iso}^{GS}$)

ω (a.u.)	0.0	0.005	0.02	0.0428
		Gas phase		
α_{xx}	47.88	47.89	47.93	48.09
α_{yy}	155.24	155.63	157.08	165.24
α_{zz}	97.80	97.85	98.80	103.88
α_{iso}	100.31	100.37	101.27	105.74
$\Delta\alpha_{iso}$	6.12	6.16	6.72	
		Dioxane		
α_{xx}	52.02	52.02	52.07	52.25
α_{yy}	101.68	101.65	101.11	99.43
α_{zz}	108.72	108.77	109.50	115.98
α_{iso}	87.47	87.44	87.56	89.22
$\Delta\alpha_{iso}$	−23.2	−21.43	−21.88	−22.41
		Acetonitrile/nonEq		
α_{xx}	51.48	51.49	51.53	51.71
α_{yy}	89.36	89.30	88.36	84.83
α_{zz}	106.00	106.04	106.74	112.68
α_{iso}	82.28	82.28	82.21	83.07
$\Delta\alpha_{iso}$	−55.22	−27.02	−27.52	−28.99
		Acetonitrile/Eq		
α_{xx}	51.46	51.46	51.51	51.69
α_{yy}	47.98	47.82	45.35	35.14
α_{zz}	107.35	107.40	108.15	113.10
α_{iso}	68.93	68.34	68.34	66.64
$\Delta\alpha_{iso}$	−68.57	−40.41	−41.39	−45.42

well as for all the excited state static and dynamic polarizabilities a nonequilibrium response is always assumed.

We note that Jonsson et al. did not observe this decrease in the polarizability upon excitation from the ground to the excited state of the solvated molecules [16]. Although the reasons for these different results require a careful analysis, they are most likely due to the different computational approaches used to obtain the excited-state polarizability, in our approach by using linear response theory from the state-optimized wave function, and in the case of Ref. [16] as the double residue of the cubic response function.

We do not discuss the excited-state polarizabilities of the $1B_1$ state, since the effects of the solvent on the excited-state polarizability are here much smaller. The excited-state polarizabilities of this state have been discussed in Ref. [17].

4.3. The excited-state polarizability of nitrobenzene

In Table 3 we have collected the polarizabilities for the charge-transfer state $2A_1$ of nitrobenzene. We note that the same reduction of the dipole-axis component of the polarizability upon excitation observed in pNA is present also for the solvated nitrobenzene molecule, although the effect is less pronounced for the isotropic polarizability of this molecule due to a larger relative increase in the two perpendicular components of the polarizability. As such it is only for acetonitrile in the non-equilibrium scheme that an effective reduction in the polarizability upon excitation from the ground to the $2A_1$ state is observed.

Table 3
CASSCF(10/10) values of the components of the dynamic polarizability tensor $\alpha(-\omega\ \omega)/a.u.$ of the singlet excited state 2^1A_1 of nitrobenzene in gas and in solution, and ground-to-excited state change of the isotropic polarizabilities ($\Delta\alpha_{iso} = \alpha_{iso}^{2^1A_1} - \alpha_{iso}^{GS}$)

ω (a.u.)	0.0	0.005	0.02	0.0428
		Gas phase		
α_{xx}	42.87	42.88	42.92	43.16
α_{yy}	140.04	140.17	142.10	151.41
α_{zz}	115.11	116.31	186.29	84.86
α_{iso}	99.34	99.79	123.77	93.14
$\Delta\alpha_{iso}$	+19.16	+19.59	+43.33	+11.74
		Dioxane		
α_{xx}	46.05	46.06	46.10	46.27
α_{yy}	124.64	124.71	125.76	130.95
α_{zz}	113.16	113.29	115.99	150.03
α_{iso}	94.62	94.69	95.95	109.08
$\Delta\alpha_{iso}$	+6.53	+7.56	+8.50	+20.35
		Acetonitrile/nonEq		
α_{xx}	45.54	45.54	45.58	45.74
α_{yy}	117.67	117.73	118.63	123.11
α_{zz}	108.73	108.82	110.76	125.20
α_{iso}	90.65	90.70	91.66	98.02
$\Delta\alpha_{iso}$	−11.08	+4.45	+5.09	+10.18
		Acetonitrile/Eq		
α_{xx}	45.29	45.29	45.33	45.49
α_{yy}	97.95	97.98	98.41	100.98
α_{zz}	103.94	103.85	106.46	109.43
α_{iso}	82.39	82.37	83.40	85.30
$\Delta\alpha_{iso}$	−19.33	−3.88	−3.16	−2.54

It is worthwhile to compare the ground-to-excited state polarizabilities in the acetonitrile solvent in the equilibrium and non-equilibrium schemes as summarized in Table 3. We recall that as for pNA also for nitrobenzene we have placed the molecule on the yz plane with its dipole axis along the y direction.

The importance of taking the dynamical aspects of the solvent reorganization into account is clearly illustrated for this solvent, as we for the frequency-dependent polarizabilities observe a difference in sign in the ground-to-excited state polarizability. Indeed, the equilibrium scheme underestimates the change in the α_{yy} (in particular) and α_{zz} components of the excited state polarizability upon solvation, leading to a difference in sign in the change of the isotropic polarizability upon excitation from the ground to the $2A_1$ state.

We also note that for the acetonitrile-solvated nitrobenzene molecule the large ground-to-excited state change in the static polarizability can be explained in the same way as for pNA (see above): only in this case in fact the ground state polarizability is computed with a complete equilibrium response.

In Table 4 we have collected the excited-state polarizabilities for the $1B_1$ state, and the change in the polarizability upon excitation from the ground electronic state to this excited state. In contrast to the CT $2A_1$ state, only minor changes in the isotropic polarizability relative to the ground-state polarizability is observed both in gas phase and in the non-polar solvent. This is due to the fact that this excited state involves litte reorganization of the electronic distribution relative to the ground state, which is also evident from the small differences in the ground and excited state dipole moments (see Table 1). As we switch to the more polar acetonitrile solvent, a much larger solvent effect is observed for the difference

Table 4
CASSCF(10/10) values of the components of the dynamic polarizability tensor $\alpha(-\omega\ \omega)$/a.u. of the singlet excited state 1^1B_1 of nitrobenzene NB in gas and in solution, and ground-to-excited state change of the isotropic polarizabilities ($\Delta\alpha_{iso} = \alpha_{iso}^{1^1B_1} - \alpha_{iso}^{GS}$)

ω (a.u.)	0.0	0.005	0.02	0.0428
		Gas phase		
α_{xx}	43.41	43.41	43.45	43.59
α_{yy}	110.86	110.93	112.09	117.81
α_{zz}	97.04	97.08	97.81	101.20
α_{iso}	83.77	83.81	84.45	87.53
$\Delta\alpha_{iso}$	+3.59	+3.61	+4.01	+6.12
		Dioxane		
α_{xx}	47.06	47.06	47.10	47.27
α_{yy}	139.42	139.58	142.21	156.18
α_{zz}	115.16	115.24	116.45	122.26
α_{iso}	100.55	100.63	101.92	108.57
$\Delta\alpha_{iso}$	+12.46	+13.50	+14.47	+19.84
		Acetonitrile/nonEq		
α_{xx}	46.48	46.49	46.53	46.69
α_{yy}	159.52	159.82	164.74	193.88
α_{zz}	113.74	113.82	115.04	120.88
α_{iso}	106.58	106.71	108.77	120.48
$\Delta\alpha_{iso}$	+4.86	+20.46	+22.21	+32.64
		Acetonitrile/Eq		
α_{xx}	46.39	46.40	46.43	46.59
α_{yy}	180.69	181.17	189.03	240.99
α_{zz}	114.44	114.52	115.78	121.85
α_{iso}	113.84	114.03	117.08	136.48
$\Delta\alpha_{iso}$	+12.12	+27.78	+30.52	+48.64

between the ground and excited-state polarizability, which may be due to the increased stabilization of more polar excited states that enter into the sum-over-states expression for the polarizability relative to the stabilization of this $1B_1$ state.

4.4. *The non-polar molecule benzene*

We will finally present some results for the non-polar molecule benzene. The (free) energies and excitation energies calculated as differences between the (free) energies of the excited and ground electronic states are reported in Table 5. Our calculated CASSCF excitation energies for benzene in the gas phase is in good agreement with earlier CASSCF studies [12,61], which indicates that our basis set is sufficiently large to adequately describe these excited states with the chosen active space. We note that further extensions of the active space have been shown to change the excitation energies only little [61], but it is known that our 6-in-6 CASSCF wave function lacks significant amounts of dynamical correlation and thus overshoots the experimental excitation energies significantly [61,62]. As can be expected for $\pi \rightarrow \pi^*$ transitions in such a non-polar molecule as benzene, the solvent hardly affects the calculated excitation energies, even in a fairly polar solvent such as acetonitrile.

The excited-state polarizabilities for the $1B_{2u}$ and $1B_{1u}$ states are given in Table 6. Before addressing the solvent effects on the excited-state polarizabilities, let us first consider the gas-phase results for the $1B_{2u}$ state. A recent Stark effect investigation found the in-plane components of the ground-state

Table 5
Energies (in a.u.) of ground, 1^1B_{1u} and 1^1B_{2u} states of benzene and ground-to-excited state energies (eV). For a description of the active space and basis sets used, see the text

	1^1A_g	1^1B_{2u}		1^1B_{1u}	
	Energy	Energy	Δ	Energy	Δ
Gas	−230.799366	−230.617430	4.95	−230.512469	7.81
Dioxane	−230.800024	−230.618011	4.95	−230.513251	7.80
Acetonitrile/NonEq	–	−230.618802	4.96	−230.514340	7.80
Acetonitrile/Eq	−230.800984	−230.618840	4.96	−230.514369	7.99

Table 6
CASSCF values of the components of the dynamic polarizability tensor $\alpha(-\omega\ \omega)$/a.u. of the singlet excited state 1^1B_{2u} and 1^1B_{1u} in gas and in solution, and ground-to-excited state change of the isotropic polarizabilities ($\Delta\alpha_{iso} = \alpha_{iso}^{EXC} - \alpha_{iso}^{GS}$)

ω (a.u.)	0.0	0.005	0.02	0.0428	0.0	0.005
	B_{2u}				B_{1u}	
	Gas phase				Gas phase	
$\alpha_\perp$	67.31	67.32	67.44	67.92	134.14	134.99
$\alpha_\parallel$	43.24	43.24	43.31	43.59	64.73	64.84
α_{iso}	59.29	59.29	59.38	59.81	111.00	111.61
$\Delta\alpha_{iso}$	−3.93	−3.94	−3.98	−4.03	47.78	48.38
	Dioxane				Dioxane	
$\alpha_\perp$	72.40	72.41	72.55	73.10	156.46	157.64
$\alpha_\parallel$	46.12	46.12	46.20	46.52	70.00	70.11
α_{iso}	63.64	63.65	63.76	64.14	127.65	128.47
$\Delta\alpha_{iso}$	−4.65	−4.65	−4.69	−5.20	59.36	60.17
	Acetonitrile/nonEq				Acetonitrile/nonEq	
$\alpha_\perp$	71.52	71.53	71.67	72.20	151.90	152.95
$\alpha_\parallel$	45.99	45.99	46.07	46.38	69.28	69.38
α_{iso}	63.01	63.02	63.13	63.59	124.32	128.42
$\Delta\alpha_{iso}$	−4.54	−4.54	−4.57	−4.67	56.77	60.86
	Acetonitrile/Eq				Acetonitrile/Eq	
$\alpha_\perp$	71.56	71.57	71.70	72.24	152.62	153.70
$\alpha_\parallel$	45.95	45.95	46.03	46.34	69.45	69.55
α_{iso}	63.02	63.03	63.14	63.60	124.90	125.66
$\Delta\alpha_{iso}$	−4.53	−4.53	−4.56	−4.66	57.35	58.10

polarizabilities to be 79 a.u. and the out-of-plane component to be 48 a.u., to be compared with our ground-state data of 73.5 a.u. and 42.7, respectively. Thus, our CASSCF wave function slightly overestimates the polarizability compared to this experimental investigation.

More interestingly, the experimental investigation found that upon excitation to the $1B_{2u}$ state, the in-plane components of the polarizability hardly change (an increase of 2 a.u.) whereas there was a significant increase in the out-of-plane component of 15 a.u. This is in marked contrast to our results where we instead observe a decrease in the in-plane components of about 5.5 a.u., and only a very moderate increase in the out-of-plane component of about 1 a.u. We note that our results are supported by an earlier theoretical investigation where the excited state polarizability were obtained both using linear response functions from an explicitely optimized state as well as from the double residue of the cubic response function [12]. It would clearly be of interest to investigate this difference at a higher correlated level such as CASPT2 or CCSD/CC3.

We note that all components of the polarizability tensor of the B_{2u} state increase quite significantly upon going from gas phase to solution, even though hardly any change could be observed in the corresponding excitation energy. However, the increase is less than that observed in the ground state, leading to a further reduction of the ground-to-excited state polarizability for this state. The changes in the ground-to-excited state polarizabilities upon solvation are, however, not very large.

Although there is a fairly substantial change in the polarizability tensor components when going from the gas-phase to a dioxane solution, only minor differences are observed between the different solvents as well as between the equilibrium and non-equilibrium scheme for acetonitrile. Furthermore, the solvent affects the ground and excited state polarizabilities fairly equal, giving much smaller solvent effect on the change in the polarizability upon excitation from the ground to the excited B_{2u} state.

For the B_{1u} state we only show results for the static polarizability and at a frequency of 0.005 a.u. This is due to the close-lying E_{1u} state in benzene, which makes the dispersion very strong and thus not properly treated by the response method, as we for the explicitely optimized B_{1u} state have the E_{1u} state less than 0.03 a.u. above in energy. Still, the presence of the E_{1u} state has a major effect on the polarizability tensor for this state, leading to a significant increase of all polarizability tensors, but in particular the in-plane components.

Due to the close-lying E_{1u} state, the solvent effects on the excited-state polarizability is also very strong. Even though there is hardly any difference in the excitation energies between the benzene molecule in the gas-phase and in dioxane, the in-plane components of the polarizability increase by about 20%, the overall change in the ground-to-excited state isotropic polarizability also increasing by about 20% when the molecule is taken from gas-phase to dioxane.

As for the B_{2u} state, the presence of the dielectric continuum is the main contributor to the change in the excited-state polarizability upon solvation, and only minor changes are observed when increasing the polarity of the solvent, an effect which is not unexpected considering that benzene is a non-polar molecule and thus will not gain much additional energetic stabilization from increasing the polarity of the solvent.

5. Concluding remarks and outlook

The field of *ab initio* calculation of ground-state polarizabilities are now fairly well developed, both with respect to methods designed to provide highly accurate results for small and medium-sized molecules, as well as methods for calculating reliable polarizabilities for large molecules.

We have also discussed the use of dielectric continuum models as an efficient tool for providing in many cases an accurate and reliable estimate and understanding of the effects of a solvent on the properties and energetics of a solute. We have in particular tried to emphasize the general applicability of this approach, and thus how it easily can be combined with any theoretical tool designed to calculate properties and energetics of molecules in the gas-phase.

Many different theoretical methods have in recent years been developed for investigating excited state properties of gas-phase molecules. The set of theoretical methods include the highly accurate coupled-cluster approaches based on response theory [33,34] or Equation-of-Motion theory [63–65], as well as methods aimed at reliable calculations on large molecules using Hartree – Fock theory [22–24] or DFT [26,27]. For fluorescence-related properties, methods based on a multiconfigurational reference wave function used to optimize the excited state explicitly will be central [19,22,25].

Due to the robust features of the dielectric continuum models, we expect extensions of all the above methods for gas-phase studies of excited state properties to the liquid phase described through a dielectric continuum to appear within the next few years. The availability of methods that will be easy to use will allow researchers to study in detail the properties and behavior of molecular excited states of molecules in solution, including the excited state polarizabilities. These studies can be expected to shed important light on the interactions of excited molecular states with their surrounding media and how this affects properties such as solvatochromic shifts of molecules in solution or glasses.

Acknowledgements

KR has been supported by the Norwegian Research Council through a Strategic University Program in Quantum Chemistry (Grant No 154011/420). This work has also received support from the European Union under the COST D26 Action. The support of MIUR (Ministero dell'Istruzione, dell'Università e della Ricerca) "COFIN-00" is acknowledged by BM and RC.

References

[1] D.M. Bishop, *Adv. Quantum Chem.* **25** (1994), 1.
[2] D.P. Shelton and J.E. Rice, *Chem. Rev.* **94** (1994), 3.
[3] Y. Luo, H. Ågren, P. Jørgensen and K.V. Mikkelsen, *Adv. Quantum Chem.* **26** (1995), 165.
[4] S. Heitz, D. Weidauer and A. Hese, *Chem. Phys. Lett.* **176** (1991), 55.
[5] S. Heitz, D. Weidauer and A. Hese, *J. Chem. Phys.* **95** (1991), 7952.
[6] S. Heitz, D. Weidauer, B. Rosenow and A. Hese, *J. Chem. Phys.* **96** (1992), 976.
[7] D.C. Rodenberger, J.R. Heflin and A.F. Garito, *Nature* **359** (1992), 309.
[8] D.C. Rodenberger, J.R. Heflin and A.F. Garito, *Phys. Rev. A, 51* (1995), 3234.
[9] Y. Luo, D. Jonsson, P. Norman, K. Ruud, O. Vahtras, B. Minaev, H. Ågren, A. Rizzo and K.V. Mikkelsen, *Int. J. Quantum Chem.* **70** (1998), 219.
[10] D. Jonsson, P. Norman, Y. Liu and H. Ågren, *J. Chem. Phys.* **105** (1996), 581.
[11] Y. Luo, P. Norman, D. Jonsson and H. Ågren, *Mol. Phys.* **89** (1996), 1409.
[12] P. Norman, D. Jonsson and H. Ågren, *Chem. Phys. Lett.* **268** (1997), 337.
[13] D. Jonsson, P. Norman and H. Ågren, *Chem. Phys.* **224** (1997), 201.
[14] E.M.S. Castanheira and J.M.G. Martinho, *J. Photochem. Photobiol. A: Chem.* **80** (1994), 151.
[15] H. Heinz, U.W. Suter and E. Leontidis, *J. Am. Chem. Soc.* **123** (2001), 11229.
[16] D. Jonsson, P. Norman, H. Ågren, Y. Luo, K.O. Sylvester-Hvid and K.V. Mikkelsen, *J. Chem. Phys.* **109** (1998), 6351.
[17] R. Cammi, L. Frediani, B. Mennucci and K. Ruud, *J. Chem. Phys* **119** (2003), 5818.
[18] J. Tomasi and M. Persico, *Chem. Rev.* **94** (1994), 2027.
[19] J. Olsen and P. Jørgensen, *J. Chem. Phys.* **82** (1985), 3235.
[20] R. McWeeny, *Methods of Molecular Quantum Mechanics*, Academic Press, 2 edition, 1992.
[21] J. Li, J. Feng and J. Sun, *Int. J. Quantum Chem.* **62** (1994), 673.
[22] H. Hettema, H.J.Aa. Jensen, P. Jørgensen and J. Olsen, *J. Chem. Phys.* **97** (1992), 1174.
[23] P. Norman, D. Jonsson, O. Vahtras and H. Ågren, *Chem. Phys. Lett.* **242** (1995), 7.
[24] P. Norman, D. Jonsson, O. Vahtras and H. Ågren, *Chem. Phys.* **203** (1996), 23.
[25] D. Jonsson, P. Norman and H. Ågren, *J. Chem. Phys.* **105** (1996), 6401.
[26] S.J.A. van Gisbergen, J.G. Snijders and E.J. Baerends, *J. Chem. Phys.* **109** (1998), 10644.
[27] P. Sałek, O. Vahtras, T. Helgaker and H. Ågren, *J. Chem. Phys.* **117** (2002), 9630.
[28] P. Jørgensen, H.J.Aa. Jensen and J. Olsen, *J. Chem. Phys.* **89** (1988), 3654.
[29] H. Ågren, O. Vahtras, H. Koch, P. Jørgensen and T. Helgaker, *J. Chem. Phys.* **98** (1993), 6417.
[30] B. Roos and P. Siegbahn, in: *Methods of Electronic Structure Theory*, H.F. Schaefer III, ed., Plenum, 1977.
[31] H.J.Aa. Jensen and P. Jørgensen, *J. Chem. Phys.* **80** (1984), 1204.
[32] T. Helgaker, P. Jørgensen and N.C. Handy, *Theor. Chim. Acta* **76** (1989), 227.
[33] H. Koch and P. Jørgensen, *J. Chem. Phys.* **93** (1990), 3333.
[34] H. Koch, H.J.Aa. Jensen, P. Jørgensen and T. Helgaker, *J. Chem. Phys.* **93** (1990), 3345.
[35] O. Christiansen, P. Jørgensen and C. Hättig, *Int. J. Quantum Chem.* **68** (1998), 1.
[36] J. Tomasi, R. Cammi, B. Mennucci, C. Cappelli and S. Corni, *Phys. Chem. Chem. Phys.* **4** (2002), 5697.
[37] S. Miertuš, E. Scrocco and J. Tomasi, *J. Chem. Phys.* **55** (1981), 117.

[38] R. Cammi and J. Tomasi, *J. Comput. Chem.* **16** (1995), 1449.
[39] E. Cancés and B. Mennucci, *J. Math. Chem.* **23** (1998), 309.
[40] E. Cancés, B. Mennucci and J. Tomasi, *J. Chem. Phys.* **107** (1997), 3032.
[41] B. Mennucci, E. Cancés and J. Tomasi, *J. Phys. Chem.* **B, 101** (1997), 10506.
[42] R. Cammi, L. Frediani, B. Mennucci, J. Tomasi, K. Ruud and K.V. Mikkelsen, *J. Chem. Phys.* **117** (2002), 13.
[43] K.V. Mikkelsen, E. Dalgaard and P. Swanstrøm, *J. Phys. Chem.* **91** (1987), 3081.
[44] K.V. Mikkelsen, H. Ågren, H.J.Aa. Jensen and T. Helgaker, *J. Chem. Phys.* **89** (1988), 3086.
[45] K.V. Mikkelsen, A. Cesar, H. Ågren and H.J.Aa. Jensen, *J. Chem. Phys.* **103** (1995), 9010.
[46] A. Willetts and J.E. Rice, *J. Phys. Chem.* **99** (1993), 426.
[47] L. Onsager, *J. Am. Chem. Soc.* **58** (1936), 1486.
[48] K.V. Mikkelsen and K.O. Sylvester-Hvid, *J. Phys. Chem.* **100** (1996), 9116.
[49] K.O. Sylvester-Hvid, K.V. Mikkelsen, D. Jonsson, P. Norman and O. Vahtras, *J. Chem. Phys.* **109** (1998), 5576.
[50] K.-O. Sylvester-Hvid, K.V. Mikkelsen, D. Jonsson, P. Norman and H. Ågren, *J. Phys. Chem.* **A, 103** (1999), 8375.
[51] D.M. Chipman, *J. Chem. Phys.* **106** (1997), 10194.
[52] E. Cancés and B. Mennucci, *J. Chem. Phys.* **115** (2001), 6130.
[53] L.S. Prabhumirashi and S.S. Kunte, *Spectrochim. Acta* **A, 44** (1988), 213.
[54] H.K. Sinha and K.J. Yates, *J. Chem. Phys.* **93** (1990), 7085.
[55] M. Takezaki, N. Hirota and M. Terazima, *J. Phys. Chem.* **A, 101** (1997), 3443.
[56] M. Okruss, R. Müller and A. Hese, *J. Chem. Phys.* **110** (1999), 10393.
[57] I. Renge, *Chem. Phys.* **167** (1992), 173.
[58] S.P. Karna, P.N. Prasad and M. Dupuis, *J. Chem. Phys.* **94** (1991), 1171.
[59] L. Frediani, R. Cammi, C.S. Pomelli, J. Tomasi and K. Ruud, *J. Comp. Chem.* **25** (2003), 375.
[60] G.U. Bublitz and S.G. Boxer, *Ann. Rev. Phys. Chem.* **48** (1997), 213.
[61] B.O. Roos, K. Andersson and M.P. Fülscher, *Chem. Phys. Lett.* **192** (1992), 5.
[62] E.N. Lassetre, A. Skerbele, M.A. Dillon and K.J. Ross, *J. Chem. Phys.* **48** (1968), 5066.
[63] J.F. Stanton and R.J. Bartlett, *J. Chem. Phys.* **98** (1993), 7029.
[64] P. Rozyczko and R.J. Bartlett, *J. Chem. Phys.* **107** (1997), 10823.
[65] C. Hättig, H. Koch and P. Jørgensen, *J. Chem. Phys.* **109** (1998), 3293.

Journal of Computational Methods in Sciences and Engineering 4 (2004) 399–409
IOS Press

Static Dipole Polarizability of *o*-, *m*- and *p*-Benzyne isomers: Ab initio, DFT and CCSD calculations

Humberto Soscún*, Carlos Toro-Mendoza, Esker Chacín and Javier Hernández
Laboratorio de Química Inorgánica Teórica, Departamento de Química, Facultad Experimental de Ciencias, La Universidad del Zulia, AP. 526, Módulo No. 2, Grano de Oro, Maracaibo, Venezuela

Abstract. We report a study of the static dipole polarizability (α) of the *o*-, m- and p-benzyne isomers in their singlet ground state. Due to the biradical character of benzynes, calculations were performed at ab initio unrestricted Hartree-Fock (UHF) level of theory. The correlation effects were accounted for with second order Møller Plesset (MP2) method, Density Functional Theory (DFT) with the BLYP and B3LYP hybrid approaches and the coupled cluster CCSD method with 6-31+G(d,p) standard basis set with optimized geometries. C_2v symmetry restriction was employed for the geometry of o- and m-benzynes (closed and open conformation) and D_2h for p-benzynes (open and closed conformation). Because the high quality of the basis set is a requirement for accurate α determination, MP2 and DFT calculations were performed with the 6-31+G(d,p), the extended 6-311++G(3d,3p) and the specialized Sadlej basis sets. The CCSD calculations were only carried out with the 6-31+G(d,p) and the Sadlej basis sets. The results indicate that the average α polarizability values for o- and m-benzynes are lower than the corresponding to the experimental value of benzene, while this property for p-benzyne is slightly higher than the benzene one. The anisotropy $\Delta\alpha$ values of benzyne isomers are always higher than the anisotropy of benzene molecule. Finally, was found that the conformations of p-C_6H_4 are the less stable isomers, and are the most polarizables and anisotropic compounds from the benzyne series.

Keywords: Dipole polarizability, ab initio, unrestricted hartree-fock, DFT, MP2, CCSD, biradical character, benzynes

1. Introduction

Benzynes are reactive intermediates derived from substituted benzenes that participate in aromatic nucleophilic substitutions, in determined cycloaddition reactions and are very useful for the synthesis of new compounds [1–4]. Three molecular isomers are known for benzyne, *ortho-benzyne* (1, o-C_6H_4, 1,2-didehydrobenzene), *meta-benzyne* (2, m-C_6H_4, 1,3-didehydrobenzene) and *para*-benzyne (3, p-C_6H_4, 1,4-didehydrobenzene). They are neutral and their main electronic feature is the biradical character dominated by through-space and through-bond interactions that change with the position of the two radical centers from o-C_6H_4 to p-C_6H_4, leading to singlet ground state configuration. These compounds have been the subject of a great variety of recent experimental and theoretical studies. Benzynes have been prepared and characterized by IR techniques in argon matrix at low-temperature [5]. Different structures have been postulated for benzyne isomers (see Fig. 1). Particularly, for o-C_6H_4 only one monocyclic structure is known that is characterized by the presence of one CC triple bond linked to a

*Corresponding author. E-mail: hsoscun@yahoo.es.

network of two CC double bonds in the same aromatic ring (1, Fig. 1) [4]. However, for *meta-benzyne* (m-C_6H_4) and *para-benzyne* (p-C_6H_4) the electron deficiency is shared between two different carbon atoms in two different bonds that can lead either to open and closed conformations. The opening of the ring gives the 2a and 3a monocyclic structures. Instead, when the ring is closed, the 2b and 3b bicyclic forms are obtained [5b]. In the closed forms, occurs a ring fusion of five and three members for m-benzyne (2b), and two of four members for p-benzyne (3b). Detailed analysis of these conformations is found in the section of geometric results.

From ab initio theoretical studies it has been concluded that the stability of these compounds in the singlet ground state follows the o-C_6H_4 > m-C_6H_4 > p-C_6H_4 order [6]. This order parallels the multireference and biradical character of benzynes [5c]. Recently, the magnetic properties of the o-, m- and p-benzyne isomers have been studied theoretically, showing that the evaluated relative aromaticities follow the same order that the energetic stability and the multireference character of benzyne series [5c]. This study, has also demonstrated that the aromaticity increases with the distance between the divalent carbon atoms in the benzyne molecules. It is clear from these statements and from the different benzyne conformations that these isomers should present significant variations in the delocalization of the π-electron system that could be reflected in the electronic properties of these molecules.

In order to characterize the electronic properties of benzyne isomers, we have carried out a theoretical research of the dipole moment μ and the static dipole polarizability α of these biradical molecules in their ground state. The α property is the intrinsic response of the molecular electronic distribution to the action of an external electric field [7], which gives relevant information about the nature of the π delocalization in aromatic molecules [8]. Also, we have addressed our investigations in the understanding of the stability of the benzyne isomer conformations. The results show that the p-C_6H_4 is the most polarizable isomer from the benzyne series and corresponds to the molecule with higher energy between the three isomers.

2. Theory

The perturbation of the electronic distribution of a molecular system by effect of an external field is reflected in the permanent molecular dipole moment μ_e according to [9].

$$\mu_e(\Im) = \mu_e(\Im = 0) + (1/2)\alpha\Im + (1/6)\beta\Im^2 + (1/24)\gamma\Im^3 + \ldots \tag{1}$$

The first linear response to the action of external electric field $\Im$ is known as the molecular dipole polarizability α. The terms of higher order are the first and second hyperpolarizabilities, respectively. Similarly, a Taylor expansion for the energy E of the system has been defined in terms of the dipole polarizabilities and the field $\Im$ as,

$$E(\Im) = E(0) - \mu\Im - (1/2)\alpha\Im^2 - (1/6)\beta\Im^3 - (1/24)\gamma\Im^4 + \ldots \tag{2}$$

where $E(0)$ is the unperturbed energy. In polarizability studies, the quantities of experimental interest are the average polarizability

$$\alpha_{\text{ave}} = \langle\alpha\rangle = 1/3(\alpha_{xx} + \alpha_{yy} + \alpha_{zz}) \tag{3}$$

and the polarizability anisotropy

$$\Delta\alpha = \left((1/2)\left[(\alpha_{xx} - \alpha_{yy})^2 + (\alpha_{xx} - \alpha_{zz})^2 + (\alpha_{yy} - \alpha_{zz})^2\right]\right)^{1/2} \tag{4}$$

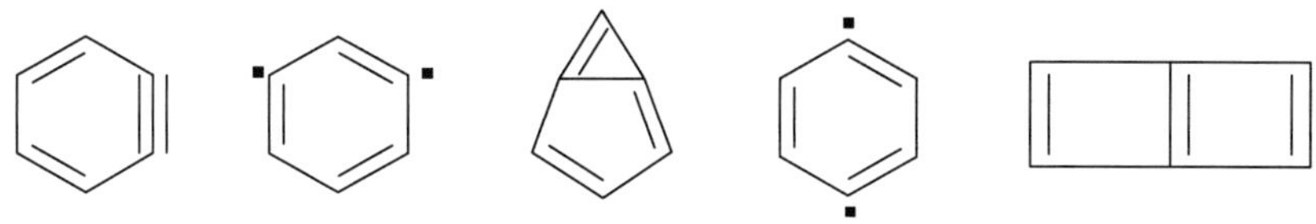

1 2a 2b 3a 3b

Fig. 1.

3. Computational details

Due to the multireference character of benzynes, an appropriate description of these molecules can be obtained either by multireference methods (MRCI, MRCC) or single-determinat approaches that consider high degree of electron correlation, such as CCSD(T) or BD(T) [10]. Despite of this, and due to the high computational cost of these higher correlated methods, density functional theory approaches based in unrestricted H-F methods (UDFT) have shown to give reasonable descriptions of the structural and electronic properties of biradical molecular species [10]. In view of these results, in the present work we have used UDFT techniques and different U-correlated methods for the calculations of the geometry and electronic properties of benzyne isomers.

Geometric structure of benzyne isomers were optimized under restriction of C_{2v} symmetry for o- and m-benzynes and D_{2h} for p-benzyne, by using gradient techniques at the unrestricted Hartree-Fock (UHF) [11], second order Møller-Plesset perturbation theory (UMP2) [12] and Density Functional Theory UDFT [13] at the UBLYP [14] and UB3LYP [15] approaches. Evaluation of electron correlation within the UHF approach was performed in order to take in consideration the biradical character of benzyne isomers. For these optimizations, we used the 6-31+G(d,p) basis set, in which the 6-31G basis function is augmented with sp diffuse functions plus a set of d and p polarized functions that have been added to the C and H atoms, respectively [16]. The dipole moment μ and the independent components ($\alpha_{xx}, \alpha_{yy}, \alpha_{zz}$) of the static dipole polarizability α were also calculated by using the analytic and direct coupled perturbed at the UCPHF method [17]. The determination of accurate theoretical values of this property require of both the accounting of the electron correlation effects and the use of extended basis sets augmented with selected sets of diffuse and polarized functions [7,8]. Also, can be used the specialized basis set of Sadlej [18]. In this context, we have also performed calculations at UHF, UMP2, UBLYP and UB3LYP levels with the 6-311++G(3d,3p) and the Sadlej basis sets for the polarizability of benzyne isomers. The good performance of these atomic functions for polarizability calculations has been pointed out elsewhere [7,8]. The calculations with the UHF and UDFT methods were carried out using direct techniques with the GAUSSIAN 98 [19] quantum chemistry package, and the UMP2 ones were performed with the Gamess program [19]. Finally, in order to obtain accurate determination of the electron correlation on the polarizability properties of the studied benzynes, calculations at UCCSD [20] level with the 6-31+G(d,p) and Sadlej basis sets with UMP2 and UB3LYP optimized geometries were performed with the Gaussian program. Finally, the dipole polarizability of benzene molecule was evaluated at the DFT and CCSD/6-31+G(d,p) levels for comparison.

4. Results

4.1. Geometry, energetic and dipole moments

In Fig. 1 are depicted the molecular notation scheme and axis orientation for the structure of benzyne isomers. The corresponding geometric parameters are shown in Table 1. The bond lengths and bond angles agree with those previously reported for these structures using similar methods and basis sets. Our results confirm a monocyclic structure for o-C_6H_4 isomer and different conformations for m-C_6H_4 and p-C_6H_4 ones. The bond distance of dehydrogenated carbons (a) in o-C_6H_4 is 1.225 Å at UHF/6-31+G(d,p) level and increases slightly with electronic correlation methods. For instance, the (a) bond lengths are 1.270 Å, 1.253 Å and 1.265 Å for UMP2, UB3LYP and UBLYP approaches, respectively. The short bond distance is indicative of the presence of a triple CC bond and agrees very well with the observed bond length for ethyne (C_2H_2 1.203 Å) [6]. In Table 1 are also reported values of optimized geometric parameters of o-C_6H_4 at the B3LYP [4] and HF and MP2 [8] levels for comparison. Our results agree very well with the reported geometries of this benzyne isomer.

The geometry of m-C_6H_4 has been the subject of debate in the literature [5,22]. In our geometry optimizations it was found that this compound is able to give two different conformations that depend of the level of theory, how has been reported [5]. For instance, at the UMP2 and UBLYP levels one biradical open structure (2a) is found, whereas one bicyclic conformation (2b) at the UHF and UB3LYP approaches there exists. The optimized geometric parameters for both conformations are shown in Table 1. These results are in agreement with previous optimizations reported recently [5,22], that are also displayed in Table 1 for comparison. It is important to note that the features between the (2a) and (2b) conformations can be depicted with the optimized values of the (a) bond distance in m-C_6H_4 isomers, where important variations are observed between the UHF and UB3LYP method values and UMP2 and UBLYP ones. At UHF level, a bicyclic structure with a bond distance (a) of 1.482 Å is formed. This bicyclic conformation is maintained when the electron correlation is accounted with the UB3LYP approach, giving a bond length that is 0.14 Å larger than the UHF one. However, the distance C-C (a) between dehydrogenated carbons at electron correlation UMP2 and UBLYP methods corresponds to a monocyclic structure for m-C_6H_4 isomer. These results are coincident with those reported by De Proft et al. with the BLYP approach [5] and by Sander et al. with the CCSD(T) level [21]. Our (a) C-C optimized distance are 2.129 Å and 2.013 Å with the UMP2 method and UBLYP hybrid approach, respectively, whereas the corresponding values at the CCSD(T) method is 2.101 Å [21]. At this point is important to mention that, based in a detailed DFT and CCSD(T) investigation of m-benzyne structure, Winkler and Sander [22] reported that the (2a) and (2b) configurations of m-benzyne are not valid representations for this system and they suggest that the best way to describe the electronic structure of this isomer is a σ-allylic monocyclic system. However, this proposal has been not verified by either calculations or experimental determinations. The fact that BLYP approach is able to give a right answer in the case of the open ring conformation of m-C_6H_4 has been recently discussed by Sander et al. [21]. In order to refine the structural details of m-benzyne, these authors have performed an investigation about the most probable conformation of m-C_6H_4 by using spectroscopic techniques and CCSD(T) high level calculations. They have verified that m-C_6H_4 possesses the biradical open structure (2a) rather than the closed conformation of two fused rings (2b) [21].

The last isomer of benzyne is the p-C_6H_4 one, which structure has been also the subject of controversy in the literature because this molecule is able to give two different conformations, one open ring structure (3a) and one bicyclic (3b) conformation (known as butalene) [5,23]. The results obtained in this work of geometry optimization, shown in Table 1, are able to find these two representations of p-C_6H_4 isomer. In the bicyclic conformation (3b) the bond distance (a) is about of 1.441 Å at HF level, and the effect of electron correlation at UMP2 increases it to a value of 1.567 Å, while this bond at the UBLYP and UB3LYP increases it to 1.584 Å and 1.568 Å, respectively. These results are consistent with the recent study of Crawford et al. [23] about the structure of this molecule by using higher correlated methods. With respect to the other conformation of p-benzyne, the corresponding optimized geometric parameters

Table 1
Bond distances in Angstrons (Å) and bond angles in degrees (°) of benzyne isomers optimized at the 6-31+G(d,p) level

	o-C_6H_4				*m*-C_6H_4				*p*-C_6H_4			
	UHF	UMP2	UBLYP	UB3LYP	UHF	UMP2	UBLYP	UB3LYP	UHF	UMP2	UBLYP	UB3LYP
a	1.225	1.270	1.265	1.253	1.482	2.129	2.013	1.617	1.441	1.567	1.584	1.568
	1.223[c]	1.269[a]		1.251[a]		2.106[a]	2.101[d]	1.600[a]	2.676[e]	2.725[a]	2.747[e]	2.727[a]
		1.268[c]				2.128[b]						2.720[e]
b	1.385	1.391	1.395	1.386	1.339	1.378	1.355	1.350	1.532	1.453	1.468	1.569
	1.383[c]	1.394[a]		1.385[a]		1.377[a]	1.375[d]		1.329[e]	1.426[a]	1.360[e]	1.348[a]
		1.389[c]				1.376[b]		1.350[a]				1.348[e]
c	1.394	1.408	1.427	1.415	1.386	1.380	1.375	1.383	1.384	1.397	1.406	1.395
	1.410[c]	1.411[a]		1.413[a]		1.383[a]	1.379[d]	1.382[a]	1.489[e]	1.382[a]	1.490[e]	1.480[a]
		1.405[c]				1.379[b]						1.482[e]
d	1.411	1.411	1.416	1.408	1.406	1.404	1.410	1.413	1.076	1.086	1.096	1.088
						1.405[a]	1.403[d]	1.413[a]	1.072[e]		1.091[e]	1.084[e]
						1.404[b]						
e	1.073	1.081	1.091	1.087	1.071	1.082	1.095	1.083				
		1.081[a]		1.085[a]		1.082[a]	1.086[d]	1.084[a]				
						1.087[b]						
f	1.076	1.084	1.095	1.083	1.077	1.087	1.095	1.086				
				1.088[a]		1.086[a]	1.082[d]	1.090[a]				
		1.084[a]				1.092[b]						
g					1.073	1.078	1.092	1.089				
						1.078[a]	1.077[d]	1.088[a]				
						1.085[b]						
ab	127.5	126.6	127.1	127.2	56.4	39.4	44.6	53.2				
	127.5[c]	122.4[a]		122.4[a]								
		126.6[c]										
bc	110.0	110.8	110.5	110.4	164.4	134.3	142.7	158.3	88.1		92.4	92.3
	110.0[c]	111.2[a]		110.4[a]		135.5[a]		159.2[a]	116.5[e]	92.3	116.5[e]	117.4[e]
		110.8[c]				134.2[b]						
be	127.1	126.8	127.0	127.1								
cd	122.5	122.6	122.4	122.5	105.7	117.9	110.7	109.2	134.8	134.5	134.5	134.3
									114.3[e]		115.2[e]	115.0[e]
ce	123.0	122.4	122.5	122.6	122.5	120.5	120.7	125.3				
cf	118.9	118.6	118.4	118.6								
df	118.6	118.8	119.1	119.0	123.7	122.8	124.6	124.2				
					67.2	101.2	90.8	73.5				
bb						99.8[a]		72.7[a]				
						101.2[b]						
bg					146.4	129.4	134.6	143.2				
ac					108.0	94.9	98.2	105.1	91.9	87.7	87.6	87.7
					112.5	114.5	110.7	111.5				
dd						114.3[a]		111.7[a]				
						114.4[b]						
de					126.8	121.6	122.9	125.5				
ad									133.3	133.2	133.3	133.4
bb									126.0[e]		124.9[e]	125.3[e]

[a]Ref. [4] B3LYP/6-31G(d,p), [b]Ref. [3] MP2/6-31G(d) and [c]Ref. [8] HF/6-31G(d,p) and MP2/6-31G(d,p); [d]CCSD(T)/6-311G(2d,2p), Ref. [21]; [e]These values correspond to the open form of the *p*-benzyne structure (3a).

for the monocyclic structure (3a) are also reported in Table 1 that corresponds to the results with the "e' superscript. In particular, the (a) bond of (3a) representation have values of 2.676 Å at the UHF,

Table 2
Total energy ($-E_t$) in Hartrees and dipole moment (μ) in Debyes for o-, m- and p-benzyne conformations. All values were calculated with standard basis set 6-31+G(d,p). Between brackets are reported the CCSD/Sadlej dipole moments

			UHF	UMP2	UBLYP	UB3LYP	UCCSD
o-Benzyne	1	$-E_t$	229.40376	230.20875	230.83744	230.98837	230.22788
		μ	1.93	1.74	1.78	1.78	2.03
							(1.91)
m-Benzyne	2a	$-E_t$	—	230.19714	230.81993	—	230.19858
		μ	—	1.36	0.86	—	1.36
							(1.25)
	2b	$-E_t$	229.39050	—	—	230.90895	230.20350[a]
		μ	0.64	—	—	0.12	0.13[a]
							(0.16[a])
p-Benzyne	3a	$-E_t$	229.29114	—	230.79123	230.86825	230.16234[a]
		μ	0.0	—	0.0	0.0	0.0[a]
	3b	$-E_t$	229.30079	230.10881	230.74544	230.83512	230.12892
		μ	0.0	0.0	0.0	0.0	0.0

[a]UCCSD calculations performed with the UB3LYP/6-31+G(d,p) optimized geometry.

2.747 Å at UBLYP and 2.720 Å at UB3LYP. These values are consistent with the structure reported by Moskaleva et al., which value at UMP2 is 2.725 Å [4], also reported in Table 1 for comparison. It is important to mention, that Andes investigated the structure of p-benzyne conformations by using unrestricted DFT methods [5], and he found that (3b) butalene, the closed form of p-benzyne, is the most probable conformation that can be isolated at low temperature.

In the following, we discuss the energetic aspects of benzyne series. Table 2 displays the calculated energies and dipole moment values for benzyne isomers at the different levels of theory. The energetic stability of these isomers follows the order o-C_6H_4 > m-C_6H_4 > p-C_6H_4. This order is independent of the structural conformations. At the UHF level of theory was not possible to find a minimum for the (2a) structure of m-benzyne. In the same context, at UMP2 level was not possible for us the determination of the (2b) and (3a) structures for m-benzyne and p-benzyne ones, respectively. Similarly, the (2b) conformation at UBLYP and the (2a) structure at the UB3LYP were not found. At UB3LYP/6-31+G(d,p) level, the m-C_6H_4 (2b) and p-C_6H_4 (3b) are 49.8 Kcal/mol and 96.2 Kcal/mol less stables than o-C6H4 one. The results of the total energy agree very well with the known order previously reported for these isomers [5]. Internally, at the UB3LYP level the bicyclic from comformation (3b) is less stable than the open one (3a). In Table 2 are also reported values of the total energies calculated at the UCCSD/6-31+G(d,p) approach for the UMP2/6-31+G(d,p) optimized structures of o-benzyne, and the (2a) and (3b) configurations of m-benzyne and p-benzyne, respectively. For the (2b) and (3a) conformations of m-benzyne and p-benzyne, respectively, the UB3LYP/6-31+G(d,p) optimized geometries were used for the UCCSD energy calculations. These results follow the same order of benzyne stability previously described by the other methods. However, internally the bicyclic (2b) conformation of m-benzyne is more stable than the open (2a) one, whereas for p-benzyne the open (3a) is more stable than the closed one (3b).

There is not report on the dipole moment (μ) of benzynes in the literature. Table 2 also collects the results obtained in the present work for this property. The analysis of the dipole moment μ of Table 2, indicate that the o-C_6H_4 isomer show values that lies in a reasonable range for all the studied methods. In particular with respect to the o-C_6H_4 UHF μ value (1.93 Debyes), this property decreases with the UMP2 (1.74 Debyes), UBLYP (1.78 Debyes) and UB3LYP (1.78 Debyes) electron correlation effects, to increase again at the UCCSD. At this level of theory, the dipole moment of this isomer is about

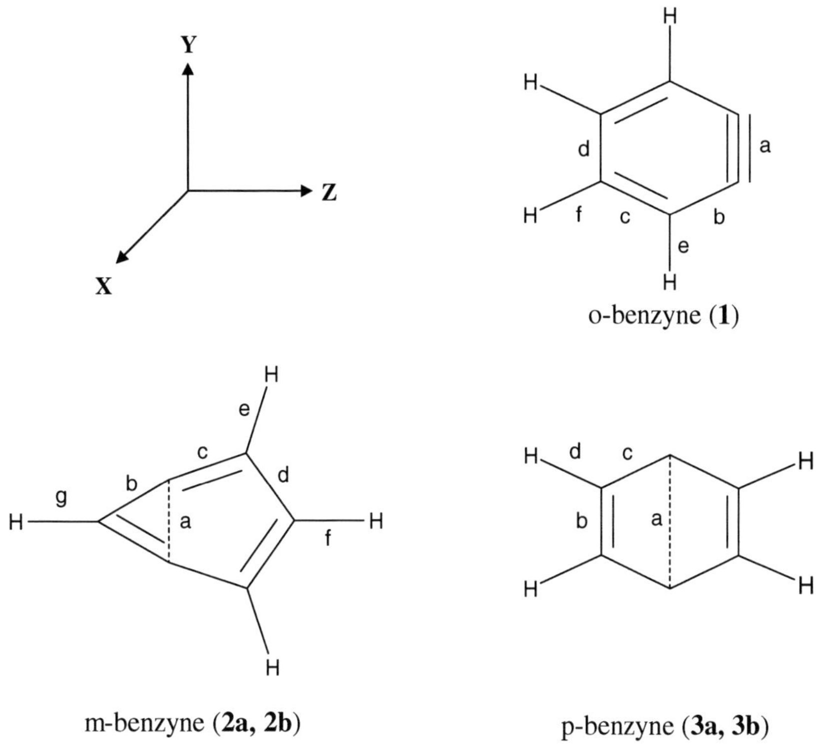

Fig. 2. Geometric structures of benzyne isomers (o-, m-, p-) showing notation scheme and molecular orientation in the coordinate axis.

2.03 Debyes with the 6-31+G(d,p), whereas with the Sadlej basis set, this property has a value of 1.91 Debyes. With respect to the m-C_6H_4, the μ values change with the level of theory employed and with the conformation of this isomer. For instance, the open representation of m-benzyne gives a structure (2a) with high dipole moment, which values oscillate from 1.36 Debyes at UMP2 and UCCSD to 0.86 Debyes at the UBLYP approach. With respect to the closed representation of this molecule (2b), the results show values for this property of 0.64 Debyes at UHF, 0.12 Debyes with the UB3LYP method and 0.13 Debyes for the UCCSD one. With the UCCSD/Sadlej methodology, the μ values for the (2a) and (2b) conformations are 1.25 Debyes and 0.16 Debyes, respectively. These results show that for the open and closed representations of m-benzyne, the dipole moment μ is randomly represented by these methodologies. Apparently, due to the lacking of multireference consideration, these approaches (MP2, BLYP, B3LYP) based in the unrestricted HF methods are not able to reproduce satisfactorily the charge distribution of the m-C_6H_4 configurations. Further work it is being developed in our laboratory in order to determine appropriate values for the dipole moment of m-benzyne.

4.2. Dipole polarizabilities

The average dipole polarizability (α_{ave}), together with the principal components ($\alpha_{xx}, \alpha_{yy}, \alpha_{zz}$) of dipole tensor, and the anisotropy of polarizability ($\Delta\alpha$) of three benzyne isomers and the experimental for benzene [24] are reported in Table 3. Calculations were performed with the UHF, UMP2, UBLYP and UB3LYP methods using the 6-31+G(d,p), 6-311++G(3d,3p) and Sadlej basis sets over optimized geometries at the 6-31+G(d,p) level of theory. The molecular orientation (Fig. 1) is such that the z

direction is the C_2 symmetry axis and zy is the molecular plane. For this scheme α follows generally the relationships $\alpha_{xx} < \alpha_{yy} < \alpha_{zz}$. Additionally, Table 3 displays for comparison the only HF and MP2 results from the literature about polarizability calculations of o-benzyne [8] and the experimental polarizability of benzene [24]. The results are organized following the features of the conformation of the benzyne isomers. The polarizability results for o-benzyne are for the unique structural representation of this molecule. For m-benzyne the UHF and UB3LYP results correspond to the closed conformation (2b), while the UMP2 and UBLYP are for the (2a) open form. For the case of p-benzyne isomer, the results are for the bicyclic form (3b), while the results with the "b" superscript in Table 3 correspond to the open (3a) conformation. At UHF and UMP2 the α_{ave} of o-benzyne and p-benzyne are almost similar, however the polarizability is increased significantly when is evaluated with the DFT approaches. Generally, the UBLYP functional give dipole polarizability values that are in the range of about 2%–8% higher than the UB3LYP method. For m-benzyne there are important variations between the UHF and UMP2 α results, mainly in the α_{yy} component, which are due to the geometry of the open and closed conformations of this isomer. These differences are not affected with the extension of the basis sets. In particular, a contraction of 13.8% for the α_{ave} at UMP2/6-311++G(3d,3p) with respect to the UHF is obtained. However, for the same open conformation with the UBLYP calculations, this property is increased in about 34.5%. It is important to note that with the UBLYP, the open form of m-benzyne is more polarizable than o-benzyne one, whereas at the UB3LYP method, the closed structure of m-benzyne has almost the same polarizability as o-benzyne. Additionally, these species are less polarizable than benzene. It should be noted, however, that at UBLYP hybrid method the yy and zz component of α are very close by symmetry as consequence of geometry variations at this level of theory in the (2a) m-C_6H_4 isomer configuration. It is assumed that the geometric differences between UBLYP and the other methods affect the linear polarizability.

Regarding to the polarizability of the conformations of p-benzyne, Table 3 shows that the open conformation of this isomer is approximately 12% more polarizable than the closed one. This difference, maintained at the different levels of theory, is because the α_{yy} and α_{zz} components of the (3a) form are higher than the corresponding in the (3b) structure. At this point, we note that the UHF method gives polarizability values that are unreliable for the (3a) conformation of p-benzyne. In general, the calculated α_{ave} values show that p-C_6H_4 in their two representations is the highest polarizable specie of the benzyne isomers, even slightly with higher polarizability than benzene molecule. At UB3LYP/6-311++G(3d,3p) level, the open p-C_6H_4 form (3a) is 14.6% and 15.9% more polarizable than o-C_6H_4 isomer and the m-C_6H_4 conformation, respectively. These results indicate that o-C_6H_4 is slightly more polarizable than m-C_6H_4. However, with the UB3LYP/Sadlej, the order is $\alpha(o\text{-}C_6H_4) < \alpha(m\text{-}C_6H_4) < \alpha(C_6H_6) < \alpha(p\text{-}C_6H_4)$. The variations in the relative differences of the polarizabilities are dependent of the method, basis set and kind of isomer.

In order to make a realistic comparison between the polarizability of the benzyne isomers, UCCSD/6-31+G(d,p) and UCCSD/Sadlej α calculations were performed with UMP2/6-31+G(d,p) optimized geometries of the (1), (2a) and (3b) benzyne conformations. Similar UCCSD calculations were carried out for the (2b) and (3a) benzyne conformations with UB3LYP/6-31+G(d,p) geometries. These results, reported in Table 4 together with polarizability values for benzene molecule (ours and from the literature [25]), show that the UCCSD/6-31+G(d,p) order of the average polarizability is o-C_6H_4 < C_6H_6 < m-C_6H_4 (2a) < p-C_6H_4 (3b). A further analysis of these results shows that the Sadlej basis increases the UCCSD/6-31+G(d,p) benzyne polarizability in the range of 6.2 to 6.6%, giving α components and α_{ave} values that are of similar quality than the calculated with the DFT methods. In particular, the UCCSD/Sadlej results give the following order for the of the benzyne isomers: o-C_6H_4

Table 3
Dipole polarizability α (au) and anisotropy of polarizability $\Delta\alpha$ (au) of benzyne isomers at the UHF, UMP2, UBLYP and UB3LYP methods using A) 6-31+G(d,p), B) 6-311++G(3d,3p) and C) Sadlej basis sets

		UHF			UMP2			UBLYP			UB3LYP		
		A	B	C	A	B	C	A	B	C	A	B	C
o-C_6H_4	α_{xx}	37.81	42.09	41.76	38.06	42.33	42.31	39.99	43.36	41.75	38.59	42.25	41.06
	α_{yy}	66.82	70.52	70.82	70.02	73.54	74.29	75.74	77.97	75.95	71.99	74.67	73.72
	α_{zz}	80.70	83.14	83.27	76.13	79.46	80.06	86.65	88.44	85.56	83.61	85.65	83.94
	α_{ave}	61.78	65.25	65.28	61.40	65.11	65.55	67.46	69.92	67.75	64.73	67.52	66.24
		61.65[a]	65.01[a]		61.54[a]								
	$\Delta\alpha$	37.90	36.42	36.90	34.72	34.55	35.22	42.27	40.86	39.88	40.48	39.08	38.79
		37.82[a]	36.44[a]		35.49[a]								
m-C_6H_4	α_{xx}	37.51	41.53	41.47	37.89	42.09	–	39.97	43.42	43.11	38.02	41.57	41.25
	α_{yy}	63.17	65.74	65.99	34.01	37.73	–	85.71	86.98	82.24	74.01	71.60	72.38
	α_{zz}	79.27	80.77	80.73	78.97	82.22	–	85.80	87.53	87.99	80.45	86.17	86.62
	α_{ave}	59.98	62.68	62.73	50.29	54.01	–	70.49	72.65	71.11	64.16	66.45	66.75
	$\delta\alpha$	34.43	34.29	34.35	27.29	42.79	–	45.79	43.84	42.30	39.60	39.39	40.19
p-C_6H_4	α_{xx}	39.75	44.32	44.30	38.84	45.24	42.94	40.67	44.32	43.38	39.46	43.10	42.76
		40.09[b]	43.80[b]					41.64[b]	44.48[b]		40.33[b]	43.45[b]	
	α_{yy}	62.70	67.37	69.10	58.37	63.47	63.77	66.41	69.58	70.20	63.81	66.82	68.68
		180.66[b]	184.35[b]					106.62[b]	107.23[b]		109.31[b]	110.07[b]	
	α_{zz}	99.05	101.75	101.60	98.59	100.66	98.83	105.50	106.94	104.80	101.59	102.18	102.52
		76.92[b]	80.29[b]					84.02[b]	86.49[b]		80.86[b]	83.60[b]	
	α_{ave}	67.17	71.15	71.67	65.27	69.79	68.51	70.86	73.61	72.79	68.29	70.70	71.32
		99.22[b]	102.81[b]					77.43[b]	79.40[b]		76.83[b]	79.04[b]	
	$\delta\alpha$	51.79	50.06	49.77	52.77		48.92	56.24	54.57	53.33	54.22	51.49	51.90
		126.25[b]	126.32[b]					57.14[b]	55.37[b]		60.04[b]	58.10[b]	
C_6H_6	α_{xx}		44.84	44.79		45.12	45.29	42.72	46.04	45.84		44.84	44.58
	α_{yy}		77.90	78.17		81.00	81.78	82.52	85.11	86.23		82.04	82.85
	α_{zz}		77.90	78.17		81.00	81.78	82.52	85.11	86.23		82.04	82.85
	α_{ave}		66.91	66.38		69.04	69.62	69.25	72.09	72.77		69.64	70.09
													67.48[c]
	$\Delta\alpha$		32.98	33.38		35.88	36.49	39.80	39.07	39.07		37.20	38.27
													35.00[c]

[a]Ref. [8] HF/6-31G(d,p), HF/6-31+G(3d,3p), MP2/6-31+G(d,p); [b]These polarizabilities values correspond to the open ring conformation of p-benzyne (3a); [c]Ref. [24] Static Rayleigh Scattering.

(65.19 au) < m-C_6H_4 [2b (65.25 au) < 2a (68.35 au)] < C_6H_6 (68.70 au) < p-C_6H_4 [3a (68.71 au) < 3b (69.40 au)]. It is worth to note that the open conformations of m-C_6H_4 (2a) and p-C_6H_4 (3a) present almost the same values for the dipole polarizability, whereas there is a significant difference between the polarizabilities of 2b and 3b benzyne bicyclic structures.

Results for $\Delta\alpha$ at UHF level show the same behavior than the α_{ave} (p-C_6H_4 > o-C_6H_4 > m-C_6H_4). However, when electron correlation is taken into account the p-C_6H_4 is the most anisotropic isomer followed by m-C_6H_4 and then o-C6H4. From Table 3, can be deduced that p-C_6H_4 is approximately 24% and 28% more anysotropic than the other isomers and benzene, respectively. From the UCCSD $\Delta\alpha$ values of Table 4 it is observed that benzene is less anisotropic than benzynes, and the anisotropy follow the order o-C_6H_4 < m-C_6H_4 (2b 2a) < p-C_6H_4 (3a < 3b).

5. Conclusions

We have studied the geometries, dipole moments and dipole polarizabilities α of o-, m- and p-benzyne isomers in their ground states. Two different structural configurations were studied for m- and p-benzyne, the open and bicyclic ones. The properties for benzene were also calculated for

Table 4
Dipole polarizability components α_{ii} (au), average polarizability α_{ave} (au) and anisotropy of polarizability $\Delta\alpha$ (au) of benzyne isomers at the UCCSD/6-31+G(d,p) and the UCCSD/Sadlej (between brackets) levels. For 1, 2a and 3b conformations were employed UMP2/6-31+G(d,p) optimized geometries and B3LYP/6-31+G(d,p) geometries for 2b and 3a structures

	Conformation	UCCSD/6-31+G(d,p) (UCCSD/Sadlej)				
	(Fig. 1)	α_{xx}	α_{yy}	α_{zz}	α_{ave}	$\Delta\alpha$
o-C_6H_4	1	37.33	67.73	78.32	61.13	36.69
		(41.58)	(72.11)	(81.88)	(65.19)	(36.41)
m-C_6H_4	2a	38.00	76.38	78.68	64.35	39.58
		(42.40)	(80.04)	(82.60)	(68.35)	(38.98)
	2b	(41.36)	(70.66)	(83.73)	(65.25)	(37.58)
p-C_6H_4	3a	(42.60)	(80.62)	(82.92)	(68.71)	(39.21)
	3b	38.05	60.64	95.86	64.85	50.46
		(42.80)	(66.40)	(99.00)	(69.40)	(48.88)
C_6H_6		39.95	74.80	74.80	63.18	34.85
					(68.70)[a]	(34.85)[a]

[a]CCSD/Sadlej calculations, Ref. [25].

comparison. These properties were evaluated by using unrestricted Hartree-Fock methods, UMP2, UBLYP and UB3LYP DFT hybrid approaches with the 6-31+G(d,p), 6-311++G(3d,3p) and Sadlej basis sets. Additionally, calculations of polarizabilities within the UCCSD/6-31+G(d,p)//UMP2/6-31+G(d,p), UCCSD/Sadlej//UMP2/6-31+G(d,p) and UCCSD/Sadlej//B3LYP/6-31+G(d,p) methodology were performed. The results of dipole moment indicate that o-benzyne is a polar molecule, which value of μ lies in the range between 1.74 Debyes (UMP2) to 1.91 Debyes (UCCSD/Sadlej). For m-benzyne, the open conformation (2a) leads to a molecule with high values of μ that lies between 0.86 Debyes (UBLYP) to 1.25 Debyes (UCCSD/Sadlej), and the close configuration gives low values of μ that ranges from 0.12 Debyes (UB3LYP), 0.16 Debyes (UCCSD/Sadlej) to 0.64 Debyes (UHF). The results of polarizability indicate that α_{ave} values for o- and m-benzynes are lower than those of benzene, while the p-benzyne α is higher than the benzene one. At UMP2 level, a random behavior is observed on the α_{ave} of the benzyne isomers. In particular, it is important to note that the polarizability values for m-benzyne with the UMP2 method are too small if are compared with the UBLYP ones. The corresponding predicted values with the UB3LYP/Sadlej method for the o-, and the closed conformation of m- and p-benzynes are 66.24 au, 66.75 au (2b), 71.32 au (3b), respectively. This tendency is in agreement with the UCCSD calculations, which corresponding values are 65.19 au, 65.25 au and 69.40 au, respectively. The UCCSD/Sadlej values for the 2a and 3a open conformations of m- and p- benzynes are 68.35 au and 68.71 au, respectively. Additionally, the $\Delta\alpha$ values for benzynes are higher than the anisotropy of benzene and follow the same tendency as the α_{ave} values.

We have also analyzed the energetic stability of the benzyne isomers. Our calculations have shown that order of the average dipole polarizability does not follow the order of stability of the benzyne isomers. However, was found that p-C_6H_4 in their two conformations is the less stable isomer and is the most polarizable and anisotropic compound from the three benzyne series.

Acknowledgments

Authors thank to CONICIT of Venezuela for Project of Apoyo a Grupos of Optica Cuántica (PG 97000593) and CONDES of La Universidad del Zulia for partial support.

References

[1] P.G.S. Dkhar and R.H. Duncan, *Proc. Indian Acad. Sci. (Chem. Sci.)* **112** (2000), 97.

[2] W. Langenaeker, F. De Proft and P. Geerlings, *J. Phys. Chem. A* (102) (1998), 5944.

[3] S.G. Wierschke, J.J. Nash and R.R. Squires, *J. Am. Chem. Soc.* **115** (1993), 11958.; P.G. Wenthold, J. Hun and R.R. Squires, *J. Am. Chem. Soc.* **118** (1996), 11865.

[4] L.V. Moskaleva, L.K. Madden and M.C. Lin, *Phys. Chem. Chem. Phys.* **1** 1999, 3967; W. Sander, *Acc. Chem. Res.* **32** (1999), 669.

[5] J.G. Radziszewski, B.A. Hess, Jr. and R. Zahradnik, *J. Am. Chem. Soc.* **114** (1992), 114; *m*-Benzyne: R. Marquard, W. Sander and E. Kraka, *Angew. Chem. Int. Ed.* **35** (1996), 746; *p*-Benzyne: R. Marquardt, A. Balster, W. Sander, E. Kraka, D. Cremer and J.G. Radziszewski, *Angew. Chem. Int. Ed.* **37** (1998), 955; B. Andes, Jr., *Eur. J. Org. Chem.* (2001), 2185; F. de Proft, P. v Ragué Schleyer, J.H. v Lenthe, F. Stahl and P. Geerlings, *Chem. Eur, J.* **8** (2002), 3402.

[6] K. Yamaguchi, A. Nishio, S. Yabushita and T. Fueno, *Chem. Phys. Lett.* **53** (1978), 109.

[7] A. Hinchliffe and H.J. Soscún, *J. Mol. Struct.* **300** (1993), 1.

[8] A. Hinchliffe and H.J. Soscún, *J. Mol. Struct. (Theochem)* **313** (1994), 265.

[9] A.D. Buckingham, *Adv. Chem. Phys.* **12** (1967), 107.

[10] J. Grafenstein, E. Kraka, M. Filatov and D. Cremer, *Int. J. Mol. Sci.* **3** (2002), 360.

[11] C.C.J. Roothaan, *Rev. Mod. Phys.* **23** (1951) 69; J.A. Pople and R.K. Nesbet, *J. Chem. Phys.* **22** (1954), 571.

[12] C. Møller and M.S. Plesset, *Phys. Rev.* **46** (1934), 618.

[13] R.G. Parr and W. Yang, *Density functional theory of atoms and molecules*, Oxford Univ, Press, Oxford, 1989; J.P. Perdew and Y. Wang, *Phys. Rev. B* **45** (1992), 13244; J.P. Perdew, J.A. Chevary, S.H. Vosko, K.A. Jackson, M.R. Pederson, D.J. Singh and C. Fiolhais, *Phys. Rev. B* **46** (1992), 6671; J.K. Labanowsky and J.W. Andzelm, eds, *Density Functional methods in chemistry*, Springer-Verlag, New York, 1991.

[14] A.D. Becke, *Phys. Rev. A* **38** (1988), 3098; C. Lee, W. Yang and R.G. Parr, *Phys. Rev. B* **37** (1988), 785; B. Miehlich, A. Savin, H. Stoll and H. Preuss, *Chem. Phys. Lett.* **157** (1989), 200.

[15] A.D. Becke, *J. Chem. Phys.* **98** (1993), 5648; C. Lee, W. Yang and R.G. Parr, *Phys. Rev. B* **37** (1988), 785; B. Miehlich, A. Savin, H. Stoll and H. Preuss, *Chem. Phys. Lett.* **157** (1989), 200.

[16] W.J. Hehre, R. Ditchfield and J.A. Pople, *J. Chem. Phys.* **56** (1977), 2257; R. Krishnan, M.J. Frisch and J.A. Pople, *J. Chem. Phys.* **72** (1980), 4244; M.J. Frisch, J.A. Pople and J.S. Binkley, *J. Chem. Phys.* **80** (1984), 3265.

[17] M.J. Frisch, M. Head-Gordon and J.A. Pople, *J. Chem. Phys.* **141** (1990), 189.

[18] A.J. Sadlej, *Coll. Czech. Chem. Commum.* **53** (1988), 1995; A.J. Sadlej, *Theor. Chim. Acta* **79** (1991), 123.

[19] Gaussian 98, Revision A.7, M.J. Frisch, G.W. Trucks, H.B. Schlegel, G.E. Scuseria, M. Robb, J.R. Cheeseman, V.G. Zakrzewski, J.A. Montgomery, Jr., R.E. Stratmann, J.C. urant, S. Dapprich, J.M. Millam, A.D. Daniels, K.N. Kudin, M.C. Strain, O. Farkas, J. Tomasi, V. Barone, M. Cossi, R. Cammi, B. Mennucci, C. Pomelli, C. Adamo, S. Clifford, J. Ochterski, G.A. Petersson, P.Y. Ayala, Q. Cui, K. Morokuma, D.K. Malick, A.D. Rabuck, K. Raghavachari, J.B. Foresman, J. Cioslowski, J.V. Ortiz, A.G. Baboul, B.B. Stefanov, G. Liu, A. Liashenko, P. Piskorz, I. Komaromi, R. Gomperts, R.L. Martin, D.J. Fox, T. Keith, M.A. Al-Laham, C.Y. Peng, A. Nanayakkara, C. Gonzalez, M. Challacombe, P.M.W. Gill, B. Johnson, W. Chen, M.W. Wong, J.L. Andres, C. Gonzalez, M. Head-Gordon, E.S. Replogle and J.A. Pople, *Gaussian, Inc.* Pittsburgh PA, 1998; Gamess, Version 2002. M.W. Schmidt, K.K. Baldridge, J.A. Boatz, S.T. Elbert, M.S. Gordon, J.H. Jensen, S.Koseki, N. Matsunaga, K.A. Nguyen, S.J. Su, T.L. Windus, together with M. Dupuis and J.A. Mongtomery, *J. Comp. Chem.* **14** (1993), 1347.

[20] G.E. Scuseria and H.F. Schaefer, III., *J. Chem. Phys.* **90** (1989), 3700.

[21] W. Sander, M. Exner, M. Winkler, A. Balster, A. Hjerpe, E. Kraka and D. Cremer, *J. Am. Chem. Soc.* **124** (2002), 13072.

[22] M. Winkler and W. Sander, *J. Phys. Chem. A* **105** (2001), 10422.

[23] T. Danield Crawford, E. Kraka, J.F. Stanton and D. Cremer, *J. Chem. Phys.* **114**(24) (2001), 10638.

[24] S.P. Karna, G.B. Talatrapa and P.N. Prassad, *J. Chem. Phys.* **95** (1991), 5873.

[25] O. Christiansen, C. Hättig and P. Jørgensen, *Spectrochimica Acta part A: Molecular and Biomolecular Spectroscopy* **55**(3) (1999), 509.

Journal of Computational Methods in Sciences and Engineering 4 (2004) 411–418
IOS Press

First hyperpolarizability of 6-vertex carboranes 2. DFT study of NH_2/NO_2 -substituted 1,2-*closo*-dicarbahexaboranes

Kyrill Yu Suponitsky[a,b,*] and Tatiana V. Timofeeva[b]
[a]*X-ray Structural Center, Institute of Organoelement Compounds, Russian Academy of Sciences, Vavilov St., 28, Moscow, 117813, Russia*
[b]*Department of Natural Sciences, New Mexico Highlands University, Las Vegas, New Mexico, 87701, USA*

Abstract. The structure and molecular first hyperpolarizability (β) of the nitro-amino-substituted 1,2-dicarba-closo-hexaboranes were investigated at B3LYP/6-31+G(d) level of theory. Conformations with different orientations of the substituents with respect to the carborane cage were considered. It is shown that two factors significantly influence the value of β: orientation of the amino- and nitro-group with respect to the cage, and their mutual orientation. The latter factor appears to be more important. The results obtained have revealed that a carborane substituted at boron atoms is characterized by higher value of the molecular first hyperpolarizability.

Keywords: Nonlinear optics, hyperpolarizability, carboranes, three-dimensional delocalization, molecular design, DFT

1. Introduction

Closo-carboranes are known to be extremely thermally and chemically stable compounds [1,2]. Different approaches have been emerged to explain their stability by three-dimensional delocalization of the electron density and aromaticity. These approaches are graph theory combined with Huckel approximation [3], resonance stabilization [4], three-dimensional Huckel theory [5], topological analysis of the electron density in terms of theory of atoms in molecules [6], nucleus-independent chemical shift [7]. Therefore it seems reasonable to use *closo*-carboranes as building blocks for the molecular design of nonlinear optical (NLO) materials. It has long been recognized that NLO-active molecule for second harmonic generation should contain conjugated system endcapped with donor (D) and acceptor (A) groups [8]. A variety of conjugated bridges has been proposed and investigated [9–12]. This has led to dramatic increase of molecular first hyperpolarizability (β) in proposed systems relative to well-known organic standards such as urea and *para*-nitroaniline (pNA). However, the efficiency-stability tradeoff is still an open question, and the search for NLO materials with high thermal stability is important. This was our motivation to study NLO-active molecules in which the conjugated system is built up of carboranes. Because of the three-dimensional delocalization of the electron density in *closo*-carboranes, they might be utilized as a part of a conjugated bridge, thereby increasing thermal stability of a compound.

*Corresponding author. Fax.: +7 095 135 5085; E-mail: kira@xrlab.ineos.ac.ru.

To our knowledge, there are only few reports related to the investigation of the NLO properties of carborane-containing molecules [13–16]. In our first paper on this subject [17], hyperpolarizabilities of 6-vertex NH_2/NO_2-substituted 1,6-dicarbaboranes were evaluated in terms of DFT method. It was shown that the hyperpolarizability significantly depends on relative orientation of the substituents, which makes it difficult to predict *a priori* their "optimal" orientation, in contrast to conjugated organic molecules (for instance, *p*NA or N,N'-dimethylaminonitrostilbene) where planar structure is always preferable.

In the present study, we have continued investigation of 6-vertex *closo*-carboranes and have considered NH_2/NO_2-substituted 1,2-*closo*-dicarbahexaboranes.

2. Calculation

To be consistent with our previous study [17], we used the B3LYP method which is the "cheapest" way to incorporate electron correlation that is necessary for reliable estimation of molecular hyperpolarizability [18]. Choice of this modest level of theory looks reasonable because we are interesting in the relative values of hyperpolarizabilities. It was shown in the literature many times [19,20], and, in particular for boron clusters [21], that the relative order of β values is adequately reproduced by B3LYP method. Our earlier results have shown that hyperpolarizability is not significantly affected by going from a 6-31+G(d) basis set to a 6-311++G(df,p) basis, therefore the 6-31+G(d) basis has been utilized both for geometry optimization and calculation of hyperpolarizability. All calculations have been carried out with the Gaussian94 program [22]. Hyperpolarizability has been estimated in terms of CPHF approach as implemented in the Gaussian program. According to this approach components of hyperpolarizability tensor are presented as

$$\beta_{ijk} = -\frac{\partial^3 W(E)}{\partial E_i \partial E_j \partial E_k}.$$

After that vectorial part of β can be calculated according to equation

$$\beta = \sqrt{\beta_x^2 + \beta_y^2 + \beta_z^2}, \text{ where } \beta_i = \beta_{ixx} + \beta_{iyy} + \beta_{izz}.$$

3. Results and discussion

As model molecules we considered NH_2/NO_2-substituted carboranes in which the substituents are not attached to the nearest boron or carbon atoms. Investigated molecules are shown in the scheme on the next page.

Orientation of the NH_2/NO_2 substituents relative to the carborane cage in these molecules might be different. As it was shown for NH_2/NO_2-1,6-dicarbahexaboranes [17] and as we will see on the next page, barrier of rotation of the amino- or nitro-group is low, hence molecular geometry in condensed phase might be significantly affected by surrounding molecules. Therefore, in order to obtained detailed information on the hyperpolarizability of molecules under consideration it is necessary to investigate different conformations even though some of them in isolated state do not correspond to the minimum on potential energy surface (PES).

For both molecules, we first optimized their structures without symmetry constraints and then considered geometries restricted by C_s symmetry. All investigated conformations for both molecules are presented in Fig. 1. Geometry optimization of molecule **I** without symmetry constraints leads to conformer

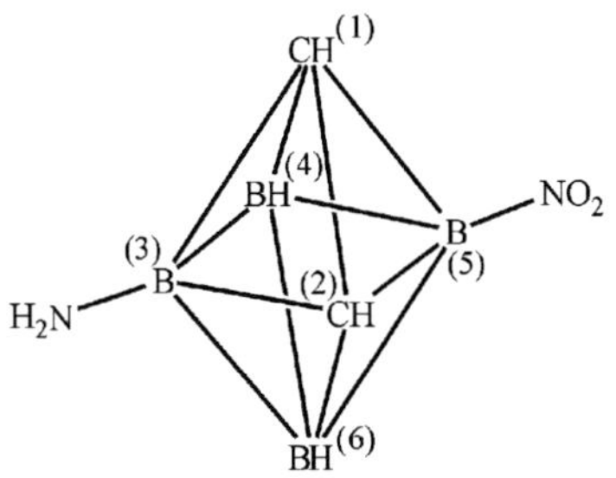

3-amino-5-nitro-1,2-
dicarba-*closo*-hexaborane
I

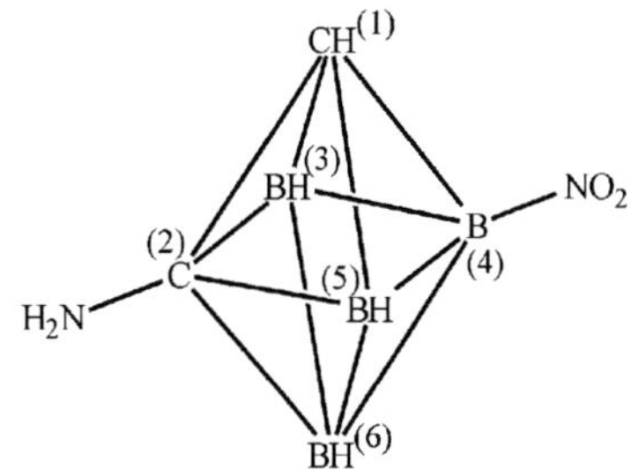

2-amino-4-nitro-1,2-
dicarba-*closo*-hexaborane
II

Table 1
Bond lengths (Å), total energies* (a.u.), relative energies† (kcal·mol^{-1}), and molecular first hyperpolarizabilities (β, a.u.) of the conformations **Ia-c** and **IIa-b**. Bond lengths in unsubstituted 1,2-carborane are listed for comparison

	Unsubstituted 1,2-$C_2B_4H_6$	**Ia**	**Ib**	**Ic**		Unsubstituted 1,2-$C_2B_4H_6$	**IIa**	**IIb**
B(3)-C(1)	1.621	1.713	1.635	1.638	C(2)-C(1)	1.543	1.596	1.577
B(3)-B(4)	1.734	1.730	1.750	1.747	C(2)-B(3)	1.621	1.598	1.613
B(3)-C(2)	1.621	1.591	1.635	1.638	C(2)-B(6)	1.630	1.673	1.647
B(3)-B(6)	1.734	1.771	1.750	1.747	C(2)-B(5)	1.621	1.598	1.613
B(5)-C(1)	1.621	1.568	1.598	1.601	B(4)-C(1)	1.630	1.582	1.601
B(5)-B(4)	1.734	1.753	1.717	1.712	B(4)-B(3)	1.734	1.741	1.721
B(5)-C(2)	1.621	1.636	1.598	1.601	B(4)-B(6)	1.708	1.668	1.681
B(5)-B(6)	1.734	1.681	1.717	1.712	B(4)-B(5)	1.734	1.741	1.721
C(1)-C(2)	1.543	1.557	1.548	1.532	C(1)-B(3)	1.621	1.630	1.622
C(1)-B(4)	1.630	1.624	1.644	1.655	B(3)-B(6)	1.734	1.747	1.750
C(2)-B(6)	1.630	1.657	1.644	1.655	B(6)-B(5)	1.734	1.747	1.750
B(4)-B(6)	1.708	1.723	1.712	1.706	C(1)-B(5)	1.621	1.630	1.622
Total energy	–	−439.0439	−439.0419	−439.0406	Total energy	–	−439.0114	−439.0117
Relative energy	–	0.0	1.3	2.1	Relative energy	–	0	0.2
β	–	808	940	238	β	–	403	238

*zero-point correction is included.
†relative energies are calculated assuming energy for **Ia** (for molecule **I**) and for **IIa** (for molecule **II**) to be zero.

Ia, which corresponds to the global minimum. In **Ia**, the oxygens of the nitro-group are approximately projected onto the C(2) and B(4) atoms of the 4-membered ring of the cage (C(1)-C(2)-B(6)-B(4)), while the lone pair (LP) of the amino-group appears in *trans*-orientation to the B(3)-B(6) bond[1] (approximately parallel to the NO_2-group). We will refer to such a conformation as an eclipsed orientation of the substituents relative to the cage and parallel orientation of the substituents relative to each other (the line connecting the projections of the two amino group hydrogens onto the C(1)-C(2)-B(6)-B(4) plane is approximately parallel with the line connecting projections of the two nitro group oxygens onto the same plane). In **Ia**, the B(5)-C(2), B(5)-B(4), B(3)-C(1) and B(3)-B(6) bonds are elongated relative to unsubstituted carborane (calculated at the same level of theory), and the elongation is more pronounced for the latter two bonds, which are located at the donor part of the molecule (Table 1). This elongation

[1]For all presented structures of molecules **I** and **II**, the amino-group can be rotated by 180°. In some cases (**Ia-c**) we were unable to locate such a conformation, while in other cases (**IIa,b**) differences between two conformations are negligible (less than 0.005 Å in the bond lengths, less than 0.1 kcal mol^{-1} in energy, and less than 1% in the value of hyperpolarizability). Therefore this is not discussed.

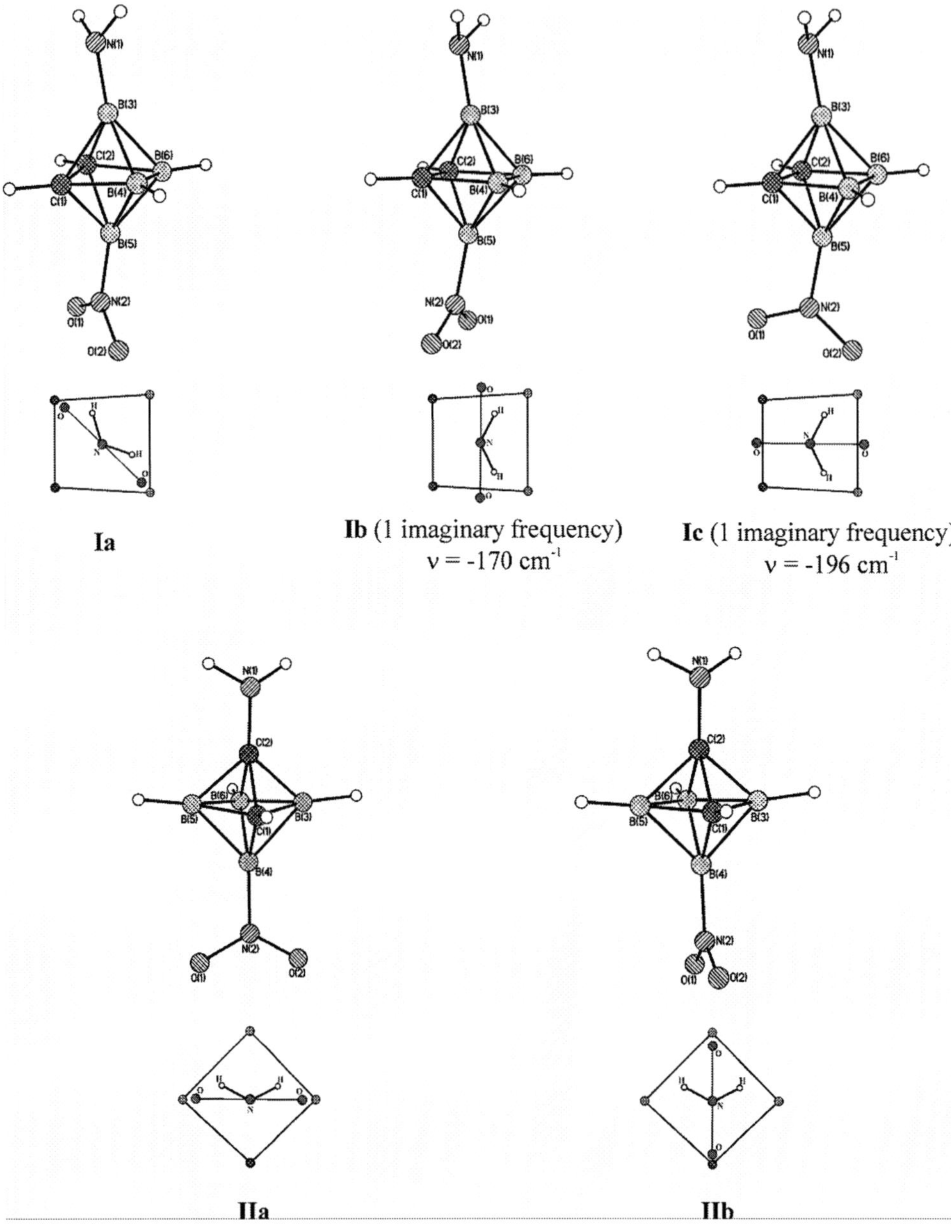

Fig. 1. Structures of the calculated conformers of the molecules **I** and **II**. Relative orientation of the substituents is shown schematically as a projection onto corresponding 4-membered ring. The presence of imaginary frequences is indicated.

might be explained by the backdonation of the electron density of the N-O π-bonds and LP of the amino-group to the corresponding antibonding MOs of the carborane cage. The other bonds at the B(3) and B(5) atoms (B(5)-C(1), B(5)-B(6), B(3)-C(2), B(3)-B(4)) are shortened relative to unsubstituted carborane. At the same time, bonds in the C(1)-C(2)-B(6)-B(4) ring are less affected by substituents. This situation is analogous to the results obtained for NH_2/NO_2-1,6-dicarbahexaboranes.

The influence of substituents on the geometry of the carborane cage in the other two conformations of molecule **I** (**Ib, Ic**), which are of C_s symmetry, is different. According to C_s symmetry, the amino-group

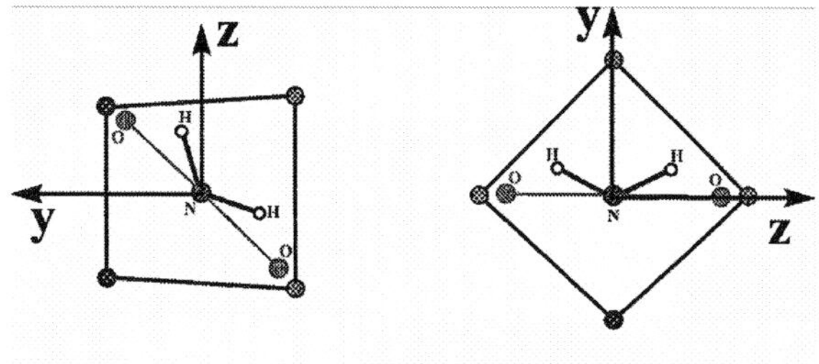

Fig. 2. Orientations of the coordinate systems for molecules **I** and **II**. Axis X is perpendicular to the plane of the figue and is directed from N(1) to N(2).

can only be in staggered orientation in which LP is located above the center of the C(1)-C(2) bond while the nitro-group is parallel (**Ib**) or perpendicular (**Ic**) to the amino-group. The bonds at the B(3) atom bearing donor substituent are longer than those in unsubstituted carborane, while bonds at the opposite B(5) atom, which is connected to the acceptor group, are shortened (Table 1). The same trends were observed in our previous study [17], as well as by X-ray study of acceptor-substituted boranes [23] and for donor/acceptor-substituted benzene derivatives [24].

Geometry optimization of the molecule **II** without symmetry constraints leads to the structure **IIa** which is of C_s symmetry. The conformation **IIa** is characterized by eclipsed orientation of the substituents which are parallel to each other. The other possible conformation of C_s symmetry **IIb** differs by rotation of the nitro-group in which conformation the amino- and nitro-groups are perpendicular (Fig. 1). The bond lengths distributions in the cages of conformers **IIa** and **IIb** are analogous to what is found for the structure **Ia**. The lengths of bonds which are in cis- or trans-orientation to LP of the amino-group or to the N-O bonds of the nitro-group are elongated, while the other bond lengths at the B(4) and C(2) atoms are shortened (Table 1). The bond lengths in the C(1)-B(3)-B(6)-B(5) ring are only slightly elongated.

From Table 1, it can be seen that interaction of the substituents with the carborane cage is more pronounced in the structures **Ia**, **IIa** and **IIb** which correspond to minima on PES. In those conformers, the substituents are in eclipsed orientation relative to the cage. From the consideration of the relative energies of the conformers, it follows that the eclipsed orientation of the substituents with respect to the carborane cage has more significant influence on energy than their mutual orientation. For instance, energy of the conformer **IIb** with perpendicular orientation of the substituents is only 0.2 $kcal{\cdot}mol^{-1}$ higher than that of **IIa** (with parallel orientation). In two conformations of the molecule **I** (**Ia** and **Ib**), the amino- and nitro-groups are parallel to each other. These conformations differ by orientation of the substituents with respect to the carborane cage being eclipsed for **Ia** and staggered for **Ib**. This leads to increase of energy of **Ib** by 1.3 $kcal{\cdot}mol^{-1}$.

Calculated hyperpolarizabilities for all structures are presented in Table 1. It can be seen that although the change of the orientation of the NH_2/NO_2 groups does not significantly influence molecular geometry, it might lead to significant change in the value of β. Influence of the orientation of the NH_2/NO_2-groups on hyperpolarizability is opposite to its influence on energy. Minor change in energy when going from **IIa** to **IIb** is accompanied by an almost two times decrease in the value of β. The hyperpolarizabilities of **Ia** and **Ib** (which are characterized by parallel orientation of substituents to each other, but differ by their orientation with respect to the cage (Fig. 1)) are close (Table 1), while difference in energy is sizable. It should be noted that less stable conformation **Ib** has higher value of β. Perpendicular orientation of the amino- and nitro-groups in **Ic** leads to a dramatic decrease in β. At the same time, difference in energy between **Ib** and **Ic** is less than that between **Ia** and **Ib**.

Table 2
Hyperpolarizability tensor components (a.u.) for the conformations **Ia-c** and **IIa-b***

	Ia	**Ib**	**Ic**	**IIa**	**IIb**
β_{xxx}	882	972	296	451	297
β_{xyy}	−46	−11	−50	−25	−40
β_{xzz}	−28	−31	−11	−30	−20
β_{yxx}	14	143	−24	51	2
β_{yyy}	1	11	8	23	29
β_{yzz}	−7	−24	−17	−4	−4
β_{zxx}	25				
β_{zyy}	16				
β_{zzz}	−12				
β_{xyz}	−9				

*Orientations of the coordinate systems are depicted in Fig. 2.

Table 2 contains all non-vanishing components of β tensor for all studied structures. Conformations with parallel relative orientations of NH_2/NO_2 groups (**Ia, Ib, IIa**) are characterized by almost one-dimensional hyperpolarizability. If the nitro-group is rotated to be perpendicular to the amino-group, it significantly decreases the interaction between donor and acceptor groups and, as a consequence, leads to a dramatic decrease in main tensor component β_{xxx}, while changes in off-diagonal components are less pronounced. Nevertheless hyperpolarizability of **Ic** and **IIb** conformations might be described as rather one-dimensional.

From the above text, it follows that the influence of the orientation of substituents in NH_2/NO_2-carboranes on hyperpolarizability is much more complicated than in π-conjugated organic molecules. To maximize the value of β, organic D-π-A molecules must adopt a planar conformation which is usually most stable [25]. It has also been shown that influence of *cis/trans* orientation between fragments in such molecules is of minor importance [12]. It makes the search for "optimal" molecular structure of organic compounds easier. Unfortunately this is not the case for D/A-substituted carboranes. Two factors, namely orientation of substituents to each other and their orientation with respect to the carborane cage, must be taken into account.

4. Conclusions

In the present study we have investigated the structures and first hyperpolarizabilities of NH_2/NO_2- substituted 1,2-*closo*-dicarbahexaboranes. In accord with our previous study on 1,6-*closo*-dicarbahexaboranes, substitution at the boron atoms of the carborane cage leads to higher values of β. Detailed analysis of different conformations has allowed us to come to following conclusion. In order to maximize the value of β, structure must adopt parallel orientation of the substituents to each other. This factor has more significant influence on β than orientation of the substituents relative to the carborane cage.

Acknowledgments

We thank Prof. N. Allinger's group (University of Georgia at Athens) for providing us with computational time for the calculations. We are also thank D. Thornburg for help with preparation of the manuscript.

References

[1] V.I. Bregadze, Dicarba-*closo*-dodecaboranes $C_2B_{10}H_{12}$ and their derivatives, *Chemical Reviews* **92** (1992), 209–223.

[2] J. Plešek, Potential applications of the boron cluster compounds, *Chemical Reviews* **92** (1992), 269–278.

[3] R.B. King and D.H. Rouvray, Chemical applications of group theory and topology. 7. A traph-theoretical interpretation of the bonding topology in polyhedral boranes, carboranes, and metal clusters, *Journal of the American Chemical Society* **99** (1977), 7834–7840.

[4] J. Aihara, Three-dimensional aromaticity of polyhedral boranes, *Journal of the American Chemical Society* **100** (1978), 3339–3342.

[5] B.M. Gimarc and M. Zhao, Three-dimensional hulckel theory for closo-carboranes, *Inorganic Chemistry* **35** (1996), 825–834.

[6] R.W.F. Bader and D.A. Legare, Properties of atoms in molecules: Structures and reactivities of boranes and carboranes, *Canadian Journal of Chemistry* **70** (1992), 657–676.

[7] P.V.R. Schleyer and K. Najafian, Stability and three-dimensional aromaticity of *closo*-monocarborane anions, $CB_{n-1}H_n^-$, and *closo*-dicarboranes, $C_2B_{n-2}H_n$, *Inorganic Chemistry* **37** (1998), 3454–3470.

[8] B.L. Davydov, L.D. Derkacheva, V.V. Dunina, M.E. Zhabotinsky, V.F. Zolin, L.G. Koreneva and M.A. Samokhina, Interrelation of charge transfer and second harmonic generation of OQG, *JETP Letters* **12** (1970), 24–26.

[9] L.-T. Cheng, W. Tam, S.H. Stevenson, G.R. Meredith, G. Rikken and S.R. Marder, Experimental investigations of organic molecular nonlinear optical polarizabilities. 1. Methods and results on benzene and stilbene derivatives, *Journal of Physical Chemistry* **95** (1991), 10631–10643; L.-T. Cheng, W. Tam, S.R. Marder, A.E. Stiegman, G. Rikken and C.W. Spangler, Experimental investigations of organic molecular nonlinear optical polarlzabllities. 2. A study of conjugation dependences, *Journal of Physical Chemistry* **95** (1991), 10643–10652.

[10] K.J. Drost, A.K.-Y. Jen and V.P. Rao, Designing organic NLO materials, *ChemTech* (1995), 16–24; V.P. Rao, A.K.-Y. Jen, J. Chandrasekhar, I.N.N. Namboothiri and A. Rathna, The important role of heteroaromatics in the design of efficient second-order nonlinear optical molecules: Theoretical investigation on push-pull heteroaromatic stilbenes, *Journal of the American Chemical Society* **118** (1996), 12443–12448.

[11] T. Verbiest, S. Houbrechts, M. Kauranen, K. Clays and A. Persoons, Second-order nonlinear optical materials: Recent advances in chromophore design, *Journal of Materials Chemistry* **7** (1997), 2175–2189.

[12] E.M. Breitung, C.-F. Shu and R.J. McMahon, Thiazole and thiophene analogues of donor-acceptor stilbenes: Molecular hyperpolarizabilities and structure-property relationships, *Journal of the American Chemical Society* **122** (2000), 1154–1160.

[13] D.M. Murphy, D.M.P. Mingos and J.M. Forward, Synthesis of icosahedral carboranes for second-harmonic generation. Part 1, *Journal of Materials Chemistry* **3** (1993), 67–76; D.M. Murphy, D.M.P. Mingos, J.L. Haggitt, H.R. Powell, S.A. Westcott, T.B. Marder, N.J. Taylor and D.R. Kanis, Synthesis of icosahedral carboranes for second-harmonic generation. Part 2, *Journal of Materials Chemistry* **3** (1993), 139–148.

[14] J. Abe, N. Nemoto, Y. Nagase, Y. Shirai and T. Iyoda, A new class of carborane compounds for second-order nonlinear optics: Ab initio molecular orbital study of hyperpolarizabilities for 1-(1',X"-Dicarba-closo-dodecaborane-1'-yl)-*closo*-dodecaborate Dianion ($X = 2, 7, 12$), *Inorganic Chemistry* **37** (1998), 172–173.

[15] B. Grüner, Z. Janoušek, B. King, J.N. Woodford, C.H. Wang, V. Všetečka and J. Michl, Synthesis of 12-substituted 1-carba-closo-dodecaborate anions and first hyperpolarizability of the 12-$C_7H_6^+$-$CB_{11}H_{11}$- Ylide, *Journal of the American Chemical Society* **121** (1999), 3122–3126.

[16] D.G. Allis and J.T. Spencer, Polyhedral-based nonlinear optical materials. Part 1. Theoretical investigation of some new high nonlinear optical response compounds involving carboranes and charged aromatic donors and acceptors, *Journal of Organometallic Chemistry* **614–615** (2000), 309–313; D.G. Allis and J.T. Spencer, Polyhedral-based nonlinear optical materials. 2. Theoretical investigation of some new high nonlinear optical response compounds involving polyhedral bridges with charged aromatic donors and acceptors, *Inorganic Chemistry* **40** (2001), 3373–3380; J. Taylor, J. Caruso, A. Newlon, U. Englich, K. Ruhlandt-Senge and J.T. Spencer, Polyhedral-based nonlinear optical materials. 3.1 Synthetic studies of cyclopentadiene- and cycloheptatriene-substituted polyhedral compounds: Synthesis of 1,12-[$(C_7H_7)C_2B_{10}H_{10}(C_5H_3Me_2)$] and related species, *Inorganic Chemistry* **40** (2001), 3381–3388.

[17] K. Yu Suponitsky and T.V. Timofeeva, Hyperpolarizability of 6-vertex carboranes, quantum chemical study, *Central European Journal of Chemistry* **1** (2003), 1–9.

[18] H.A. Kurtz and D.S. Dudis, Quantum mechanical methods for predicting nonlinear optical properties, in: *Reviews in Computational Chemistry*, K.B. Lipkowitz and D.B. Boyd, eds, Wiley-VCH, John Wiley and sons: NY, 1998, pp. 241–279.

[19] A.J. Cohen, N.C. Handy and D.J. Tozer, Density functional calculations of the hyperpolarizabilities of small molecules, *Chemical Physics Letters* **303** (1999), 391–398.

[20] J.A. van Gisbergen, J.G. Snijders and E.J. Baerends, *Journal of Chemical Physics* **109** (1998), 10657–10668.

[21] H. Reis, M.G. Papadopoulos and I. Boustani, DFT calculations of static dipole polarizabilities and hyperpolarizabilities for the boron clusters B_n (n =3–8,10), *Int. J. Quant. Chem.* **78** (2000), 131–135.

[22] M.J. Frisch, G.W. Trucks, H.B. Schlegel, P.M.W. Gill, B.G. Johnson, M.A. Robb, J.R. Cheeseman, T. Keith, G.A. Petersson, J.A. Montgomery, K. Raghavachari, M.A. Al-Laham, V.G. Zakrzewski, J.V. Ortiz, J.B. Foresman, J. Cioslowski, B.B. Stefanov, A. Nanayakkara, M. Challacombe, C.Y. Peng, P.Y. Ayala, W. Chen, M.W. Wong, J.L. Andres, E.S. Replogle, R. Gomperts, R.L. Martin, D.J. Fox, J.S. Binkley, D.J. Defrees, J. Baker, J.P. Stewart, M. Head-Gordon, C. Gonzales and J.A. Pople, *Gaussian 94*, Revizion E.2; Gaussian Inc.: Pittsburgh, PA, 1995.

[23] A. Franken, W. Preetz, M. Rath and K-F. Hesse, Preparation of mononitropentahydrohexaborate(2-) and crystal structure of $M_2[B_6H_5(NO_2)]$, $M = K$, Cs, *Zeitschrift fur naturforschung, Teil B* **48** (1993), 1727–1731; C. Drewes and W. Preetz, Synthesis and spectroscopic characterization of methylnitro-closo-hexaborates and crystal structures of cis-$(Ph_4As)_2[B_6H_4(CH_3)(NO_2)]$, fac-$(Ph_4As)_2[B_6H_3(CH_3)(NO_2)_2]{\cdot}CH_3CN$ and mer-$(Ph_4P)_2[B_6H_3(CH_3)(NO_2)c_2]$, *Zeitschrift fur Naturforschung, Teil B* **54** (1999), 349–356; C. Drewes and W. Preetz, Synthesis, spectra and crystal structure of cis-monobenzylmononitrotetrahydro-closo-hexaborate(2-), *Zeitschrift fur Naturforschung, Teil B* **54** (1999), 1219–1221.

[24] A. Domenicano, in: *accurate molecular structures: Their determination and importance*, A. Domenicano and I. Hargittai, eds, Oxford University Press: Oxford, New York, 1992.

[25] M. Barzoukas, A. Fort, G. Klein, A. Boeglin, C. Serbutoviez, L. Oswald and J.F. Nicoud, Conformational dependence of the quadratic hyperpolarizabilities of a series of push-pull diaryl acetylenes: An experimental and computational investigation, *Chemical Physics* **153** (1991), 457–464; M. Barzoukas, A. Fort, G. Klein, A. Boeglin, C. Serbutoviez, L. Oswald and J.F. Nicoud, Conformational dependence of the quadratic hyperpolarizabilities of a series of push-pull stilbenes: characterization and investigation of empirical correlations, *Chemical Physics* **164** (1992), 395–406.

Journal of Computational Methods in Sciences and Engineering 4 (2004) 419–425
IOS Press

Polarizabilities of amino acid residues

Marcel Swart[a,*], Jaap G. Snijders[b,**] and Piet Th. van Duijnen[b]
[a]*Organische en Anorganische Chemie, Vrije Universiteit Amsterdam, De Boelelaan 1083, 1081 HV Amsterdam, The Netherlands*
[b]*Theoretische Chemie, Rijksuniversiteit Groningen, Nijenborgh 4, 9747 AG Groningen, The Netherlands*

Abstract. Over the last couple of years, it has been shown that Time Dependent Density Functional Theory (TD-DFT) is able to predict accurately and efficiently the polarizability of molecules, when using appropriate exchange-correlation potentials and (large) basis sets. In a previous paper, we compared the accuracy of the predicted mean polarizabilities of 15 organic molecules with experiment, and with two other computational methods: the Restricted Hartree-Fock (RHF) method and the Direct Reaction Field (DRF) approach, the first of which is ignored in this paper. The (empirical) DRF approach however was shown to give comparable accuracies to TD-DFT with the values computed in just a few seconds. In this paper, we use TD-DFT to compute molecular polarizabilities of the twenty amino acid residues, and compare them with the results obtained with the DRF approach. Although the mean absolute deviation of the DRF values from the TD-DFT values is reasonable (7%), it is more than two times the accuracy normally found with the DRF approach. Therefore we decided to optimize the atomic parameters for these systems, and found after optimization, a good agreement with the TD-DFT values (mean absolute deviation 1.0%). As the TD-DFT calculations were necessarily obtained with two additional hydrogens to saturate the backbone bonds, the molecular value of the polarizability of the amino acid residues is overestimated by the TD-DFT calculations. Therefore, the DRF approach (with the newly optimized atomic parameters) has been used to get the actual polarizabilities of the amino acid residues.

Keywords: Amino acid residues, polarizabilities, polarizable force field, (Time Dependent) Density Functional Theory

1. Introduction

Over the last couple of years, many papers have appeared regarding the reliability, accuracy and efficiency of Time Dependent Density Functional Theory (TD-DFT) [1–3] with respect to the computation of molecular polarizabilities [4–13]. It has been shown [14] that if a sufficiently large basis set is used (for instance a triple zeta valence plus double polarization functions plus diffuse functions, e.g. TZ2P+) with an appropriate exchange-correlation potential (like the van Leeuwen-Baerends potential, usually referred to as LB94 [15]), the deviation from experiment of the computed polarizabilities is comparable to, or better than, the uncertainty associated with the experimental value [14]. Although the polarizabilities of polymers seemed to be an exception to the rule regarding the success of TD-DFT [16,17], recent progress using Time Dependent Current Density Functional Theory [18,19] has overcome also this shortcoming of standard TD-DFT.

In a previous paper [14], we have shown for a set of 15 organic molecules that the average absolute deviation from experiment, when using the LB94 potential in a large (TZ2P++) basis set, is only 3.0%,

*Corresponding author: E-mail: swart@chem.vu.nl.
**In loving memory of Jaap Snijders who passed away unexpectedly on August 13, 2003

which is comparable to the estimated experimental uncertainty of 2–4%. In the same paper, we compared the TD-DFT results with the values obtained with Restricted Hartree-Fock (RHF) calculations and the Direct/Discrete Reaction Field (DRF) [20–26] approach.[1] Using a smaller TZP basis set, in order to make a meaningful comparison between the RHF and TD-DFT results, the average absolute deviation of TD-DFT increased to 8% [14]; however, this is still acceptable in comparison with the RHF results, which showed a deviation of 25%. For that reason, the RHF method is ignored in the present paper. The DRF approach on the other hand was shown to give results comparable to TD-DFT, in only a few seconds. Therefore, the DRF approach has been included also in this paper.

The molecular polarizability is an entity which may be used for polarizable force fields, such as the Direct/Discrete Reaction Field approach [20–26]. Applications of such force fields are scarce, especially when looking at biomolecules, where usually polarization effects are either completely ignored or mimicked by effective two-body potentials. Although some recent progress has been reported on polarizable force fields for biomolecules [27], the well defined molecular polarizability property has not yet been recognized and/or utilized for the development of such a force field. Moreover, molecular polarizability values of the amino acid residues have until now, to the best of our knowledge, not yet been reported.

Therefore, we computed the molecular polarizability tensors of all twenty amino acid residues[2] using TD-DFT. For these calculations, we used the same geometries as employed in another paper where a new and accurate charge analysis [28] has been reported. The residues consist therefore of **H**NHCαHRCO**H** (with R the side chain), i.e. apart from the **bold** hydrogen atoms which are needed to satisfy the valency of the system, the normal amino acid residues as they are found in proteins and enzymes. After checking the accuracy of the DRF approach in comparison with these computed TD-DFT values, we can then easily obtain the *actual* molecular polarizability of the amino acid residues (i.e. the actual residues without the **bold** hydrogens) by ignoring the **bold** hydrogens in the DRF calculations.

2. Computational details

The polarizability of a molecule can be obtained from a Taylor expansion of the energy U about the electric field strength E:

$$U = U_0 + \mu_i \cdot E_i + \frac{1}{2}\alpha_{jk} \cdot E_j \cdot E_k; \quad i,j,k \in \{x,y,z\}$$

Details of the specific TD-DFT formulas for obtaining the molecular polarizabilities can be found elsewhere [1–4,10,29], but as has been stated in our previous paper, the formulas are comparable to the Coupled Perturbative Hartree Fock (CPHF) equations [30]. For clarity, the expressions used in the DRF approach are reported here again, as they are less well-known than the TD-DFT or CPHF equations.

Consider a system of N polarizable points, which are placed in an external field E_{ext}; the induced dipole moment in point p is then a function of the external field as well as the induced dipole moments

[1]Note that due to the recent coupling of the DRF approach with Density Functional Theory, the approach has been renamed as Discrete Reaction Field.

[2]In fact, the total number of molecules considered is 21, as for His the proton has been placed at either the N-delta or N-epsilon position.

in the other polarizable points q:

$$\mu_p = \alpha \left(E_{\text{ext}} + \sum_{q \neq p} T_{pq}\mu_q \right)$$

This can be rewritten in matrix notation ($\boldsymbol{M} = \boldsymbol{A}(\boldsymbol{E} + \boldsymbol{T}\boldsymbol{M})$), and solved to give $\boldsymbol{M} = (\boldsymbol{A}^{-1} - \boldsymbol{T})^{-1}\boldsymbol{E}$, which gives the molecular response to an external field. The $3N \times 3N$ $(\boldsymbol{A}^{-1} - \boldsymbol{T})^{-1}$-matrix can then be reduced to a normal 3×3 polarizability tensor to give the molecular polarizability tensor. Without screening for overlapping charge densities this equation could lead to the so-called polarization catastrophe, which has been "repaired" by Thole in 1981 [31]. In the Thole model [20,31,32], the interactions between polarizabilities are screened for overlapping charge densities by using a simple exponential model (where u is a reduced distance due to the screening, and the f_V, f_E and f_T factors screen the normal electrostatic potential, field and dipole field respectively):

$$u = r_{ij}/(\alpha_i\alpha_j)^{1/6} \qquad f_V = 1 - (\frac{1}{2}v + 1)e^{-v}$$

$$v = au \qquad f_E = f_V - \left(\frac{1}{2}v^2 + \frac{1}{2}\right)e^{-v}$$

$$\rho(u) = \frac{a^3}{8\pi}e^{-au} \qquad f_T = f_E - \frac{1}{6}v^3 e^{-v}$$

All TD-DFT calculations have been performed with the Amsterdam Density Functional (ADF) [33] program (version 2002.02), where 4–5 nodes have been used in parallel (LAM-MPI) jobs to speed up the calculations. As sulphur had not been included in the previously used TZ2P++ basis set, the largest standard available basis set in the 2002.02 version of ADF (QZ4P ZORA basis set, which does include sulphur) has been used for all molecules, using the LB94 exchange-correlation potential; as already indicated this basis set should be and has been used using scalar relativistic calculations using the ZORA Hamiltonian [34]. This Hamiltonian provides an efficient way to include relativistic corrections in the calculations [33] and has negligible effects when used on molecules containing no heavy atoms (as in this case). For systems containing heavy atoms this should be the preferred option anyway, while in this case no harm is being done by adding the (in this case negligible) relativistic correction total CPU time needed (in these parallel runs) ranges then from less than 2 hours for Gly to almost 27 hours for Trp on a Pentium-Linux cluster.

3. Results

Before reporting the molecular polarizability values of the amino acid residues, we should get an idea of the effect of using the QZ4P basis set instead of the previously used TZ2P++ basis. Therefore we reran the calculations on the previously used 15 molecules with the same geometries, but now with the QZ4P basis. No direct comparison can be made regarding the CPU time, as the current jobs were performed on a Pentium-Linux cluster, while the previous jobs were performed on a SGI Powerchallenge; nevertheless, the total time needed for these jobs is only 22 hours of walltime (on 2 nodes of the Pentium-Linux cluster).

The LB94/QZ4P results (see Table 1) show a small improvement over the LB94/TZ2P++ results; the average deviation decreases from 0.9 to 0.2%, while the average absolute deviation decreases slightly (from 3.0 to 2.7%). Also the standard deviations of these averages are smaller for the QZ4P basis (3.5%

Table 1
Comparison of the computed polarizabilities (a.u.) in the QZ4P basis with the previously computed TZ2P++ values [14] and experimental data [14]; LB94 results for 15 organic molecules used in previous study

	exp [14]	TZ2P++	QZ4P		exp [14]	TZ2P++	QZ4P
acetamide	40.5	41.2	41.2	methylcyanide	29.7	30.8	30.0
acetylene	22.5	22.4	21.6	neopentane	69.0	68.6	68.1
benzene	70.1	69.7	69.0	propane	42.4	42.1	43.1
chlorine	31.1	29.9	31.1	TCFM[a]	57.5	59.3	58.1
cyclohexanol	79.9	80.7	79.8	TCMC[a]	70.5	73.7	72.5
dimethylether	35.0	35.6	35.3	TFM[a]	19.0	18.6	18.2
formaldehyde	16.5	17.8	17.6	water	9.94	9.2	9.2
hydrogen	5.3	5.6	5.6				

a) TCFM trichlorofluoromethane; TCMC trichloromethylcyanide; TFM trifluoromethane.

Table 2
Computed polarizabilities (a.u.) of the amino acid residues by TD-DFT with the LB94 XC-potential in the QZ4P basis, with the standard DRF parameters [32], the optimized DRF parameters, and the true amino acid residue value

	sidechain R	*TD-DFT*	$DRF_{\text{std a}}$	$DRF_{\text{a opt}}$	$DRF_{\text{b actual}}$
ALA	CH_3	52.78	50.62	52.13	50.16
ARG	$(CH_2)_3NHCNHNH_2$	117.11	109.59	117.13	114.81
ASN	CH_2CONH_2	75.41	72.38	75.39	73.15
ASP	CH_2COOH	70.05	69.11	71.31	69.09
CYS	CH_2SH	76.88	66.74	76.89	74.99
GLN	$(CH_2)_2CONH_2$	88.94	85.36	90.70	88.79
GLU	$(CH_2)_2COOH$	85.08	82.19	86.62	84.67
GLY	H	39.93	38.54	38.87	36.66
HISD	$CH_2C_3N_2H_3$	102.28	91.42	101.31	99.35
HISE	$CH_2C_3N_2H_3$	101.61	92.03	101.62	99.65
ILE	$CH(CH_3)CH_2CH_3$	92.08	87.52	93.18	91.21
LEU	$CH_2CH(CH_3)_2$	93.06	87.67	93.63	91.60
LYS	$(CH_2)_4NH_2$	102.62	97.21	104.42	101.73
MET	$(CH_2)_2SCH_3$	105.06	91.60	104.28	102.31
PHE	$CH_2C_6H_5$	123.90	108.57	123.40	121.43
PRO	$(CH_2)_3$#	74.06	70.75	75.72	73.47
SER	CH_2OH	56.62	56.08	55.96	53.82
THR	$CH(OH)CH_3$	70.41	67.60	68.98	66.46
TRP	$CH_2C_2H_2NC_6H_4$	156.94	134.96	155.00	153.06
TYR	$CH_2C_6H_4OH$	130.60	114.27	128.17	126.19
VAL	$CH(CH_3)_2$	77.85	74.23	77.85	76.09
average deviation			−6.89%	−0.03%	
average absolute deviation			6.89%	0.96%	

a) with NH2CHRCOH representan of amino acid residue
b) with true NHCHRCO representation of amino acid residue
#) in PRO, the N and C-alpha atoms form together with the sidechain a ring.

for the average deviation, 2.2% for average absolute deviation) than for the TZ2P++ basis (3.7% and 2.4% respectively).

Now that we have established that the results may be obtained as accurately in the QZ4P basis as in the TZ2P++ basis, we move on to the actual amino acid residues. The molecular polarizability values in the QZ4P basis with the LB94 xc-potential are given in the first column of Table 2. To the best of our knowledge, there are no experimental data to compare the computed results with. The

Table 3
Atomic polarizability values (a.u.) used in DRF approach (with a-factor 2.1304)

	standard [32]	*optimized*
H	2.7927	1.1772
C	8.6959	12.8161
N	6.5565	7.1288
O	5.7494	4.0539
S	16.6984	25.7699

computed polarizability values are found in the range between 40 (Gly) and 157 (Trp), and exhibit the normal features when comparing between two similar residues; i.e. the difference between glycine and alanine (replacement of a hydrogen by a methyl group) is around 13 a.u., while the subsequent change to valine (another two hydrogens replaced by methyl groups) adds another 26 a.u. Also the increase of the polarizability of cysteine (76.9 a.u.) relative to serine (56.6 a.u.) indicates the larger atomic polarizability of sulphur relative to oxygen. This is reflected already in the atomic polarizabilities as used in the DRF approach (see Table 3), which show a much larger value for sulphur than for oxygen.

The standard DRF parameters are shown to provide amino acid polarizabilities with reasonable accuracy, if the TD-DFT values are taken as reference; the polarizability values are underestimated by some 7% (as indicated by both the average deviation and average absolute deviation). Although this is only slightly larger than the experimental uncertainty of polarizability values (which has been estimated at 2–4%) [14], it is more than twice as large as previously found (for organic molecules) [14,32]. Therefore we decided to optimize the atomic polarizability values based on the TD-DFT molecular polarizabilities for the amino acid residues (in this optimization the a-factor was kept constant at 2.1304). The resulting atomic polarizability values are given in Table 3, while the molecular values from them are given in Table 2.

By optimizing the atomic polarizabilities we obtain a very good reproduction of the TD-DFT molecular polarizabilities; the average deviation is effectively zero, while the average absolute deviation is about 1%. This is a significant reduction from the values obtained with the standard DRF parameters, and a clear improvement over the values obtained usually with the DRF approach in comparison with experimental data (which have been shown to give an average absolute deviation of some 3–4%). The same patterns as for the TD-DFT data are observed when comparing different amino acid residue, i.e. the replacement of a hydrogen by a methyl group (Gly $\rightarrow$ Ala) is accompanied by an increase of the molecular polarizability of some 13 a.u., while the subsequent double replacement of two hydrogens by two methyl groups (Ala $\rightarrow$ Val) gives an increase of the molecular polarizability of roughly 26 a.u. Also the more polarizable nature of the Cys residue in comparison with the Ser residue (featuring the difference in polarizability of the the sulphur vs. oxygen atoms) is well reflected by the optimized DRF parameters. These two molecules form a typical example for the underestimation observed by the standard DRF parameters; while the TD-DFT molecular polarizability changes from 57 to 77 a.u. between the Ser and Cys amino acid residues, the difference in atomic polarizability of the oxygen and sulphur atoms is only about 10 a.u. in the standard set. With the optimized DRF parameters, the oxygen value is slightly reduced to around 4 a.u., while the sulphur value is increases significantly to almost 26 a.u., which gives rise, among other effects, to the very good agreement between the molecular polarizabilities from the DRF optimized set and the TD-DFT calculations.

As we have a very good agreement between the TD-DFT and the optimized DRF parameter set, we can apply the latter now to find molecular polarizability values for the true amino acid residues (NHCHRCO), i.e. the amino acid residues as they are found in protein/enzyme structures without the *additional* hydrogen atoms that were needed for satisfying the valency of the atoms (replacing the backbone connecting atoms). The molecular polarizability values for the true amino acid residues NHCHRCO are given in the last column of Table 2, and show on average a reduction compared to the optimized DRF values by some 2 a.u. This may not come as a surprise as the atomic polarizability of a hydrogen atom is 1.2 a.u. within the optimized DRF parameter set, but without using the DRF approach it would have been very hard to find a value for it. This is also one of the major advantages/achievements of the DRF approach, as it provides molecular polarizabilities that are based on physical grounds, easily understood and easily obtainable. As the molecular values are obtained in less than 1 sec, the reported atomic polarizability values may be useful for the development of future polarizable force fields that include amino acid residues.

4. Conclusions

We have performed high-level Time Dependent Density Functional Theory (TD-DFT) calculations to calculate the molecular polarizability tensors of the amino acid residues, and calibrated them with the Direct Reaction Field (DRF) approach. The standard atomic polarizabilities used in the latter are shown to give an average absolute deviation of ca. 7%, which is very reasonable if the time needed is taken into consideration. However, as this deviation is more than twice as large as obtained usually with the DRF approach, we decided to optimize the atomic polarizabilities. After optimization, the average deviation is nihil, while the average absolute deviation is reduced to ca. 1%. With this very good agreement kept in mind, the true molecular polarizability values of the amino acid residues (as represented by NHCHRCO instead of HNHCHRCOH) have been obtained with the optimized DRF parameter set, and showed a reduction of approximately 2 a.u.

References

[1] S.J.A. van Gisbergen, V.P. Osinga, O.V. Gritsenko, R. van Leeuwen, J.G. Snijders and E.J. Baerends, *J. Chem. Phys* **105** (1996), 3142–3151.
[2] S.J.A. van Gisbergen, J.G. Snijders and E.J. Baerends, *J. Chem. Phys.* **103** (1995), 9347.
[3] S.J.A. van Gisbergen, J.G. Snijders and E.J. Baerends, *Chem. Phys. Lett.* **259** (1996), 599.
[4] S.J.A. van Gisbergen, J.G. Snijders and E.J. Baerends, *Comput. Phys. Commun.* **118** (1999), 119–138.
[5] B. Champagne, E.A. Perpete, S.J.A. van Gisbergen, E.J. Baerends, J.G. Snijders, C. Soubra-Ghaoui, K.A. Robins and B. Kirtman, *J. Chem. Phys.* **110** (1999), 11664–11664.
[6] B. Champagne, E.A. Perpete, D. Jacquemin, S.J.A. van Gisbergen, E.J. Baerends, C. Soubra-Ghaoui, K.A. Robins and B. Kirtman, *J. Phys. Chem. A* **104** (2000), 4755–4763.
[7] A.J. Cohen, N.C. Handy and D.J. Tozer, *Chem. Phys. Lett.* **303** (1999), 391–398.
[8] M. Gruning, O.V. Gritsenko, S.J.A. van Gisbergen and E.J. Baerends, *J. Chem. Phys.* **116** (2002), 9591–9601.
[9] P.R.T. Schipper, O.V. Gritsenko, S.J.A. van Gisbergen and E.J. Baerends, *J. Chem. Phys.* **112** (2000), 1344–1352.
[10] S.J.A. van Gisbergen, C.F. Guerra and E.J. Baerends, *J. Comput. Chem.* **21** (2000), 1511–1523.
[11] S.J.A. van Gisbergen, V.P. Osinga, O.V. Gritsenko, R. van Leeuwen, J.G. Snijders and E.J. Baerends, *J. Chem. Phys.* **105** (1996), 3142–3151.
[12] S.J.A. van Gisbergen, J.G. Snijders and E.J. Baerends, *J. Chem. Phys.* **103** (1995), 9347–9354.
[13] S.J.A. van Gisbergen, J.M. Pacheco and E.J. Baerends, *Phys. Rev. A* **6306** (2001), art. no.-063201.
[14] M. Swart, P.Th. van Duijnen and J.G. Snijders, *J. Mol. Str. (THEOCHEM)* **458** (1999), 11–17.
[15] R. van Leeuwen and E.J. Baerends, *Phys. Rev. A* **49** (1994), 2421–2431.
[16] O.V. Gritsenko, S.J.A. van Gisbergen, P.R.T. Schipper and E.J. Baerends, *Phys. Rev. A* **6201** (2000), art. no.-012507.
[17] S.J.A. van Gisbergen, P.R.T. Schipper, O.V. Gritsenko, E.J. Baerends, J.G. Snijders, B. Champagne and B. Kirtman, *Phys. Rev. Lett.* **83** (1999), 694–697.
[18] M. van Faassen, P.L. de Boeij, R. van Leeuwen, J.A. Berger and J.G. Snijders, *Phys. Rev. Lett.* **88** (2002), 186401-1-4.

[19] M. van Faassen, P.L. de Boeij, R. van Leeuwen, J.A. Berger and J.G. Snijders, *J. Chem. Phys.* **118** (2003), 1044–1053.
[20] A.H. de Vries, P.Th. van Duijnen, R.W.J. Zijlstra and M. Swart, *J. El. Spectr. Rel. Phen.* **86**(1–3) (1997), 49–56.
[21] A.H. de Vries, P.Th. van Duijnen, A.H. Juffer, J.A.C. Rullmann, J.P. Dijkman, H. Merenga and B.T. Thole, *J. Comput. Chem.* **16** (1995), 37–55; 1445–1446.
[22] B.T. Thole and P.Th. van Duijnen, *Chem. Phys.* **71** (1982), 211–220.
[23] P.Th. van Duijnen, A.H. Juffer and J.P. Dijkman, *J. Mol. Str. (THEOCHEM)* **260** (1992), 195–205.
[24] P.Th. van Duijnen, Embedding in quantum chemistry: the direct reaction field approach, in: *New challenges in computional quantum chemistry*, R. Broer, P.J.C. Aerts and P.S. Bagus, eds, Dept. of Chemical Physics and Material Science, 1994.
[25] P.Th. van Duijnen and A.H. de Vries, *Int. J. Quant. Chem.* **60** (1996), 1111–1132.
[26] L. Jensen, P.Th. van Duijnen and J.G. Snijders, *J. Chem. Phys.* **118** (2003), 514–521.
[27] G.A. Kaminski, H.A. Stern, B.J. Berne, R.A. Friesner, Y.X.X. Cao, R.B. Murphy, R.H. Zhou and T.A. Halgren, *J. Comput. Chem.* **23** (2002), 1515–1531.
[28] M. Swart, P.Th. van Duijnen and J.G. Snijders, *J. Comput. Chem.* **22** (2001), 79–88.
[29] S.J.A. van Gisbergen, C.F. Guerra and E.J. Baerends, *Abstr. Pap. Am. Chem. Soc.* **221** (2001), 65–COMP.
[30] M. Dupuis, A. Farazdel, S.P. Karma and S.A. Maluendes, HONDO: a general atomic and molecular electronic structure system, in: *MOTECC-90*, E. Clementi, ed., ESCOM, 1990, pp. 277–342.
[31] B.T. Thole, *Chem. Phys.* **59** (1981), 341–350.
[32] P.Th. van Duijnen and M. Swart, *J. Phys. Chem. A* **102** (1998), 2399–2407.
[33] G. te Velde, F.M. Bickelhaupt, E.J. Baerends, C. Fonseca Guerra, S.J.A. van Gisbergen, J.G. Snijders and T. Ziegler, *J. Comput. Chem.* **22** (2001), 931–967.
[34] E. van Lenthe, E.J. Baerends and J.G. Snijders, *J. Chem. Phys.* **99** (1993), 4597–4610.

Journal of Computational Methods in Sciences and Engineering 4 (2004) 427–438
IOS Press

Are polarizabilities useful as aromaticity indices? Tests on azines, azoles, oxazoles and thiazoles

Robert J. Doerksen**, Valerie J. Steeves and Ajit J. Thakkar*
Department of Chemistry, University of New Brunswick, Fredericton, NB E3B 6E2, Canada

Received 14 April 2003
Accepted 28 June 2003

Abstract. Recent interest in quantitative aromaticity indices has focussed on structural, magnetic, and energetic criteria. In this work, aromaticity indices based on polarizabilities are compared with indices based upon bond orders for benzene and 12 azines, pyrrole and 9 azoles, furan and 9 oxazoles, and thiophene and 8 thiazoles. The best polarizability-based index of aromaticity we find is the polarizability anisotropy of the π-electrons. However, none of the indices constructed from polarizabilities seem to be entirely suitable as measures of aromaticity. The comparison of the structural and polarizability scales enables us to formulate three critical tests that can be used to eliminate quickly unsuitable aromaticity scales for this set of heterocycles.

Keywords: Aromaticity indices, heterocyclic rings, polarizabilities, MP2, uncoupled Hartree-Fock

1. Introduction

Aromaticity is both a theoretical and practical concept in organic chemistry. Since its introduction in the mid-19th century [1–3], it has grown tremendously in importance as attested to by the voluminous literature on aromaticity that has, in turn, spawned several monographs [4–8] and too many review articles to list here. The most recent reviews [9–28] on aromaticity offer a variety of different perspectives of this dynamic field.

There has been much recent work on quantifying aromaticity. Quantitative measures would enable one to determine the relative aromaticity of any particular set of molecules, and perhaps to gain insight into the fundamental meaning of the term. However, aromaticity is to some extent ill-defined, and this naturally leads to difficulties when attempts are made to quantify it. The many quantitative aromaticity indices that have been used to date often disagree with each other even in their prediction of relative aromaticity. Nevertheless, the indices thus far devised have been able to provide some insight into various practical chemical problems such as the planarity of cyclopalladated rings [29], and the choice of suitable heterocyclic bridges for tuning the hyperpolarizability of donor-acceptor chromophores [30].

*Corresponding author. E-mail: ajit@unb.ca.
**Current address: Center for Molecular Modeling, Department of Chemistry and Laboratory for Research on the Structure of Matter, University of Pennsylvania, Philadelphia, PA 19104-6323.

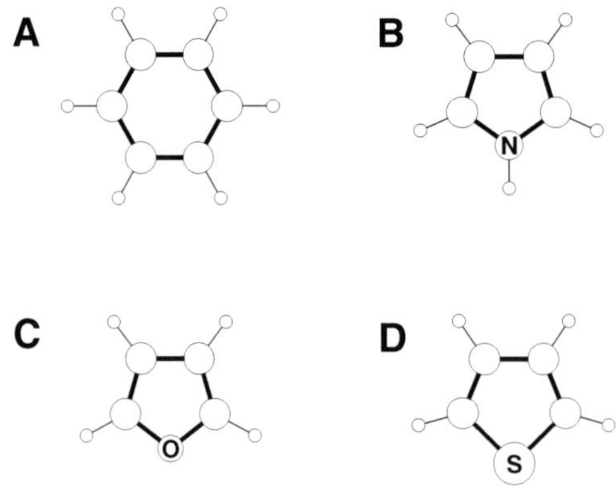

Fig. 1. Benzene (A), pyrrole (B), furan (C) and thiophene (D) shown to scale.

Various criteria of aromaticity have been proposed, and corresponding quantitative indices have developed around them. The main criteria have been categorized as energetic, geometric, and magnetic [8, 31,32,22,23,33–40,9,41]. Electronic criteria of aromaticity have also been considered [42–45,34,46–56, 24,57–59]. There is some evidence both for [34–37,40] and against [38,39] the idea that the different criteria describe different types of aromaticity. A recent paper written jointly by proponents [60] of many of the aromaticity criteria comes to the conclusion that when a large set of compounds is considered together there are statistically significant correlations among the various criteria, and aromaticity is a one-dimensional phenomenon in this sense. The same study finds that when comparisons are made for restricted sets of compounds, as in practical applications, the quality of the correlations deteriorates noticeably. In this sense, aromaticity is regarded as statistically multidimensional [60].

Katritzky et al. [34] think that the polarizability captures "a mixture of magnetic and classical aromaticity". But there have been only a few attempts to quantify aromaticity in terms of polarizability [42, 34,46,48–52,55]. Katritzky et al. [35] stated that such attempts were hampered by the paucity of accurate polarizability data. By contrast, indices based on the molecular geometry or related quantities have been widely used as indicators of aromaticity [33,37,61–72].

Recent advances in computational methods and hardware have made polarizabilities available for many aromatic molecules. In this work, we use a consistent set of ab initio calculations of polarizabilities to generate aromaticity scales for 42 heteroaromatic molecules. We treat benzene, pyrrole, furan, and thiophene (see Fig. 1) because they occupy a fundamental role in aromatic heterocyclic chemistry [73]. Further we treat the molecules obtained from them by aza-substitution (replacement of C-H by a nitrogen atom). Specifically, we focus on benzene and 12 azines [74], pyrrole and 9 azoles [49], furan and 9 oxazoles [51], and thiophene and 8 thiazoles [75]. The polarizability-based indices of aromaticity are compared with several structural indices of aromaticity. The comparison of the structural and polarizability scales leads us to formulate some tests for the rapid elimination of aromaticity scales unsuitable for this set of heterocycles.

Various aromaticity indices are defined in Section 2, the data we use is described in Section 3, and the polarizability and structural indices for 42 heteroaromatic 5- and 6-membered rings are compared in Section 4. Atomic units are used throughout.

2. Aromaticity indices

2.1. Indices based on bond orders

Early geometry-based aromaticity indices were functions of bond lengths [61]. Since bond lengths vary with atom type in heterocyclic compounds, it is common to compensate for this by using bond orders. Several aromaticity indices based on bond orders have been proposed [33,62,64–70]. The indices all adopt the convention that a lower value means the molecule is less aromatic.

Jug [64] introduced the ring current index given by the lowest bond order of all ring bonds, which is supposedly a measure of the magnitude of a ring current. We do not consider the Jug index further in this article.

Let n be the number of ring bonds (usually $n = 5$ or 6) in the cyclic molecule under consideration, and N_i be the bond order of the ith ring bond. The indices we consider all include some measure of the differences among the $\{N_i\}_1^n$ with increased equalization of bond orders implying increased aromaticity. Bird's index [65,66] uses a measure of variance V_n given by

$$V_n = \frac{100}{N_\mathrm{m}} \left[\frac{1}{n} \sum_{i=1}^{n} (N_\mathrm{m} - N_i)^2 \right]^{1/2} \tag{1}$$

in which N_m is the mean ring bond order

$$N_\mathrm{m} = \frac{1}{n} \sum_{i=1}^{n} N_i. \tag{2}$$

The variance V_n is then shifted and scaled to yield an index that is expected to have a $[0, 100]$ range

$$I_n = 100\left(1 - V_n/V_n^K\right) \tag{3}$$

in which V_n^K refers to the variance in the corresponding Kekulé structure [65]. Later, Bird [66] generalized the index so that it applies to both six- and five-membered rings by defining

$$I_A = I_6 = 1.235 I_5 \tag{4}$$

where 1.235 is the ratio of the Hückel delocalization energy of the cyclopentadienyl anion to that of benzene. We refer to I_A as Bird's index of aromaticity.

The Pozharskii index [33] is the average of all possible differences of the ring bond orders defined, in its unscaled form, by

$$P_u = \frac{2}{n(n-1)} \sum_{i<j}^{n} |N_i - N_j| \tag{5}$$

in which $n(n-1)/2$ is simply the number of distinct ring bond pairs. The index is then scaled and shifted

$$P = 100\,(1 - P_u/0.49) \tag{6}$$

where 0.49 is the P_u value for the cyclopentadiene anion in the Hückel model. In the aromaticity literature, P_u and P are often referred to as POZ and POZPC respectively.

There is the bond-order version of Krygowski's HOMA or harmonic oscillator model of aromaticity [62]. It is placed on a percentage scale and written as

$$H_p = 100 - \frac{360}{n}\sum_{i=1}^{n}(N_\mathrm{B} - N_i)^2 \tag{7}$$

where N_B is the C-C bond order in benzene. In the original work, Hückel π-bond orders were used so that $N_\mathrm{B} = 2/3$. We use a total bond order for benzene, $N_\mathrm{B} = 1.44$, calculated at the Hartree-Fock level. The aromaticity literature refers to H_p as HOMA$_p$. Note that H_p is quadratic in bond order differences, and that the reference bond order is fixed in H_p whereas it varies with molecule in the Bird index I_A. More recently, Krygowski has developed a new index [67], called HOMA$_d$ or simply HOMA, that uses bond lengths instead of bond orders. The only structural indices of aromaticity that we consider in this work are those based on bond orders.

2.2. *Indices based on polarizabilities*

The static dipole polarizability is a symmetric second-rank tensor with six independent components in general. Aromaticity indices should be based upon properties of this tensor that are invariant to the choice of the coordinate system. The eigenvalues of the polarizability tensor, or linear combinations thereof, are suitable quantities. Number the eigenvalues of the polarizability tensor in increasing order: $\alpha_1 \leqslant \alpha_2 \leqslant \alpha_3$. For the planar heteroaromatic molecules under consideration, the principal axis of lowest polarizability, corresponding to the smallest eigenvalue α_1, is perpendicular to the molecular plane. Thus we can define the perpendicular polarizability as

$$\alpha_\perp = \alpha_1 \tag{8}$$

and the in-plane polarizability as

$$\alpha_\parallel = (\alpha_2 + \alpha_3)/2. \tag{9}$$

The mean polarizability is the trace of the tensor and can be written as

$$\bar{\alpha} = (\alpha_1 + \alpha_2 + \alpha_3)/3 = (\alpha_\perp + 2\alpha_\parallel)/3, \tag{10}$$

and the polarizability anisotropy as

$$\Delta\alpha = \alpha_\parallel - \alpha_\perp. \tag{11}$$

Each of the quantities $\alpha_\parallel$, $\alpha_\perp$, $\bar{\alpha}$ and $\Delta\alpha$ has been suggested as an index of aromaticity [42,46,48–52]. Four other polarizability indices use only the contribution of the π-electrons to these quantities. The latter indices are less well-defined because the decomposition of the polarizability into σ and π components is not well defined except at relatively crude levels of electronic structure theory.

3. Geometry, polarizability and bond order data

Ab initio data, much of it previously published, was used for all 42 molecules. All the geometries were computed with second-order Møller-Plesset (MP2) perturbation theory [76] using a Pople-type split-valence plus polarization basis set [77]. In particular, MP2/6-31G(d) geometries were used for pyrrole

Table 1
Uncoupled Hartree-Fock (UCHF/C1) polarizabilities for thiophene and the thiazoles at MP2/6-31G(d,p) geometries

Molecule	$\bar{\alpha}$	$\alpha_{\perp}$	$\alpha_{\parallel}$	$\bar{\alpha}^{\pi}$	$\alpha_{\perp}^{\pi}$	$\alpha_{\parallel}^{\pi}$	$\bar{\alpha}^{\pi}/\bar{\alpha}$	$\alpha_{\perp}^{\pi}/\alpha_{\perp}$	$\alpha_{\parallel}^{\pi}/\alpha_{\parallel}$
Thiophene	64.08	47.49	72.37	34.88	30.47	37.08	0.544	0.642	0.512
1,2-Thiazole	59.23	42.77	67.45	31.23	26.39	33.65	0.527	0.617	0.499
1,3-Thiazole	58.75	43.32	66.46	30.56	26.74	32.47	0.520	0.617	0.488
1,2,3-Thiadiazole	57.15	39.72	65.86	29.33	23.66	32.16	0.513	0.596	0.488
1,2,4-Thiadiazole	53.39	38.71	60.73	26.55	22.79	28.43	0.497	0.589	0.468
1,2,5-Thiadiazole	55.42	38.74	63.76	29.19	23.04	32.26	0.527	0.595	0.506
1,3,4-Thiadiazole	54.39	39.46	61.85	26.78	23.35	28.49	0.492	0.592	0.461
1,2,3,4-Thiatriazole	52.03	35.88	60.10	25.14	20.37	27.52	0.483	0.568	0.458
1,2,3,5-Thiatriazole	52.15	35.61	60.42	26.48	20.31	29.56	0.508	0.570	0.489

and the azoles [49], furan and the oxazoles [51], and benzene and the azines [74]. MP2/6-31G(d,p) geometries were used for thiophene and the thiazoles [75]. The polarizabilities were obtained by finite-field MP2 calculations using the custom [6s5p2d/5s3p2d/4s2p] basis sets of contracted Gaussian-type functions, C and C1, described in earlier work [49,51,75]. These polarizabilities have been published in the same papers as the geometries [49,51,74,75].

A σ-π separation of the polarizability is not well-defined at the coupled Hartree-Fock or higher levels of electronic structure theory. However, such a separation is possible with the uncoupled Hartree-Fock (UCHF) method [78–82] in which the field is treated by perturbation theory using the field-free Hartree-Fock wave function as the unperturbed wave function. In the UCHF theory, the components of the polarizability for a closed-shell system can be expressed as a sum of contributions from the occupied orbitals as follows:

$$\alpha_{ab} = \sum_{j \in \mathrm{occ}} \alpha_{ab,j} \quad \text{for} \quad a, b \in \{x, y, z\} \tag{12}$$

in which $\alpha_{ab,j}$, the contribution of the jth doubly-occupied orbital to α_{ab}, is given by

$$\alpha_{ab,j} = 4 \sum_{s \in \mathrm{vir}} \frac{\langle j|r_a|s\rangle \langle s|r_b|j\rangle}{\varepsilon_s - \varepsilon_j} \tag{13}$$

where r_a is the a component of the electronic position vector, and the summation extends over the virtual orbitals. The π-electron contribution is then simply the sum of the $\alpha_{ab,j}$ over the occupied π-orbitals. Since UCHF polarizabilities can be in error by as much as 25% [49,51,74], we use the UCHF method only to obtain the fraction of the polarizability arising from the π-electrons. Our values of the π-electron contributions to the polarizabilities are obtained by scaling our MP2 polarizabilities by the UCHF π-fractions in the following manner:

$$\alpha_a^{\pi} \approx (\alpha_a)_{\mathrm{MP2}} \, (\alpha_a^{\pi}/\alpha_a)_{\mathrm{UCHF}} \quad \text{for} \quad a = \parallel, \perp. \tag{14}$$

The π contributions to the polarizabilities have been reported previously for benzene and the azines [74], pyrrole and the azoles [49], and furan and the oxazoles [51]. The values for thiophene and the thiazoles were computed especially for this work and are listed in Table 1.

In the aromaticity literature, it is common to estimate bond orders from bond lengths using an empirical relationship due to Gordy [83,84]. However, we have shown [85] that the subtlety of a bond order cannot be captured by a smooth function of the bond length. Hence, we calculated Mulliken-Mayer bond orders [86–88,85] at the Hartree-Fock level using basis sets consisting of a [6s4p/4s2p/2s] double-zeta substrate [89,90] plus [1d/1d/1p] polarization functions [91,92]. The order or index for the bond between atoms A and B is simply given by

$$I_{\mathrm{AB}} = \sum_{k\in \mathrm{A},\ell\in \mathrm{B}} (\mathcal{PS})_{k\ell}(\mathcal{PS})_{\ell k} \tag{15}$$

in which $\mathcal{P}$ and $\mathcal{S}$ are the one-electron density and overlap matrices respectively, and the summations are restricted to the basis functions centered on the atoms A and B.

4. Polarizability and aromaticity in 42 heterocycles

Table 2 collects our computed values for the Pozharskii index P, Bird's index I_A, the H_p or HOMA$_p$ index, the parallel polarizability $\alpha_{\parallel}$ and polarizability anisotropy $\Delta\alpha$ and their π-electron counterparts $\alpha_{\parallel}^{\pi}$ and $\Delta\alpha^{\pi}$. Both $\bar{\alpha}$ and $\alpha_{\perp}$ can be obtained from $\alpha_{\parallel}$ and $\Delta\alpha$; we have made the unusual choice of the latter because we anticipated that they are the most useful for aromaticity indices. Fuller information about the polarizability tensors of these molecules is available elsewhere [49,51,74,75].

Of the eight different polarizability scales mentioned, the shortcomings of seven of them can be illustrated with the help of two simple observations. As noted previously by us [49,51,74], each aza-substitution decreases $\bar{\alpha}$, $\alpha_{\parallel}$, $\alpha_{\perp}$ and $\Delta\alpha$ in a systematic, step-wise manner. The same trend is followed by $\bar{\alpha}^{\pi}$, $\alpha_{\parallel}^{\pi}$ and $\alpha_{\perp}^{\pi}$, except that $\alpha_{\parallel}^{\pi}$ is lower for s-triazine than it is for two tetrazines. Thus, if used as indices of aromaticity, these seven quantities predict aromaticity to decrease as the amount of aza-substitution increases. However, this differs from the predictions of the Pozharskii, Bird and HOMA$_p$ indices. For example, Table 2 shows that these three structural indices all predict 1,2,3-triazole and 1,2,3,5-tetrazole to be more aromatic than pyrrole, and 1,2,3-thiadiazole and 1,2,3,5-thiatriazole to be more aromatic than thiophene. One may disagree with any specific prediction of the structural indices, but it is reasonable to accept that aromaticity does not decrease smoothly with increasing aza-substitution. After all, the location of the aza-substitution should have an important effect on the aromaticity. The only polarizability index, of the eight we examined, that does not decrease smoothly with increasing aza-substitution is $\Delta\alpha^{\pi}$. For example, Table 2 shows that $\Delta\alpha^{\pi}$ exactly matches the predictions of the structural indices for 1,2,3-triazole, 1,2,3,5-tetrazole, 1,2,3-thiadiazole, and 1,2,3,5-thiatriazole discussed above.

The second point [49,51,74] to note is that the values of $\alpha_{\parallel}$, $\alpha_{\perp}$, $\bar{\alpha}$ and $\Delta\alpha$ show little variation among a set of isomers like 1,2,3-triazine, 1,2,4-triazine and 1,3,5-triazine. However, the structural indices show that the aromaticity of a set of isomers can vary widely. For example, Table 2 shows that the Pozharskii index is 84, 65 and 100 respectively for these triazines. Wide variation of aromaticity in a set of isomers is difficult to obtain if we use any one of $\alpha_{\parallel}$, $\alpha_{\perp}$, $\bar{\alpha}$ and $\Delta\alpha$ as an aromaticity index. Table 2 shows that the π parts of these polarizabilities have more variation among a set of isomers than the sum of the σ and π contributions. The best polarizability index from this perspective is $\Delta\alpha^{\pi}$ which has values of 15, 16 and 14 for these triazines.

Given the findings of the previous two paragraphs, we thought it more promising to take a detailed look at $\Delta\alpha^{\pi}$ than any of the other seven polarizability indices discussed above. There are two factors that cause us trepidation in doing so. One is that Fowler and Steiner [52] considered but discarded $\Delta\alpha^{\pi}$ as an aromaticity index because of the strange result that $\Delta\alpha^{\pi} \approx 0$ for the cyclopentadienyl anion ($C_5H_5^-$). The other is Occam's (or Ockham's) razor [93,94] which makes us ask whether we should use an ill-defined concept like σ-π separation of polarizabilities to examine another ill-defined concept like

Table 2
Aromaticity scales for 42 heterocycles

Molecule	P	I_A	H_p	$\alpha_{\parallel}$	$\Delta\alpha$	$\alpha_{\parallel}^{\pi}$	$\Delta\alpha^{\pi}$
Benzene	100	100	100	81.33	36.16	45.62	17.63
Pyridine	85.9	88.8	98.8	75.39	34.31	41.33	17.01
1,2-Diazine	63.5	71.7	93.4	69.58	31.77	38.32	16.85
1,3-Diazine	88.0	90.3	98.7	69.26	31.82	36.53	15.58
1,4-Diazine	84.8	86.6	98.0	69.93	32.59	38.35	17.19
1,2,3-Triazine	83.7	86.5	98.3	64.29	29.31	34.37	15.41
1,2,4-Triazine	65.3	73.3	93.9	64.03	29.43	34.22	15.65
1,3,5-Triazine	100	100	98.7	62.89	28.69	31.37	13.62
1,2,3,4-Tetrazine	67.9	74.0	94.2	58.99	26.65	31.41	14.71
1,2,3,5-Tetrazine	84.8	86.7	98.5	58.63	26.40	30.35	14.09
1,2,4,5-Tetrazine	79.3	81.3	97.1	58.70	26.20	31.72	15.05
Pentazine	86.9	88.7	98.3	53.95	23.47	28.55	13.66
Hexazine	100	100	97.1	49.16	20.18	26.18	12.63
Pyrrole	49.8	76.4	86.9	62.52	22.84	33.26	6.88
Pyrazole	43.7	73.7	85.3	56.50	21.75	29.02	7.19
Imidazole	42.0	71.6	84.1	56.47	21.96	28.84	7.13
1,2,3-Triazole	61.6	88.4	91.9	51.36	20.56	26.16	7.67
1,2,4-Triazole	38.4	69.9	83.0	50.54	20.37	24.58	6.97
1,2,5-Triazole	49.8	77.1	87.2	50.89	20.37	25.97	7.76
1,3,4-Triazole	27.3	52.9	70.9	50.71	20.40	24.44	6.63
1,2,3,4-Tetrazole	46.1	73.2	84.9	45.88	18.72	22.12	6.98
1,2,3,5-Tetrazole	59.2	87.1	91.9	45.72	18.70	22.76	7.63
Pentazole	64.5	89.7	91.9	41.18	16.86	20.07	7.21
Furan	27.8	53.7	71.9	55.22	20.13	28.23	5.73
1,2-Oxazole	12.7	42.6	62.6	50.34	19.49	24.31	5.86
1,3-Oxazole	17.1	47.1	66.2	49.45	18.92	24.19	5.79
1,2,3-Oxadiazole	22.4	51.2	69.4	45.95	18.39	22.18	6.38
1,2,4-Oxadiazole	6.5	38.5	58.7	44.62	17.80	20.10	5.39
1,2,5-Oxadiazole	8.2	32.9	53.8	45.55	18.37	21.57	6.32
1,3,4-Oxadiazole	5.3	30.4	50.0	43.87	17.13	20.07	5.21
1,2,3,4-Oxatriazole	8.6	37.1	57.8	40.12	15.99	17.72	5.15
1,2,3,5-Oxatriazole	13.9	41.4	61.5	40.85	16.57	19.06	6.21
Oxatetrazole	15.5	37.3	57.6	36.01	14.16	16.05	5.24
Thiophene	48.6	73.1	84.9	73.27	28.53	37.54	8.84
1,2-Thiazole	42.9	69.9	83.0	67.79	27.68	33.81	9.07
1,3-Thiazole	40.0	68.6	82.3	67.23	26.88	32.84	7.93
1,2,3-Thiadiazole	58.0	85.3	90.5	63.36	26.59	30.94	9.03
1,2,4-Thiadiazole	38.4	68.1	82.0	61.57	25.34	28.82	7.49
1,2,5-Thiadiazole	47.3	72.0	84.3	62.12	26.16	31.43	10.05
1,3,4-Thiadiazole	20.8	46.6	65.8	61.67	24.94	28.41	6.68
1,2,3,4-Thiatriazole	39.6	66.3	80.5	57.68	24.25	26.41	7.43
1,2,3,5-Thiatriazole	58.8	86.6	91.5	57.40	24.17	28.08	9.13

aromaticity! However, taking heart from the reminder [95,96] that rigid adherence to Occam's razor is a path to sterility, we press on with $\Delta\alpha^{\pi}$.

Before we proceed with a more detailed comparison of $\Delta\alpha^{\pi}$ and the structural indices, we compare the structural indices among themselves to see what level of correlation can be expected. Figure 2 shows that there is a reasonably good correlation between the Pozharskii index P and the HOMA index H_p. The correlation is not linear and has a distinct quadratic quality perhaps reflecting the quadratic dependence of H_p on bond order differences as opposed to the linear dependence of P on bond order differences.

Figure 3 shows a coarse linear relationship between P and the Bird index I_A with a correlation coefficient $R^2 = 0.92$. However, a closer examination of Fig. 3 shows that there are, in fact, two linear

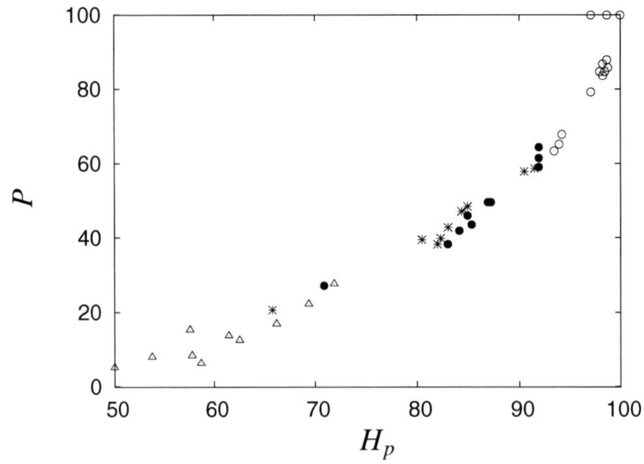

Fig. 2. Correlation between the Pozharskii index P and the H_p index. Benzene and the azines (○), pyrrole and the azoles (●), furan and the oxazoles (△), thiophene and the thiazoles (∗).

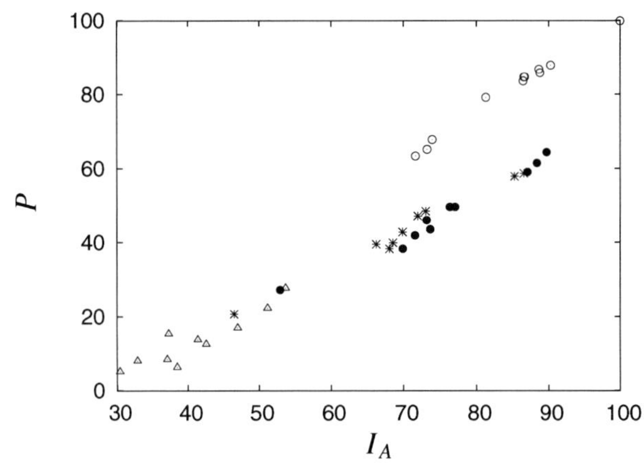

Fig. 3. Correlation between the Pozharskii index P and the Bird I_A index. Benzene and the azines (○), pyrrole and the azoles (●), furan and the oxazoles (△), thiophene and the thiazoles (∗).

relationships, one for each ring size $r = 6$ and $r = 5$. A correlation coefficient of $R^2 = 0.995$ is obtained for the equation

$$P = -29.6 + 24.6\delta_{6r} + 1.035 I_A \tag{16}$$

in which δ_{6r} is a Kronecker delta. Perhaps this difference between 5- and 6-rings is due to the manner in which Bird unified his indices for rings of differing size. The discussion above establishes that there are reasonable correlations among the P, I_A and H_p indices. We have made further comparisons among these and other structural indices of aromaticity elsewhere [97–99].

We now turn to correlations between the structural indices and $\Delta\alpha^\pi$. Figure 4 shows that there is a correlation between the Pozharskii index P and the polarizability index $\Delta\alpha^\pi$ but that it has much more scatter than the ones between P and the other structural indices H_p and I_A. The correlation improves if we treat each of the four families of molecules separately. The correlation between $\Delta\alpha^\pi$ and the other structural indices is no better. It is clear that the correlation between the polarizability index $\Delta\alpha^\pi$ and

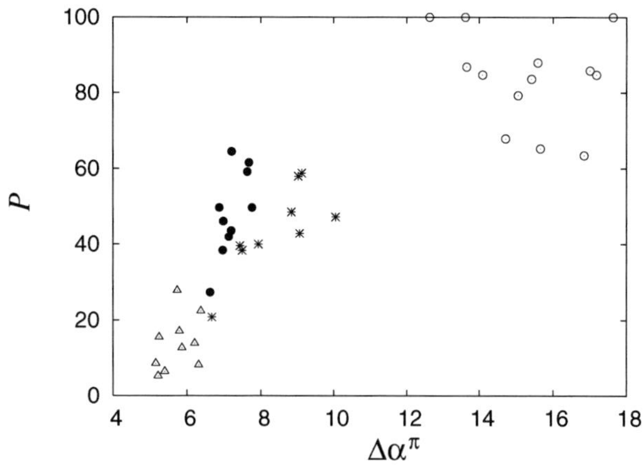

Fig. 4. Correlation between the Pozharskii index P and the $\Delta\alpha^\pi$ index. Benzene and the azines (○), pyrrole and the azoles (•), furan and the oxazoles (△), thiophene and the thiazoles (∗).

the structural indices is worse than the correlations among the latter.

The structural aromaticity scales P, I_A and H_p measure equalization of bond orders, which should be a measure of electron delocalization as it is found in molecules in the ground state and at their equilibrium geometry. The polarizability scale $\Delta\alpha^\pi$ ostensibly measures delocalization by quantifying the greater ease with which the π-electrons can be polarized in the molecular plane as opposed to perpendicular to it. It is not evident that the structural and polarizability scales measure the same thing. Thus, the relatively poor correlation seen in Fig. 4 could either be another manifestation of the multidimensional nature of aromaticity [60], or an indication that $\Delta\alpha^\pi$ has shortcomings as an index of aromaticity.

Another blow against the suitability of $\Delta\alpha^\pi$ is struck by consideration of the effect of changing the heteroatom on aromaticity. Most indices predict that aromaticity decreases in the sequence pyrrole, thiophene and furan. This is supported by the structural indices P, I_A and H_p based on bond orders as reported in this work. Schleyer et al. [38,31] obtained the same ordering from nucleus-independent chemical shifts (NICS), magnetic susceptibility exaltations (Λ), aromatic stabilization energies (ASE), and the Julg index [61] based on bond lengths. The same ordering was also predicted by the HOMA$_d$ index [70], and the Hess-Schaad resonance energy per electron (REPE) [8, p. 20]. However, our polarizability indices predict that aromaticity decreases in the sequence thiophene, pyrrole and furan. Only experimental Λ values quoted by Pozharskii [33] and Bird [39], and Bird's hardness [53] agree with the polarizability indices.

We are led to conclude that none of the polarizability-based indices of aromaticity considered in this paper are entirely suitable. In hindsight, this may be rationalized by observing that aromaticity, as measured by stabilization energies, is a probe of the electron distribution close to the nuclei whereas the polarizability is a probe of the outer regions of the electron density.

This effort to test aromaticity indices has led us to formulate a set of stringent tests for indices proposed in the future.

1. A good index of aromaticity should predict that aromaticity can vary with both the degree and location of aza-substitution in families of heterocycles like the azines, azoles, oxazoles, and thiazoles.
2. A good index of aromaticity should be able to predict differing aromaticity for isomers like the three triazines.
3. A good index should predict that aromaticity decreases in the sequence pyrrole, thiophene and furan.

5. Concluding remarks

Aromaticity indices based on polarizabilities were compared with the Bird I_A, Pozharskii P, and H_p structural indices based upon bond orders for benzene and 12 azines, pyrrole and 9 azoles, furan and 9 oxazoles, and thiophene and 8 thiazoles. The best of the polarizability indices we considered was $\Delta\alpha^{\pi}$. However, it too has its shortcomings. We conclude that none of the polarizability-based indices of aromaticity considered in this paper are entirely suitable. The comparison of the structural and polarizability scales enabled us to formulate three simple, critical tests that can be used to eliminate quickly unsuitable aromaticity scales for this set of heterocycles.

Acknowledgments

One of us (AJT) thanks Saul Wolfe for kindling his interest in aromaticity, and for encouraging him to pursue a scientific career more than thirty years ago. We thank Noureddin El-Bakali Kassimi for useful discussions and for the coordinates of the structures of thiophene and the thiazoles reported in Ref. [75]. This work was supported in part by the Natural Sciences and Engineering Research Council of Canada.

References

[1] A. Kekulé, *Bull. Soc. Chim. Paris* **3** (1865), 98–110.
[2] A. Kekulé, *Liebigs Ann. Chem.* **137** (1866), 129–196.
[3] E. Erlenmeyer, *Liebigs Ann. Chem.* **137** (1866), 327–359.
[4] G.M. Badger, *Aromatic Character and Aromaticity*, Cambridge University Press, Cambridge, 1969.
[5] E. Clar, *The Aromatic Sextet*, Wiley, London, 1972.
[6] D. Lewis and D. Peters, *Facts and Theories of Aromaticity*, Macmillan, London, 1975.
[7] P.J. Garratt, *Aromaticity*, Wiley, New York, 1986.
[8] V.I. Minkin, M.N. Glukhovtsev and B.Y. Simkin, *Aromaticity and Antiaromaticity: Electronic and Structural Aspects*, Wiley, New York, 1994.
[9] T.M. Krygowski, M.K. Cyrański, Z. Czarnocki, G. Häfelinger and A.R. Katritzky, *Tetrahedron* **56** (2000), 1783–1796.
[10] P.v.R. Schleyer, *Chem. Rev.* **101** (2001), 1115–1118.
[11] R.B. King, *Chem. Rev.* **101** (2001), 1119–1152.
[12] M. Bühl and A. Hirsch, *Chem. Rev.* **101** (2001), 1153–1184.
[13] R.V. Williams, *Chem. Rev.* **101** (2001), 1185–1204.
[14] J.R. Bleeke, *Chem. Rev.* **101** (2001), 1205–1228.
[15] L. Nyulászi, *Chem. Rev.* **101** (2001), 1229–1246.
[16] V.I. Minkin and R.M. Minyaev, *Chem. Rev.* **101** (2001), 1247–1266.
[17] M.D. Watson, A. Fechtenkötter and K. Müllen, *Chem. Rev.* **101** (2001), 1267–1300.
[18] R.H. Mitchell, *Chem. Rev.* **101** (2001), 1301–1316.
[19] K.B. Wiberg, *Chem. Rev.* **101** (2001), 1317–1332.
[20] A.D. Allen and T.T. Tidwell, *Chem. Rev.* **101** (2001), 1333–1348.
[21] J.A.N.F. Gomes and R.B. Mallion, *Chem. Rev.* **101** (2001), 1349–1384.
[22] T.M. Krygowski and M.K. Cyrański, *Chem. Rev.* **101** (2001), 1385–1420.
[23] A.R. Katritzky, K. Jug and D.C. Oniciu, *Chem. Rev.* **101** (2001), 1421–1450.
[24] F. De Proft and P. Geerlings, *Chem. Rev.* **101** (2001), 1451–1464.
[25] L.J. Schaad and B.A. Hess, *Chem. Rev.* **101** (2001), 1465–1476.
[26] K. Jug, P.C. Hiberty and S. Shaik, *Chem. Rev.* **101** (2001), 1477–1500.
[27] S. Shaik, A. Shurki, D. Danovich and P.C. Hiberty, *Chem. Rev.* **101** (2001), 1501–1540.
[28] S.W. Slayden and J.F. Liebman, *Chem. Rev.* **101** (2001), 1541–1566.
[29] A. Crispini and M. Ghedini, *J. Chem Soc., Dalton Trans.* **1997** (1997), 75–80.
[30] I.D.L. Albert, T.J. Marks and M.A. Ratner, *J. Am. Chem. Soc.* **119** (1997), 6575–6582.
[31] P.v.R. Schleyer and H. Jiao, *Pure Appl. Chem.* **68** (1996), 209–218.

[32] M.N. Glukhovtsev, *J. Chem. Ed.* **74** (1997), 132–136.
[33] A.F. Pozharskii, *Chem. Heterocycl. Compounds* **21** (1985), 717–749.
[34] A.R. Katritzky, P. Barczynski, G. Musumarra, D. Pisano and M. Szafran, *J. Am. Chem. Soc.* **111** (1989), 7–15.
[35] A.R. Katritzky, V. Feygelman, G. Musumarra, P. Barczynski and M. Szafran, *J. Prakt. Chem.* **332** (1990), 853–869.
[36] A.R. Katritzky and P. Barczynski, *J. Prakt. Chem.* **332** (1990), 885–897.
[37] K. Jug and A.M. Köster, *J. Phys. Org. Chem.* **4** (1991), 163–169.
[38] P.v.R. Schleyer, P.K. Freeman, H. Jiao and B. Goldfuss, *Angew. Chem. Int. Ed. Engl.* **34** (1995), 337–340.
[39] C.W. Bird, *Tetrahedron* **52** (1996), 9945–9952.
[40] G. Subramanian, P.v.R. Schleyer and H. Jiao, *Angew. Chem. Int. Ed. Engl.* **35** (1996), 2638–2641.
[41] M.K. Cyrański, P.v.R. Schleyer, T.M. Krygowski, H. Jiao and G. Hohlneicher, *Tetrahedron* **59** (2003), 1657–1665.
[42] S.B. Bulgarevich, T.A. Yusman and O.A. Osipov, *Zh. Obshch. Khim.* **54** (1984), 1603–1607.
[43] A. Moyano and J.C. Paniagua, *J. Org. Chem.* **51** (1986), 2250–2257.
[44] Z. Zhou, R.G. Parr and J.F. Garst, *Tetrahedron Lett.* **29** (1988), 4843–4846.
[45] Z. Zhou and R.G. Parr, *J. Am. Chem. Soc.* **111** (1989), 7371–7379.
[46] P. Lazzeretti and J.A. Tossell, *J. Mol. Struct.* (*Theochem*) **236** (1991), 403–410.
[47] A. Moyano and J.C. Paniagua, *J. Org. Chem.* **56** (1991), 1858–1866.
[48] E.F. Archibong and A.J. Thakkar, *Mol. Phys.* **81** (1994), 557–567.
[49] N.E.-B. Kassimi, R.J. Doerksen and A.J. Thakkar, *J. Phys. Chem.* **99** (1995), 12790–12796.
[50] A. Hinchliffe and H.J. Soscún M. *J. Mol. Struct.* (*Theochem*) **331** (1995), 109–125.
[51] N.E.-B. Kassimi, R.J. Doerksen and A.J. Thakkar, *J. Phys. Chem.* **100** (1996), 8752–8757.
[52] P.W. Fowler and E. Steiner, *J. Phys. Chem. A* **101** (1997), 1409–1413.
[53] C.W. Bird, *Tetrahedron* **53** (1997), 13111–13118.
[54] S.T. Howard and T.M. Krygowski, *Can. J. Chem.* **75** (1997), 1174–1181.
[55] S. Millefiori and A. Alparone, *J. Mol. Struct.* (*Theochem*) **431** (1998), 59–78.
[56] G.P. Bean, *J. Org. Chem.* **63** (1998), 2497–2506.
[57] N. Sadlej-Sosnowska, *J. Org. Chem.* **66** (2001), 8737–8743.
[58] J. Poater, X. Fradera, M. Duran and M. Solà, *Chem. Eur. J.* **9** (2003), 400–406.
[59] J. Poater, X. Fradera, M. Duran and M. Solà, *Chem. Eur. J.* **9** (2003), 1113–1122.
[60] M.K. Cyrański, T.M. Krygowski, A.R. Katritzky and P.v.R. Schleyer, *J. Org. Chem.* **67** (2002), 1333–1338.
[61] A. Julg and P. François, *Theor. Chim. Acta.* **8** (1967), 249–259.
[62] J. Kruszewski and T.M. Krygowski, *Tetrahedron Lett.* **13** (1972), 3839–3842.
[63] T.M. Krygowski and T. Wieckowski, *Croat. Chim. Acta* **54** (1981), 193–202.
[64] K. Jug, *J. Org. Chem.* **48** (1983), 1344–1348.
[65] C.W. Bird, *Tetrahedron* **41** (1985), 1409–1414.
[66] C.W. Bird, *Tetrahedron* **48** (1992), 335–340.
[67] T.M. Krygowski, *J. Chem. Inf. Comput. Sci.* **33** (1993), 70–78.
[68] T.M. Krygowski, A. Ciesielski, C.W. Bird and A. Kotschy, *J. Chem. Inf. Comput. Sci.* **35** (1995), 203–210.
[69] L. Nyulászi, P. Várnai and T. Vezprémi, *J. Mol. Struct.* (*Theochem*) **358** (1995), 55–61.
[70] T.M. Krygowski and M. Cyrański, *Tetrahedron* **52** (1996), 10255–10264.
[71] C.W. Bird, *Tetrahedron* **53** (1997), 3319–3324.
[72] S.I. Kotelevskii and O.V. Prezhdo, *Tetrahedron* **57** (2001), 5715–5729.
[73] D.I. Davies, *Aromatic Heterocyclic Chemistry*, Oxford University Press, New York, 1991.
[74] R.J. Doerksen and A.J. Thakkar, *Int. J. Quantum Chem.* **60** (1996), 1633–1642. Some entries for 1,2-diazine in Table 2 are incorrect. The correct values are $\alpha_2 = 67.78$, $\bar{\alpha} = 58.99$, $\Delta_1\alpha = 31.77$, $\Delta_2\alpha = 31.92$ and $\Delta_3\alpha = 3.12$.
[75] N.E.-B. Kassimi and Z. Lin, *J. Phys. Chem. A* **102** (1998), 9906–9911.
[76] C. Møller and M.S. Plesset, *Phys. Rev.* **46** (1934), 618–622.
[77] W.J. Hehre, L. Radom, P.v.R. Schleyer and J.A. Pople, *Ab Initio Molecular Orbital Theory*, Wiley, New York, 1986.
[78] A. Dalgarno and J.M. McNamee, *J. Chem. Phys.* **35** (1961), 1517–1518.
[79] A. Dalgarno, *Adv. Phys.* **11** (1962), 281–315.
[80] T.C. Caves and M. Karplus, *J. Chem. Phys.* **50** (1969), 3649–3661.
[81] S.P. Liebmann and J.W. Moskowitz, *J. Chem. Phys.* **54** (1971), 3622–3631.
[82] H. Nakatsuji and J.I. Musher, *J. Chem. Phys.* **61** (1974), 3737–3741.
[83] W. Gordy, *J. Chem. Phys.* **14** (1946), 305–320.
[84] W. Gordy, *J. Chem. Phys.* **15** (1947), 305–310.
[85] R.J. Doerksen and A.J. Thakkar, *Int. J. Quantum Chem.* **90** (2002), 534–540.
[86] M. Giambiagi, M.S. de Giambiagi, D.R. Grempel and C.D. Heymann, *J. Chim. Phys.* **72** (1975), 15–22.
[87] I. Mayer, *Chem. Phys. Lett.* **97** (1983), 270–274.
[88] A.B. Sannigrahi, *Adv. Quantum Chem.* **23** (1992), 301–351.

[89] A.J. Thakkar, T. Koga, M. Saito and R.E. Hoffmeyer, *Int. J. Quantum Chem. Symp.* **27** (1993), 343–354.
[90] T. Koga, M. Saito, R.E. Hoffmeyer and A.J. Thakkar, *J. Mol. Struct.* (*Theochem*) **112** (1994), 249–260.
[91] T.H. Dunning, Jr. *J. Chem. Phys.* **90** (1989), 1007–1023.
[92] D.E. Woon and T.H. Dunning, *J. Chem. Phys.* **98** (1993), 1358–1371.
[93] W.M. Thorburn, *Mind* (*New Ser.*) **24** (1915), 287–288.
[94] W.M. Thorburn, *Mind* (*New Ser.*) **27** (1918), 345–353.
[95] R. Hoffman, V.I. Minkin and B.K. Carpenter, *Bull. Soc. Chim. Fr.* **133** (1996), 117–130.
[96] R. Hoffman, V.I. Minkin and B.K. Carpenter, *HYLE Int. J. Phil. Chem.* **3** (1997), 3–28.
[97] V.J. Steeves, *A Look at Classical Aromaticity Indices*, B.Sc. thesis, University of New Brunswick, Fredericton, Canada, 1997.
[98] R.J. Doerksen, *Geometries, Polarizabilities, and Aromaticity of Ring Molecules*, Ph.D. thesis, University of New Brunswick, Fredericton, Canada, 1998.
[99] R.J. Doerksen, V.J. Steeves and A.J. Thakkar, Structural indices of aromaticity. To be published.

Journal of Computational Methods in Sciences and Engineering 4 (2004) 439–450
IOS Press

Molecular polarizability of Si/Ge/GaAs semiconductors clusters

Francisco Torrens
Institut Universitari de Ciència Molecular, Universitat de València, Dr. Moliner 50, E-46100 Burjassot, València, Spain
Tel.: +34 96 35 43 182; Fax: +34 96 35 43 156; E-mail: Francisco.Torrens@uv.es

Abstract. The interacting induced dipole polarization model implemented in our program for the calculation of molecular polarizabilities (POLAR) is used for the calculation of the molecular dipole–dipole polarizability $\overline{\overline{\alpha}}$. POLAR is tested with Si_n, Ge_n and Ga_nAs_m small clusters. The polarizability is an important quantity for the identification of clusters with different numbers of atoms and even for the separation of isomers. The results for the polarizability are in agreement with reference calculations performed with our version of the program PAPID (polarisabilités atomiques par interactions dipolaires) and with reference computations from Dr. J. R. Chelikowsky. The bulk limit for the polarizability is estimated from the Clausius–Mossotti relationship. The polarizability trend for the clusters as a function of size is different from what one might have expected. The clusters are more polarizable than what one might have inferred from the bulk polarizability. Previous experimental work yielded the opposite trend for somewhat larger clusters. At present, the origin of this difference is problematic. One might argue that smaller clusters need not behave like those of intermediate size. The high polarizability of small clusters is attributed to dangling bonds at the surface of the cluster. Indeed, most of the atoms within small clusters reside on the surface. In this respect, semiconductor clusters resemble *metallic* clusters.

Keywords: Polarization, polarizability, dipole moment, quadrupole moment, semiconductor cluster, isomer, nanostructure

PACS Codes: 36.40.-c; 36.40.Cg; 61.46.+w

1. Introduction

The role of size in modifying the properties of a material has not been exploited until recently [1–6]. One example of experimental technique allows clusters to be deposited on an inert substrate, and then probed with photons [7]. Another experimental procedure is to produce a cluster beam, and to perform photoemission measurements on the clusters within the beam. In both cases, extracting sufficient a *signal-to-noise* ratio is a real problem [8].

On the basis of interatomic potentials, the structure of Si_{13} is special [9]. The predicted ground state is a close-packed icosahedron. Unfortunately, studies based on *ab initio* methods yielded a structure, which is a capped antiprism structure [10]. It was proposed that current electronic structure theory is not capable of handling the contributions of exchange and correlation in this system [11]. Jackson et al. used density functional theory (DFT) to calculate the polarizabilities for several low-energy geometries of Si_n $(10 \leqslant n \leqslant 20)$ clusters [12]. The calculations indicated that the polarizability per atom for Si clusters approaches the bulk limit from above as a function of size. Deng et al. calculated the polarizabilities of Si_n $(9 \leqslant n \leqslant 28)$ clusters using DFT [13,14]. They used geometries carefully selected by energy

1472-7978/04/$17.00

minimization. The polarizability showed fairly irregular variation with cluster size, but all calculated values were higher than the bulk one.

Hohm et al. deduced an experimental value of 116.7 ± 1.1 a.u. for the polarizability of As_4 from the analysis of refractivity measurements in arsenic vapour [15]. This was in close agreement with the *ab initio* finite-field many-body perturbation theory and coupled-cluster calculation result of 119.5 ± 3.6 a.u.

Nanocrystalline powders can be used to synthesize materials with physical processing such as sintering [16,17]. It was demonstrated for rutile that such processing can reduce sintering temperatures by hundreds of degrees without sacrificing desirable mechanical properties [18]. Quantum confinement can be used to engineer the excitation energy [19]. Polarizability measurements were performed for small semiconductor clusters [20].

The present author has tried to develop a scheme for the calculation of molecular polarizabilities (POLAR) that is less demanding in time and resources than are first-principle calculations such as DFT. A good starting point is program PAPID (polarisabilités atomiques par interactions dipolaires) [21,22]. PAPID was designed to fit the coupled Hartree-Fock polarizability of a molecule as a function of atomic contributions. Then, PAPID uses an interacting induced dipole polarization formulation to estimate the polarizability of this (or a similar) molecule. The obtained molecular polarizability is analyzed in *effective* (coupled) atomic contributions, which are different from the starting (uncoupled) atomic polarizabilities. For a given atomic number Z, PAPID differentiates several classes depending on the functional group in which the atom is included. At present, there are 6 classes for H, 7 for C, 3 for N, 5 for O, 1 for F, 1 for Cl, 1 for Br and 1 for I. A possible limitation is the short number of classes. For instance, all hydrocarbons are limited to one only class for H and one only class for C. On the other hand, POLAR solves this limitation including the effect of the net charge, lone pairs and connectivity of each atom. As an example, POLAR differentiates the atomic polarizabilities of central and ending atoms in O_3. The final aim of POLAR will be its application to large molecules and assemblies. The price to be paid in both PAPID and POLAR is then the necessity of semiempirical calibrations. However, POLAR does not require a particular calibration for each new molecule. In previous articles, fullerenes, Sc clusters and Sc-cluster endohedral fullerenes were studied [23]. This work was extended to graphite models and to Sc *hexagonal close packing* [24–26]. In this report, the following semimetallic clusters have been calculated: $Si_n, Ge_n (n \leqslant 10)$ and $Ga_n As_m (n, m \leqslant 4)$.

In the next section, the improved interacting induced dipole polarization model for molecular polarizabilities is presented. A description of programs PAPID and POLAR is given in the two following sections. Following that, results are presented and discussed. The last section summarizes the conclusions.

2. Improved interacting induced dipole polarization model

Computational programs have been used for the calculation of the dipole – dipole polarizability $\overline{\overline{\alpha}}$, with the interacting induced dipole polarization model [27–29], which calculates tensor effective anisotropic point polarizabilities, generalized for polyatomic molecules by Applequist et al. [30,31]. The molecular polarizability, $\alpha_{\alpha\beta}^{\text{mol}}$, is defined as the linear response to an external electric field,

$$\mu_{\alpha}^{\text{ind}} = \alpha_{\alpha\beta}^{\text{mol}} E_{\beta}^{\text{ext}} \tag{1}$$

where μ_α^{ind} is the induced molecular dipole moment. Considering a set of N interacting atomic polarizabilities, the atomic induced dipole moment has a contribution also from the other atoms,

$$\mu_{p,\alpha}^{\text{ind}} = \alpha_{p,\alpha\beta}\left(E_\beta^{\text{ext}} + \sum_{q\neq p}^{N} T_{pq,\beta\gamma}^{(2)}\mu_{q,\gamma}^{\text{ind}}\right) \tag{2}$$

where $T_{pq,\beta\gamma}^{(2)}$ is the interaction tensor

$$T_{pq,\alpha\beta}^{(2)} = \frac{3r_{pq,\alpha}r_{pq,\beta}}{r_{pq}^5} - \frac{\delta_{\alpha\beta}}{r_{pq}^3} \tag{3}$$

where r_{pq} is the distance between atoms p and q, and δ represents the Kronecker delta function: $\delta(\alpha,\beta) = 1$ if $\alpha = \beta$, and $\delta(\alpha,\beta) = 0$ if $\alpha \neq \beta$. The molecular polarizability can then be written as

$$\alpha_{\alpha\beta}^{\text{mol}} = \sum_{q,p}^{N} B_{pq,\alpha\beta} \tag{4}$$

where B is the relay matrix defined as (in a supermatrix notation)

$$\mathbf{B} = (\alpha^{-1} - T^{(2)})^{-1} \tag{5}$$

The atomic parameters α_p have previously been fitted to the full polarizability tensors of quantum-mechanical computed molecular polarizabilities of a series of 115 substituted aliphatic and aromatic molecules containing the elements H, C, N, O, F and Cl.

The following improvements have been implemented in the model.

1. In order to build up the relay matrix $\mathbf{B}$ the atomic polarizability tensors given by $\overline{\overline{\alpha}}^i = \overline{\overline{\alpha}}_\sigma^i + \overline{\overline{\alpha}}_\pi^i$ have been used instead of the scalar polarizability α^i.
2. Damping functions have been used in the calculation of the interaction tensor to prevent the polarizability from going to infinity in the models by Applequist and Birge (Eq. (3)). Thole [32] proposed

$$T_{pq,\alpha\beta}^{(2)} = \frac{3v_{pq}^4 r_{pq,\alpha}r_{pq,\beta}}{r_{pq}^5} - \frac{(4v_{pq}^3 - 3v_{pq}^4)\delta_{\alpha\beta}}{r_{pq}^3} \tag{6}$$

   where $v_{pq} = r_{pq}/s_{pq}$ if $r_{pq} < s_{pq}$; otherwise $v_{pq} = 1$. The term s is defined as $(\Phi_p\Phi_q)^{1/4}$ where Φ_p is a fitting parameter proportional to the atomic second-order moment [33]. The atomic parameters Φ_p have been fitted to the full polarizability tensors of quantum-mechanical computed molecular polarizabilities. Miller [34,35] proposed

$$T_{pq,\alpha\beta}^{(2)} = \left[\frac{3r_{pq,\alpha}r_{pq,\beta}}{r_{pq}^5} - \frac{\delta_{\alpha\beta}}{r_{pq}^3}\right]\left\{1 - \exp\left[-\left(\frac{r_{pq}}{0.7\rho_{pq}}\right)^{10}\right]\right\} \tag{7}$$

   with $\rho_{pq} = \rho_p + \rho_q$ (sum of the van der Waals radii).
3. The interaction between bonded atoms and atoms with a distance lying in an interval defined by $[r^{\text{inf}}, r^{\text{sup}}]$ has been neglected [36]. The starting values for this interval are $[0, 10^{30}]$ and r^{inf} is incremented if resonance conditions are detected.

4. The following tests indicating a resonance condition have been implemented [21,22]: (a) test whether matrix $\mathbf{B}^{-1}$ is singular; (b) test whether matrix $\mathbf{B}^{-1}$ is not defined positive; (c) test whether matrix $\mathbf{B}$ is not defined positive; and (d) test whether some effective matrix $\mathbf{B}_p$ is not defined positive.

3. Description of program PAPID

PAPID [21,22] allows the following polarization options: (1) model of atomic dipoles μ^i without interaction, i.e., at each atom a dipole moment is induced by the electric field created by only the permanent multipole moments of the remaining atoms, (2) model of μ^i with optimization of isotropic atomic polarizabilities α^i, (3) iterative Applequist model of interacting μ^i with isotropic α^i, i.e., at each atom a dipole moment is induced by the electric field created by the permanent and induced multipole moments of the remaining atoms [30,31], (4) iterative Applequist model with optimization of α^i, i.e., the inverse of matrix $\mathbf{B}^{-1}$ is calculated analitically, (5) non-iterative Applequist model, i.e., the inverse of matrix $\mathbf{B}^{-1}$ is calculated by fourth-order perturbation theory, (6) non-iterative Applequist model with optimization of α^i, (7) iterative Birge model of interacting μ^i with anisotropic α^i [37], (8) iterative Birge model with optimization of α^i, (9) iterative Birge model with optimization of α^i and atomic anisotropy constants ζ^i, (10) iterative Birge model with optimization of α^i and one-centre valence state electron repulsion integrals g^i, (11) non-iterative Birge model, (12) non-iterative Birge model with optimization of α^i, (13) non-iterative Birge model with optimization of α^i and ζ^i and (14) non-iterative Birge model with optimization of α^i and g^i.

The optimization of the atomic parameters allows selecting between the Fletcher [38] and Lagrange three-point extrapolation methods. The Fletcher algorithm permits selecting the precision on the stability of the minimum obtained from the optimization of the atomic parameters among a vast extrapolation of surfaces associated to a weak stability of the minima, an intermediate value, and a good stability of the obtained minimum associated to a weak modification of the parameters at each iteration of the optimization process.

PAPID allows whether to use the vibrational correction of atomic polarizabilities to average molecular vibrational stretching modes. The modified atomic polarizabilities are calculated as

$$\alpha_j = \alpha_j^o + \sum_n \frac{d\alpha_j}{dR_n} R_n \tag{8}$$

where R_n is the displacement of the n-th bond length from its rest value, and α_j^o is the atom polarizability at the rest configuration of the molecule [39]. Optimizing the ratios $d\alpha_j/dR_n$ to best represent the experimental values of the derivative of the molecular polarizability tensor, it is possible to model the molecular polarizability variations during bond vibrations.

The optimization of the atomic parameters permits selecting between the *fitting* of the molecular dipole–dipole polarizabilities α_{ab}, and the simultaneous fitting of α_{ab} and all the non-zero components $A_{a,bc}$ of the dipole–quadrupole polarizability matrix $\overline{\overline{\overline{A}}}$. Applequist [39] estimates that the variations of quadrupole moments can be represented using combinations of induced dipoles, what allows calculating certain molecular polarizabilities of upper order as

$$A_{\alpha,\beta\gamma} = \sum_{i,j} \left[\frac{3}{2}(1 + S_{\beta\gamma}) B_{\beta\gamma}^{ij} r_\gamma^j - \delta_{\beta\gamma} B_{\alpha\varepsilon}^{ij} r_\varepsilon^j \right] \tag{9}$$

where $S_{\beta\gamma}$ is the permutation operator on indices β and γ, and $\delta_{\beta\gamma}$ is the Kronecker's δ function.

The inversion of the relay matrix **B** allows selecting between the analytic and perturbative methods. The analytic calculation permits selecting the test level on the obtained results among no test, test on the stability of the determinant of matrix **B**, and test on the stability of the determinant and eigenvalues of matrix **B**. The perturbative treatment calculates up to the fourth order perturbation. PAPID allows whether to take into account the dipolar interactions between bonded atoms. Moreover, the dipolar interaction between atoms i and j can be cut off by an upper radius describing the outer limit of the interaction sphere, centred on unit i. Furthermore, their dipolar interaction can be cut off by the radius of a sphere centred on i and containing no interaction.

PAPID contains a database of optimized polarization parameters for the following atomic classes: H [alkane, alcohol, aldehyde, amide, amine and amine ion$^{(+)}$], C [alkane, carbonyl, alcohol, amine, amide, carboxylic acid/ester and carboxylate$^{(-)}$], N [amide, amine and amine ion$^{(+)}$], O [alcohol, ether, carbonyl, carboxylic acid/ester and carboxylate$^{(-)}$], F (halomethanes), Cl (halomethanes), Br (halomethanes) and I (halomethanes).

The following improvements have been implemented in PAPID.

1. The calculation can start from a predefined value of r^{inf} and increment r^{inf} by one atomic unit, each time resonance conditions are detected, as far as r^{sup}.
2. Three damping strategies are allowed for the expression of the interaction tensor: (1) Böttcher dipole interaction model (no damping), (2) Thole modification (Eq. 6) and (3) Miller modification (Eq. 7).
3. The atomic polarizability parameters can be selected between optimized *ab initio* and experimental values.
4. The atomic classes can be either read or assigned by the program.
5. The molecule can be reoriented by its principal axes of inertia.
6. The atomic polarization parameters have been included in the database for He, Li, Be, B, Ne, Na, Mg, Al, Si, P, S, Ar, K, Ca, Sc, Ti, V, Cr, Mn, Fe, Co, Ni, Cu, Zn, Ga, Ge, As, Se, Kr, Rb, Sr, Y, Zr, Nb, Mo, Tc, Ru, Rh, Pd, Ag, Cd, In, Sn, Sb, Te and Xe.
7. The atomic polarization parameters from Thole and Miller damping models have been included in the database.
8. The optimized experimental polarization parameters have been included in the database for the following atomic classes: H (alkane, alcohol, aldehyde, amide, amine and aromatic), C (alkane, carbonyl, nitrile and aromatic), N (amide, amine and nitrile), O (alcohol, ether and carbonyl), F (halomethanes), Cl (halomethanes), Br (halomethanes) and I (halomethanes) [39].

4. Description of program POLAR

POLAR allows whether to re-orientate the molecule by its principal axes of inertia. In describing the partial charge method developed for the Mulliken scale [40], Huheey mentioned that most elements approximately double their electronegativities as the partial charge approaches +1, whereas their electronegativities essentially disappear as the partial charge approaches −1 [41]. The Mulliken and Pauling scales are roughly proportional, so Huheey's observation may be expressed in Pauling units as

$$X_{eq} = X_A + \Delta_A X_A \tag{10}$$

Here, X_{eq} is the electronegativity equalized through Sanderson principle, X_A is the initial, pre-bonded electronegativity of a particular atom A, and Δ_A is the σ partial charge on A [42]. Charge conservation leads to a general expression for X_{eq}:

$$X_{eq} = \frac{N + q}{\sum_{\text{atoms}} \frac{V_A}{X_A}} \tag{11}$$

where $N = \sum v$ equals the total number of atoms in the species formula and q is the σ molecular charge. The σ partial charge Δ_A on atom A can be generalized as:

$$\Delta_A = \sum_{\text{bonds}} \frac{X_{eq,b} - X_A}{X_A} \tag{12}$$

and the electronegativity equalized for bonds is given as

$$X_{eq,b} = \frac{2 + \frac{q}{m}}{\frac{1}{X_A} + \frac{1}{X_B}} \tag{13}$$

where m is the number of bonds in the molecule.

The π-atomic charges and polarizabilities have been evaluated with the Hückel molecular orbital (HMO) method. The π-conjugation vanishes for perpendicular structures, e.g., biphenyl. HMO β parameter can be evaluated, in first approximation, between p_z orbitals twisted from co-planarity by an angle θ as

$$\beta = \beta_o \cos\theta \tag{14}$$

where β_o is equal to the β parameter for benzene [43]. Joachim et al. evaluated the electronic coupling V_{ab} of the binuclear mixed-valence M^{II}–L–M^{III} complex $[(NH_3)_5$Ru-bipyridyl-Ru$(NH_3)_5]^{5+}$ [44]. When a pyridine ring rotates around the ligand axis, $\pi - \pi V_{ab}(\theta)$ can be best fitted by a function $\cos^{1.15}\theta$. From this observation, the β function is assumed universal with this same form:

$$\beta = \beta_o \cos^{1.15}\theta \tag{15}$$

The dipole and tensor quadrupole moments have been calculated from the point distribution of net charges. The dipole moment vector is calculated as

$$\mu_a = \sum_i q^i r_a^i \tag{16}$$

and the quadrupole moment tensor is calculated as

$$\Theta_{ab} = \Theta_{ba} = \frac{1}{2}\sum_i q^i \left[3r_a^i r_b^i - (r^i)^2\delta_{ab}\right] \tag{17}$$

where q^i is the i-th element of charge at the point $\vec{r}^i$ relative to an origin fixed at the centre of mass in the molecule. The δ_{ab} stands for the Kronecker δ function. The subscripts $a, b\ldots$ denote vector or tensor components and can be equal to the Cartesian components x, y, z. Only the first, nonvanishing moment is independent of the choice of origin. Thus for an ion like OH^-, the dipole and quadrupole moments vary with the origin. Similarly, in HF, Θ depends on origin. In order to avoid these dependences, the origin is taken at the centre of mass and the molecule is brought into its principal internal coordinate system, e.g., linear molecules on the x-axis and planar molecules on the xy plane.

It can be shown that for every square matrix $\overline{\overline{\Theta}}$ there exists a matrix $\overline{\overline{V}}$ and its inverse $\overline{\overline{V}}^{-1}$, which transforms $\overline{\overline{\Theta}}$ to a diagonal matrix $\overline{\overline{D}}$ [45]:

$$\overline{\overline{D}} = \overline{\overline{V}}^{-1}\overline{\overline{\Theta}}\,\overline{\overline{V}} = \mathrm{diag}(\Theta_1, \Theta_2, \Theta_3) \tag{18}$$

where scalar Θ_i are the eigenvalues of matrix $\overline{\overline{\Theta}}$. The right *eigenvectors* of $\overline{\overline{\Theta}}$ are non-zero column vectors $\overline{x}$ such that satisfy the following equation:

$$\overline{\overline{\Theta}}\,\overline{x} = \Theta_i \overline{x} \tag{19}$$

The eigenvectors are independent of the choice of origin and are easily compared with other values. Moreover, the reduction from nine Θ_{ab} matrix elements (actually six because $\overline{\overline{\Theta}}$ is symmetric) to three facilitates the comparison with other results. The *trace* of matrix $\overline{\overline{\Theta}}$ is defined as the sum of its diagonal elements:

$$Tr(\overline{\overline{\Theta}}) = \Theta_{xx} + \Theta_{yy} + \Theta_{zz} = \Theta_1 + \Theta_2 + \Theta_3 \tag{20}$$

$Tr(\overline{\overline{\Theta}})$ is independent of the choice of origin and can be calculated in Cartesian coordinates or from the eigenvalues. The *mean* $\overline{\overline{\Theta}}$ of the matrix is defined as:

$$\Theta = \frac{Tr(\overline{\overline{\Theta}})}{3} = \frac{\Theta_{xx} + \Theta_{yy} + \Theta_{zz}}{3} = \frac{\Theta_1 + \Theta_2 + \Theta_3}{3} \tag{21}$$

Sanderson principle [42] allows the calculation of the σ atomic polarizability as:

$$\alpha_A = \frac{\partial \Delta_A}{\partial X_A} = \sum_{\text{bonds}} \frac{f_A s_A (1 - \Delta_A)\left(2 - \frac{q}{m}\right) X_B}{(2 - \Delta_A - \Delta_B)(X_A + X_B)^2} \tag{22}$$

where the coefficients f_A and s_A have been introduced to take into account the effects of the atomic charge (Δ_A), internal sub-shells and lone pairs on α_A [36]. The f_A is calculated as:

$$f_A = 1 - 1.5\Delta_A + 0.5\Delta_A^2 \tag{23}$$

and s_A is obtained as:

$$s_A = C_A + 0.15 L_A \tag{24}$$

where C_A is related to the number of internal sub-shells [$C_A = 1$ (H–Ne), 3 (Na–Ar), 4 (K–Zn), 5 (Ga–Kr), 6 (Rb–Cd) and 7 (In–Xe)] and L_A is the number of lone pairs on atom A [36]. The σ tensor atomic polarizability is calculated according to:

$$\overline{\overline{\alpha}}_\sigma = \sum_{\text{bonds}} \frac{3\alpha_\sigma}{\alpha^{\|} + 2\alpha^{\perp}} \begin{bmatrix} \alpha^{\perp} & 0 & 0 \\ 0 & \alpha^{\perp} & 0 \\ 0 & 0 & \alpha^{\|} \end{bmatrix} = \sum_{\text{bonds}} \frac{3\alpha_\sigma}{3.676} \begin{bmatrix} 1 & 0 & 0 \\ 0 & 1 & 0 \\ 0 & 0 & 1.676 \end{bmatrix} \tag{25}$$

where the z-axis is defined as the bond direction for each bond [36]. The diagonal form of $\overline{\overline{\alpha}}_\sigma$ has two distinct components noted as $\alpha^{\|}$ and $\alpha^{\perp}$, parallel and perpendicular, respectively, to the bond axis. The parameter $\alpha^{\|}/\alpha^{\perp} = 1.676$ has been obtained fitting the isotropic bonding polarizabilities of Vogel [46]. The bonding polarizabilities have been implemented in the database of program SIBFA [47].

The π tensor atomic polarizability is calculated as:

$$\overline{\overline{\alpha}}_{\pi} = \sum_{\text{bonds}} \frac{3\alpha_{\pi}}{2\alpha^{\|} + \alpha^{\perp}} \begin{bmatrix} \alpha^{\|} & 0 & 0 \\ 0 & \alpha^{\|} & 0 \\ 0 & 0 & \alpha^{\perp} \end{bmatrix} = \sum_{\text{bonds}} \frac{3\alpha_{\pi}}{3.741} \begin{bmatrix} 1 & 0 & 0 \\ 0 & 1 & 0 \\ 0 & 0 & 1.741 \end{bmatrix} \tag{26}$$

where the xy plane is the σ-plane [36]. The $\alpha^{\|}$ and $\alpha^{\perp}$ are parallel and perpendicular, respectively, to the σ-plane. The parameter $\alpha^{\perp}/\alpha^{\|} = 1.741$ has been obtained fitting the experimental polarizabilities of aromatic molecules.

POLAR performs the calculation of molecular polarizablities and allows selecting between both iterative Applequist and Birge models [36]. POLAR permits whether to take into account the dipolar interactions between bonded atoms. Moreover, the dipolar interaction between a pair of atoms can be cut off by lower and upper radii describing the inner and outer limits of the interaction sphere.

A fully operative version of POLAR, including the whole interacting induced dipoles polarization model, has been implemented in the molecular mechanics program MM2 [48,49], and in the empirical conformational energy program for peptides ECEPP2 [50–52]. The new versions are called MMID [53, 54] and ECEPPID [55], respectively. MMID has been extended and parametrized for transition metal complexes. The new version is called MMIDX [56–58].

5. Calculation results and discussion

In this work, the geometries of Si_n, Ge_n $(n \leqslant 10)$ and $Ga_nAs_m(n, m \leqslant 4)$ clusters have been optimized with MMID [53,54]. Two different structures have been obtained for Si_6 and two others for Ge_6. The polarizability $\overline{\overline{\alpha}}$ has been calculated for small semiconductor clusters. The variation of the computed polarizability values with the number of atoms is illustrated in Fig. 1. Notice that the results for Si_1, Si_4, Si_6 (cf. Fig. 1a), Ge_1, Ge_5, Ge_6, Ge_7, Ge_8, Ge_9 (Fig. 1b), Ga_1, As_1, Ga_1As_2, Ga_2As_1, Ga_2As_2, Ga_2As_3, Ga_3As_2, Ga_3As_4 and Ga_4As_3 (Fig. 1c) are superposed. On varying the number of atoms, the clusters show numbers indicative of particularly polarizable structures. The polarizability results agree with reference calculations carried out with PAPID (Ref.-1), and with those from Chelikowsky [59] performed with DFT (Ref.-2) [60]. As another reference, the bulk limit for the polarizability has been calculated from the Clausius–Mossotti relationship:

$$\alpha = \frac{3(\varepsilon - 1)v}{4\pi(\varepsilon + 2)} \tag{27}$$

where v is the elementary volume per atom in the crystalline state and ε is the bulk dielectric constant. In this work, α is calculated as 3.75Å^3 (Si) and 4.50Å^3 (Ge) per atom, and $\alpha(Ga_nAs_m)/(n+m)$ is taken as $\alpha(GaAs)/2 = 4.15$Å^3.

The polarizability trend for the clusters as a function of size is different from what one might have expected. The clusters calculated with POLAR and PAPID are, in general, more polarizable than what one might have inferred from the bulk polarizability. Previous experimental work [20] yielded the opposite trend for somewhat larger clusters, i.e., in this work the polarizability of clusters tends to be lower than the bulk limit and approaches this limit from below. At present, the origin of this difference

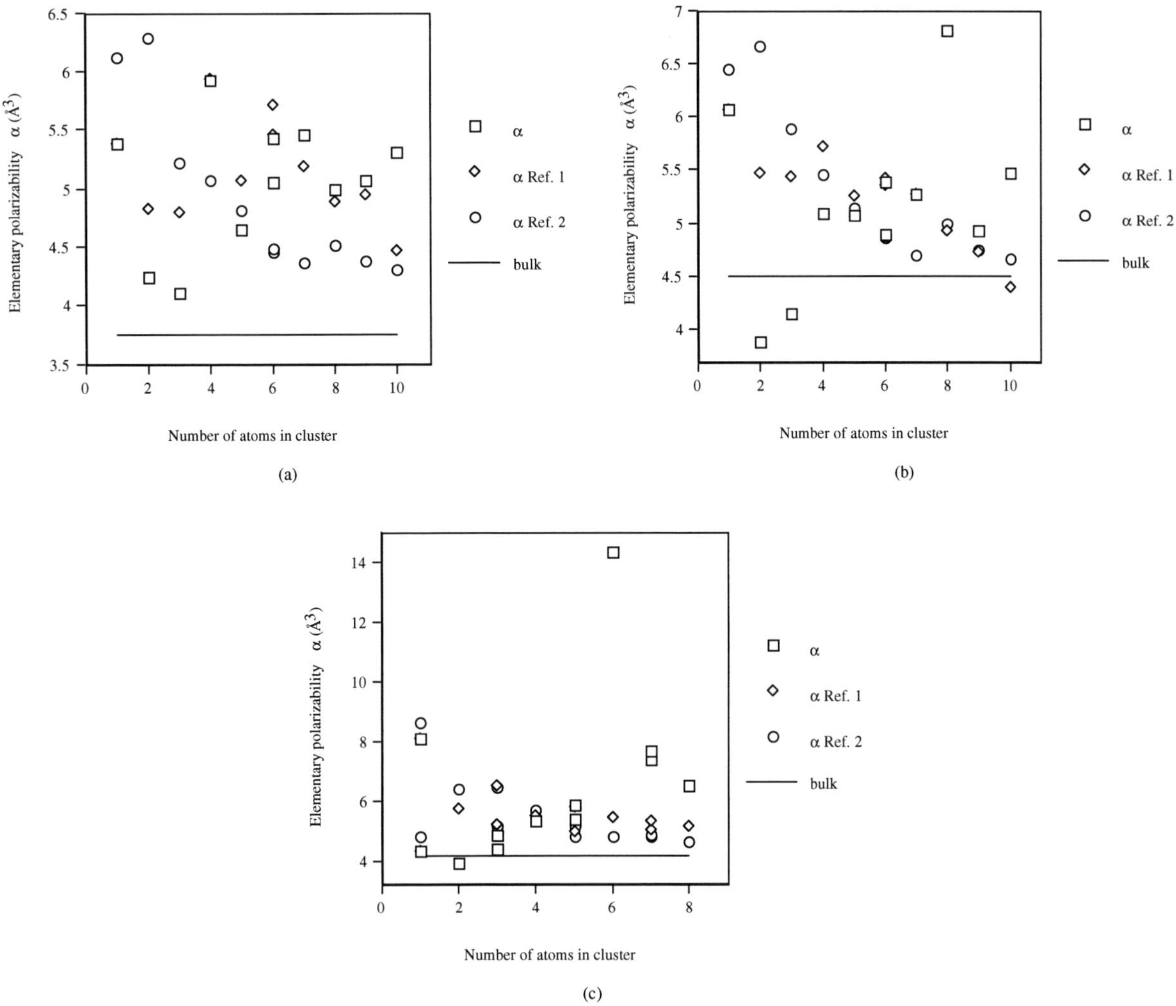

Fig. 1. Elementary polarizability of (a) Si_n, (b) Ge_n and (c) Ga_nAs_m clusters vs. cluster size. Ref. 1: PAPID calculation. Ref. 2: DFT calculation taken from Reference 59.

is problematic. One might argue that smaller clusters need not behave like those of intermediate size. In addition, the error bars in the experiments are fairly large.

The high polarizability of small clusters is attributed to dangling bonds at the surface of the cluster. Indeed, most of the atoms within small clusters reside on the surface. In this respect, semiconductor clusters resemble *metallic* clusters. They tend to have higher coordination numbers than those in the crystalline state. In fact, these structures are thought more closely related to the high-pressure metallic phases than to the diamond structure [2]. For example, it was shown that the polarizabilities of alkali metal clusters significantly exceed the bulk limit and tend to decrease with increasing cluster size [61, 62].

The results presented in this report are not consistent with experiments. (1) Calculated cluster polarizabilities are rather sensitive with respect to the geometry of the cluster provided by the force field, because of the damping function and neglect of interaction between atoms closer than r^{inf}, needed to avoid resonance conditions. (2) The experiments were carried out in a static electric field. Therefore, so-called pure vibrational contributions to the polarizability are important [63].

6. Conclusions

From the preceding results the following conclusions can be drawn:

1. On varying the number of atoms, the clusters show numbers indicative of particularly polarizable structures. The polarizability is an important quantity for the identification of clusters with different numbers of atoms and even for the separation of isomers.
2. The polarizability trend for the clusters as a function of size is different from what one might have expected. The clusters are more polarizable than what one might have inferred from the bulk polarizability.
3. The high polarizability of small clusters is attributed to dangling bonds at the surface of the cluster. In this respect, semiconductor clusters resemble metallic clusters.

Acknowledgement

The author acknowledges financial support of the Dirección General de Enseñanza Superior of the Spanish MCT (Plan Nacional I+D+I, Project No. BQU2001-2935-C02-01).

References

[1] R.P. Andres, R.S. Averback, W.L. Brown, L.E. Brus, W.A. Goddard III, A. Kaldor, S.G. Louie, M. Moscovits, P.S. Peercy, S.J. Riley, R.W. Siegel, F. Spaepen and Y. Wang, Research opportunities on clusters and cluster-assembled materials: A Department of Energy, Council on Materials Science, *J. Mater. Res.* **4** (1989), 704–736.
[2] M.F. Jarrold, Nanosurface chemistry on size-selected silicon clusters, *Science* **252** (1991), 1085–1092.
[3] L.A. Bloomfield, R.R. Freeman and W.L. Brown, Photofragmentation of mass-resolved Si^+_{2-12} clusters, *Phys. Rev. Lett.* **54** (1985), 2246–2249.
[4] P. Jena, B.K. Rao and S.N. Khanna, eds, *Physics and Chemistry of Small Clusters*, Plenum, New York, 1987.
[5] J.R. Chelikowsky and S.G. Louie, eds, *Quantum Theory of Real Materials*, Kluwer, New York, 1996.
[6] J.M. Alford, R.T. Laaksonen and R.E. Smalley, Ammonia chemisorption studies on silicon cluster ions, *J. Chem. Phys.* **94** (1990), 2618–2630.
[7] E.C. Honea, A. Ogura, C.A. Murray, K. Raghavachari, W.O. Sprenger, M.F. Jarrold and W.L. Brown, Raman spectra of size-selected silicon clusters and comparison with calculated structures, *Nature (London)* **366** (1993), 42–44.
[8] O. Cheshnovsky, S.H. Yang, C.L. Pettiette, M.J. Craycraft, Y. Liu and R.E. Smalley, Ultraviolet photoelectron spectroscopy of semiconductor clusters: Silicon and germanium, *Chem. Phys. Lett.* **138** (1987), 119–124.
[9] J.R. Chelikowsky, K.M. Glassford and J.C. Phillips, Interatomic force fields for silicon microclusters, *Phys. Rev. B* **44** (1991), 1538–1545.
[10] J.C. Grossman and L. Mitáß, Quantum Monte Carlo determination of electronic and structural properties of Si_n clusters ($n \leqslant 20$), *Phys. Rev. Lett.* **74** (1995), 1323–1326.
[11] J.C. Phillips, Electron-correlation energies and the structure of Si13, *Phys. Rev. B* **47** (1993), 14132–14135.
[12] K. Jackson, M. Pederson, C.-Z. Wang and K.-M. Ho, Calculated polarizabilities of intermediate-size Si clusters, *Phys. Rev. A* **59** (1999), 3685–3689.
[13] K. Deng, J. Yang and C.T. Chan, Calculated polarizabilities of small Si clusters, *Phys. Rev. A* **61** (2000), 025201–1-4.
[14] K. Deng, J. Yang, L. Yuan and Q. Zhu, Hybrid density-functional study of Si13 clusters, *Phys. Rev. A* **62** (2000), 045201–1-4.
[15] U. Hohm, D. Goebel, P. Karamanis and G. Maroulis, Electric dipole polarizability of As4, a challenging problem for both experiment and theory, *J. Phys. Chem. A* **102** (1998), 1237–1240.
[16] L. Brus, Electronic wave functions in semiconductor clusters: Experiment and theory, *J. Phys. Chem.* **90** (1986), 2555–2560.

[17] Y. Kayanuma, Quantum-size effects of interacting electrons and holes in semiconductor microcrystals with spherical shape, *Phys. Rev. B* **38** (1988), 9797–9805.
[18] R.W. Siegel, S. Ramasamy, H. Hahn, Z. Li, T. Lu and R. Gronsky, Synthesis, characterization, and properties of nanophase titanium dioxide, *J. Mater. Res.* **3** (1988), 1367–1372.
[19] A.P. Alivisatos, Semiconductor clusters, nanocrystals, and quantum dots, *Science* **271** (1996), 933–937.
[20] R. Schäfer, S. Schlecht, J. Woenckhaus and J.A. Becker, Polarizabilities of isolated semiconductor clusters, *Phys. Rev. Lett.* **76** (1996), 471–474.
[21] C. Voisin, A. Cartier and J.-L. Rivail, Computation of accurate electronic molecular polarizabilities, *J. Phys. Chem.* **96** (1992), 7966–7971.
[22] C. Voisin and A. Cartier, Determination of distributed polarizabilities to be used for peptide modeling, *J. Mol. Struct. (Theochem)* **286** (1993), 35–45.
[23] F. Torrens, Molecular polarizability of Sc_n, C_n, and endohedral $Sc_n@C_m$ clusters, *Microelectron. Eng.* **51–52** (2000), 613–626.
[24] F. Torrens, Molecular polarizability of Sc and C (*fullerene* and *graphite*) clusters, *Molecules* **6** (2001), 496–509.
[25] F. Torrens, Molecular polarizability of Sc_n, C_n and endohedral $Sc_n@C_m$ clusters, *Nanotechnology* **13** (2002), 433–438.
[26] F. Torrens, Molecular polarizability of fullerenes and endohedral metallofullerenes, *J. Phys. Org. Chem.* **15** (2002), 742–749.
[27] L. Silberstein, Molecular refractivity and atomic interaction, *Philos. Mag.* **33** (1917), 92–128.
[28] L. Silberstein, Dispersion and the size of molecules of hydrogen, oxygen, and nitrogen, *Philos. Mag.* **33** (1917), 215–222.
[29] L. Silberstein, Molecular refractivity and atomic interaction. II, *Philos. Mag.* **33** (1917), 521–533.
[30] J. Applequist, J.R. Carl and K.-K. Fung, An atom dipole interaction model for molecular polarizability. Application to polyatomic molecules and determination of atom polarizabilities, *J. Am. Chem. Soc.* **94** (1972), 2952–2960.
[31] J. Applequist, Atom charge transfer in molecular polarizabilities. Application of the Olson-Sundberg model to aliphatic and aromatic hydrocarbons, *J. Phys. Chem.* **97** (1993), 6016–6023.
[32] B.T. Thole, Molecular polarizabilities calculated with a modified dipole interaction, *Chem. Phys.* **59** (1981), 341–350.
[33] L. Jensen, P.-O. Åstrand, K.O. Sylvester-Hvid and K.V. Mikkelsen, Frequency-dependent molecular polarizability calculated within an interaction model, *J. Phys. Chem. A* **104** (2000), 1563–1569.
[34] K.J. Miller, Calculation of the molecular polarizability tensor, *J. Am. Chem. Soc.* **112** (1990), 8543–8551.
[35] K.J. Miller, Additivity methods in molecular polarizability, *J. Am. Chem. Soc.* **112** (1990), 8533–8542.
[36] F. Torrens, J. Sánchez-Marín, I. Nebot-Gil, Interacting induced dipoles polarization model for molecular polarizabilities. Reference molecules, amino acids and model peptides, *J. Mol. Struct. (Theochem)* **463** (1999), 27–39.
[37] R.R. Birge, Calculation of molecular polarizabilities using an anisotropic atom point dipole interaction model which includes the effect of electron repulsion, *J. Chem. Phys.* **72** (1980), 5312–5319.
[38] R. Fletcher and M.J.D. Powell, A rapidly converging descent method for minimization, *Comput. J.* **6** (1963), 163–168.
[39] J. Applequist, An atom dipole interaction model for molecular optical properties, *Acc. Chem. Res.* **10** (1977), 79–85.
[40] R.S. Mulliken, A new electroaffinity scale; together with the data on valence states and on valence ionization potentials and electron affinities, *J. Chem. Phys.* **2** (1934), 782–793.
[41] J.E. Huheey, The electronegativity of groups, *J. Phys. Chem.* **69** (1965), 3284–3291.
[42] R.T. Sanderson, An interpretation of bond lengths and classification of bonds, *Science* **114** (1951), 670–672.
[43] R.S. Mulliken, C.A. Rieke, D. Orloff and H. Orloff, Formulas and numerical tables for overlap integrals, *J. Chem. Phys.* **17** (1949), 1248–1267.
[44] C. Joachim, G. Treboux and H. Tang, in: Molecular Electronics: Science and Technology, *AIP Conference Proceedings*, Vol. 262, AIP, New York, 1992, pp. 107–117.
[45] H. Margenau and G.M. Murphy, eds, *The Mathematics of Physics and Chemistry*, van Nostrand-Reinhold, New York, 1964.
[46] A.I. Vogel, Physical properties and chemical constitution. XXIII. Miscellaneous compounds. Investigation of the so-called coördinate or dative link in esters of oxy acids and in nitro paraffins by molecular refractivity determinations. Atomic, structural, and group parachors and refractivities, *J. Chem. Soc.* (1948), 1833–1855.
[47] N. Gresh, P. Claverie and A. Pullman, Intermolecular interactions: Reproduction of the results of ab initio supermolecule computations by an additive procedure, *Int. J. Quantum Chem., Symp.* **13** (1979), 243–253.
[48] N.L. Allinger, Conformational analysis. 130. MM2. A hydrocarbon force field utilizing V_1 and V_2 torsional terms, *J. Am. Chem. Soc.* **99** (1977), 8127–8134.
[49] N.L. Allinger and Y.H. Yuh, MM2, QCPE Program No. 395, 1980.
[50] G. Némethy, M.S. Pottle and H.A. Scheraga, Energy parameters in polypeptides. 9. Updating of geometrical parameters, nonbonded interactions, and hydrogen bond interactions for the naturally occurring amino acids, *J. Phys. Chem.* **87** (1983), 1883–1887.
[51] G. Némethy, M.S. Pottle and H.A. Scheraga, ECEPP2, QCPE Program No. 454, 1983.

[52] M.J. Sippl, G. Némethy and H.A. Scheraga, Intermolecular potentials from crystal data. 6. Determination of empirical potentials for O-H... O=C hydrogen bonds from packing configurations, *J. Phys. Chem.* **88** (1984), 6231–6233.

[53] F. Torrens, M. Ruiz-López, C. Cativiela, J.I. García and J.A. Mayoral, Conformational aspects of some asymmetric Diels-Alder reactions. A molecular mechanics + polarization study, *Tetrahedron* **48** (1992), 5209–5218.

[54] F. Torrens, J. Sánchez-Marín and J.-L. Rivail, Interacting induced dipoles polarization in a force field for dipeptide models (glycine derivative), *An. Fís. (Madrid)* **90** (1994), 197–204.

[55] F. Torrens, Polarization force fields for peptides implemented in ECEPP2 and MM2, *Mol. Simul.* **24** (2000), 391–410.

[56] F. Torrens, Nature of Fe^{III}–O_2, Fe^{II}–CO and FeIII–CN complexes of hemoprotein models, *Polyhedron* **22** (2003), 1091–1098.

[57] F. Torrens, Nature of O_2, CO and CN binding to hemoprotein models, *Int. J. Quantum Chem* **99** (2004), 963–971.

[58] F. Torrens, Nature of Fe^{III}–O_2, FeII–CO and Fe^{III}–CN complexes of hemoprotein models, *Int. J. Mol. Sci.*, submitted for publication.

[59] J.R. Chelikowsky, *The electronic and structural properties of semiconductor clusters and nanostructures*, University of Minnesota, Supercomputing Institute, Research Report 97/132, 1997.

[60] A. Stathopoulos, S. Ögüt, Y. Saad, J. Chelikowsky and H. Kim, Parallel methods and tools for predicting material properties, *Comput. Sci. Eng.* **2**(4) (2000), 19–32.

[61] W.A. de Heer, The physics of simple metal clusters: Experimental aspects and simple models, *Rev. Mod. Phys.* **65** (1993), 611–675.

[62] M. Brack, The physics of simple metal clusters: Self-consistent jellium model and semiclassical approaches, *Rev. Mod. Phys.* **65** (1993), 677–732.

[63] D.M. Bishop, Molecular vibrational and rotational motion in static and dynamic electric fields, *Rev. Mod. Phys.* **62** (1990), 343–374.

Journal of Computational Methods in Sciences and Engineering 4 (2004) 451–459
IOS Press

First-principle study on polarizability and hyperpolarizability of a transition metal cluster, $[Mo_2S_3(C_6H_{11})_3(CO)_6]\cdot N(C_2H_3)_4$

Kechen Wu*, Ping Liu, Zhangfeng Zhou, Chensheng Lin and Botao Zhuang
State Key Laboratory of Structural Chemistry, Fujian Institute of Research on the Structure of Matter, Chinese Academy of Sciences, Fuzhou, Fujian 350002, China

Abstract. We reported the first-principle study of linear and nonlinear optical polarizability of a di-nuclear transition metal cluster, $[Mo_2S_3(CO)_6(C_6H_{11})_3]\cdot N(C_2H_3)_4$, by using finite-field method. The calculations were performed at an HF/Lanl2dz level. The spatial average value of the first hyperpolarizability along z direction was very large, $\beta_z = 13.8 \times 10^{-30}$ esu, indicating the potential applications of this cluster for IR nonlinear optical processes. Our results showed that the charge transfer inside the metal core cluster, $[Mo_2S_3(CO)_6]$, played an important role in nonlinear optical response while the joint effect of metal core and organic ligands $(C_6H_{11})_3$ also contributed to nonlinear optical polarizability. In addition, the organic ionic group, $[N(C_2H_3)_4]^+$, situated in z direction enhanced the β_z value. The study may help us to understand the nonlinear optical nature of the transition metal clusters and provided an example to show the computational efficiency of finite-field method in the first-principle theoretical study of the complex system containing transition-metal element.

Keywords: Hyperpolarizability, finite-field method, transition metal cluster

PACS: 31.15.Ar, 42.70.Mp, 72.80.G

1. Introduction

The study of novel nonlinear optical (NLO) materials for laser device applications is one of the most important areas of electro-optics [1]. Along the way of the discovery of various kinds of materials for NLO applications, from the inorganic mineral oxides represented by potassium titanyl phosphate (KTP), beta-barium borate (BBO) and lithium triborate (LBO), to organic molecular crystals such as 3-methoxy-4-hydroxy-benzaldehyde (MHBA), p-nitroaniline (PNA) and m-dinitrobenzene (MDNB) to organometallic complexes exampled by ferrocenyl derivatives, most recently transition-metal cluster become a new area for this purpose, especially for the second-harmonic generation (SHG) materials applied in IR region. The development of the novel IR SHG materials are motivated by the great demands in some high-technique fields such as optical communication.

In the past decade, a great improvement has been made in computational NLO material science. The in-depth understanding of the relationship between the microscopic structure and NLO properties, the predesigned tailing and modifying at atomic and molecular level and the modeling and simulation of

*Corresponding author. Fax: +86 591 3714946; E-mail: wkc@ms.fjirsm.ac.cn.

specified NLO properties become the key and efficient way in developing novel NLO materials. However, the high-level computational study of the NLO property of the inorganic transition-metal clusters remains less unexplored, especially within the first-principle theoretical framework. The special characteristics of a transition-metal cluster include the large size of structural configuration, large number of electrons and nonnegligible relativistic effect. The traditional high-level accuracy computational methods such as coupled-cluster method (CC) and analytical coupled-perturbated HF method (CPHF) would encounter great difficulties in calculating the electronic polarizability and hyperpolarizability of transition-metal cluster due to the limited computer resources, particularly when the cluster has no symmetry. But in screening potential NLO materials the efficiency and timesaving are critically required. To this end the finite-field (F-F) approach, which may track back to the Buckingham expansion [2], shows its value. The electronic polarizability and hyperpolarizability could be obtained by calculating the differences of total energies or dipole moments in the weak external electric fields. Relatively speaking, the computational requirements for F-F calculations is not that critical as CC or CPHF methods because it may perform single-point total energy calculation at SCF level (with external electric field being applied). Consequently, F-F method is a suitable choice for a complex system such as transition-metal cluster, which is usually with low symmetry or even without any symmetry. In our previous report, we did some work in the formulation of F-F methods for NLO polarizability calculations of transition-metal clusters [3]. In this paper, we will applied the method to a new synthesized di-nuclear homometallic cluster compound, $[Mo_2S_3(CO)_6(C_6H_{11})_3]\cdot N(C_2H_3)_4$, with the emphasis on the first hyperpolarizability. Our results showed the efficiency of F-F method in evaluating the NLO property of a complex system containing transition-metal elements. This new synthesized cluster was estimated to have very large second-order NLO property, which may lead to the discovery of a new material.

In Section 2 of this paper, we presented the detail of F-F method used in the (hyper)polarizability calculations and the computational model clusters we used in the calculations. In the Section 3, we presented the calculated results of the static molecular (hyper)polarizability of the subject cluster and discussed the structural origin of the NLO properties. Finally, we in the last section summarized the conclusions and evaluate the potential NLO application of the subject cluster compound.

2. Computational method and theoretical models

According to the F-F method, the total energy of an isolated system in the presence of a sufficiently weak electric field can be expressed by the Buckingham type expansion [4]:

$$E(\vec{F}) = E(0,0,0) - \mu_i F_i - \frac{1}{2}\alpha_{ij}F_iF_j - \frac{1}{6}\beta_{ijk}F_iF_jF_k - \frac{1}{24}\gamma_{ijkl}F_iF_jF_kF_l - \ldots (i,j,k = x,y,z)$$

where $\mu_i, \alpha_{ij}, \beta_{ijk}$ and γ_{ijkl} are the dipole moment, polarizability, first hyperpolarizabilities and the second hyperpolarizability, respectively and $\vec{F} = F_x\vec{x} + F_y\vec{y} + F_z\vec{z},$. We has summarized, in our previous paper [3], a general formulation classified by symmetry,

$$\alpha_{xx} = \frac{1}{(\sigma^2-\sigma^4)F^2}\{\sigma^4[\Delta E(F,0,0) + \Delta E(-F,0,0)] - [\Delta E(\sigma F,0,0) + \Delta E(-\sigma F,0,0)]\}$$

$$\alpha_{yy} = \frac{1}{(\sigma^2-\sigma^4)F^2}\{\sigma^4[\Delta E(0,F,0) + \Delta E(0,-F,0)] - [\Delta E(0,\sigma F,0) + \Delta E(0,-\sigma F,0)]\}$$

$$\alpha_{zz} = \frac{1}{(\sigma^2 - \sigma^4)F^2}\{\sigma^4[\Delta E(0,0,F) + \Delta E(0,0,-F)] - [\Delta E(0,0,\sigma F) + \Delta E(0,0,-\sigma F)]\},$$

$$\beta_{xxx} = \frac{3}{(\sigma^3 - \sigma)F^3}\{\sigma[\Delta E(F,0,0) - \Delta E(-F,0,0)] - [\Delta E(\sigma F,0,0) - \Delta E(-\sigma F,0,0)]\},$$

$$\beta_{yyy} = \frac{3}{(\sigma^3 - \sigma)F^3}\{\sigma[\Delta E(0,F,0) - \Delta E(0,-F,0)] - [\Delta E(0,\sigma F,0) - \Delta E(0,-\sigma F,0)]\},$$

$$\beta_{zzz} = \frac{3}{(\sigma^3 - \sigma)F^3}\{\sigma[\Delta E(0,0,F) - \Delta E(0,0,-F)] - [\Delta E(0,0,\sigma F) - \Delta E(0,0,-\sigma E)]\},$$

$$\beta_{xxy} = \frac{1}{2F^3}\{2\Delta E(0,F,0) - 2\Delta E(0,-F,0) - \Delta E(F,F,0) - \Delta E(-F,F,0) + \Delta E(F,-F,0)$$
$$+\Delta E(-F,-F,0)\},$$

$$\beta_{yyz} = \frac{1}{2F^3}\{2\Delta E(0,0,F) - 2\Delta E(0,0,-F) - \Delta E(0,F,F) - \Delta E(0,-F,F) + \Delta E(F,F,-F)$$
$$+\Delta E(0,-F,-F)\},$$

$$\beta_{zzx} = \frac{1}{2F^3}\{2\Delta E(F,0,0) - 2\Delta E(-F,0,0) - \Delta E(F,0,F) - \Delta E(F,0,-F) + \Delta E(-F,0,F)$$
$$+\Delta E(-F,0,-F)\},$$

$$\beta_{xyy} = \frac{1}{2F^3}\{2\Delta E(F,0,0) - 2\Delta E(-F,0,0) - \Delta E(F,F,0) + \Delta E(-F,F,0) - \Delta E(F,-F,0)$$
$$+\Delta E(-F,-F,0)\},$$

$$\beta_{yzz} = \frac{1}{2F^3}\{2\Delta E(0,F,0) - 2\Delta E(0,-F,0) - \Delta E(0,F,F) + \Delta E(0,-F,F) - \Delta E(0,F,-F)$$
$$+\Delta E(0,-F,-F)\},$$

$$\beta_{zxx} = \frac{1}{2F^3}\{2\Delta E(0,0,F) - 2\Delta E(0,0,-F) - \Delta E(F,0,F) + \Delta E(F,0,-F) - \Delta E(-F,0,F)$$
$$+\Delta E(-F,0,-F)\},$$

$$\beta_{xyz} = \frac{1}{(\sigma - \sigma^2)F^3}\{(2\sigma + \sigma^2 - 2)[\Delta E(F,0,0) + \Delta E(0,F,0) + \Delta E(0,0,F)]$$
$$-[\Delta E(\sigma F,0,0) + \Delta E(0,\sigma F,0) + \Delta E(0,0,\sigma F)] - (\sigma^2 + 2\sigma - 1)$$
$$\cdot[\Delta E(F,F,0) + \Delta E(0,F,F) + \Delta E(F,0,F) + [\Delta E(\sigma F,F,0) + \Delta E(0,\sigma F,F)$$
$$+\Delta E(F,0,\sigma F) + \Delta E(0,\sigma F,0) + \Delta E(0,F,\sigma) + \Delta E(\sigma F,0,F)$$
$$+(\sigma^2 + 2\sigma)\Delta E(F,F,F) - [\Delta E(\sigma F,F,F) + \Delta E(F,\sigma F,F) + \Delta E(F,F,\sigma F)]\}$$

$$\gamma_{xxxx} = \frac{12}{(\sigma^4 - \sigma^2)F^4}\{\sigma^2[\Delta E(F,0,0) + \Delta E(-F,0,0)] - [\Delta E(\sigma F,0,0) + \Delta E(-\sigma F,0,0)]\},$$

$$\gamma_{yyyy} = \frac{12}{(\sigma 64 - \sigma^2)F^4}\{\sigma^2[\Delta E(0,F,0) + \Delta E(0,-F,0)] - [\Delta E(0,\sigma F,0) + \Delta E(0,-\sigma F,0)]\},$$

$$\gamma_{zzzz} = \frac{12}{(\sigma^4 - \sigma^2)F^4}\{\sigma^2[\Delta E(0,0,F) + \Delta E(0,0,-F)] - [\Delta E(0,0,\sigma F) + \Delta E(0,0,-\sigma F)]\},$$

where $\Delta E(F,0,0) = E(F,0,0) - E(0,0,0)$ and so on. The homogenous external electric field has a value of 0.005 a.u., σ is set to 0.6. Our study on a series of tri-nuclear transition-metal cluster showed that these two value, $F = 0.005$ a.u. and $\sigma F = 0.003$ a.u., could provide reliable results of NLO polarizability [5].

The subject transition-metal cluster crystal belongs to Pna_{21}, a non-centric symmetric space group. The detail description of the crystal structure will be reported in another paper. The neutral molecular structural building unit, $[Mo_2S_3(CO)_6(C_6H_{11})_3]\cdot N(C_2H_3)_4$, was used as model cluster **I** (Fig. 1), which were consisted of a negatively charged metal cluster $[Mo_2S_3(CO)_6(C_6H_{11})_3]^-$ and the ionic organic tetraethyl amine group $[N(C_2H_3)_4]^+$. A trigonally tetrahedral structure was formed by metal cluster core $[Mo_2S_3]$, which was linked to organic ligands. The chemical environments of Mo_1 and Mo_2 were slightly different. The S atoms coordinated to Mo_1/Mo_2 atoms were not equivalent as well. There were three carboxyl ligands linked to each molybdenum atom, and there was a hexamethyl thioalcohol ligand, (C_6H_{11}) link to each S atom. We set up the molecular coordination as being shown in Fig. 1 with the z direction along the connection line of Mo_1 and Mo_2. In order to analyze the role of another building unit, ionic group $[N(C_2H_3)_4]^+$, played in the NLO response, we set another theoretical model, model cluster **II**, which was constructed only by anionic metal cluster $[Mo_2S_3(C_6H_{11})_3]^-$, with the coordinate unchanged.

The ECP/Lanl2dz basis sets [6] were used to take the relativistic effect of transition metal molybdenum into account. This ECP/Lanl2dz basis set is a small basis set but the number of basis functions for the subject model reached 525 and the primitive gaussians was 1323. The larger basis set is impractical for the subject transition metal cluster. The previous study by this basis set showed the validity in analysis the nature of NLO property of metal clusters [5]. We did not consider the electronic correlation effect (such as to MP2 level) and extended diffuse functions due to the large number of the electrons and nano-size structure (~4 nm), which would ask for the computer resources that we could not afford.

All the computations were performed by Gaussian 98 program [7] on a 10-node parallel Pentium IV PC cluster. We used the traditional electrostatic unit (esu) for NLO polarizability throughout the paper.

3. Results and discussions

The calculated static first hyperpolarizability of the model **I** and model **II** were listed in Table 1. The spatial average value of the fist hyperpolarizability in z direction, $\beta_z = \beta_{zxx} + \beta_{zyy} + \beta_{zzz}$, of the neutral model was 13.8×10^{-30} esu. This value was in the same order as the β value of the well-know NLO organic molecule, PNA molecule [8]. However, the subject transition-metal cluster was found to have a larger density with 4 molecular clusters in a unit cell. As a result, the larger macroscopic second-order NLO effect of the subject compound could be strongly expected.

Table 1
The calculated independent components of the static first hyperpolarizability ($\times 10^{-30}$ esu) and the spatial average values β_i of model **I**.

	β_{xxx}	β_{xxy}	β_{xxz}	β_{xyy}	β_{xzz}	β_{yyy}	β_{yyz}	β_{yzz}	β_{zzz}	β_x	β_y	β_z
Model **I**	−0.2	−2.6	0.2	0.3	0.2	4.5	0.2	−1.4	13.5	0.4	0.5	13.8
Model **II**	−0.2	−2.7	0.6	0.2	0.6	3.6	0.1	1.9	−2.6	0.5	2.8	−3.1

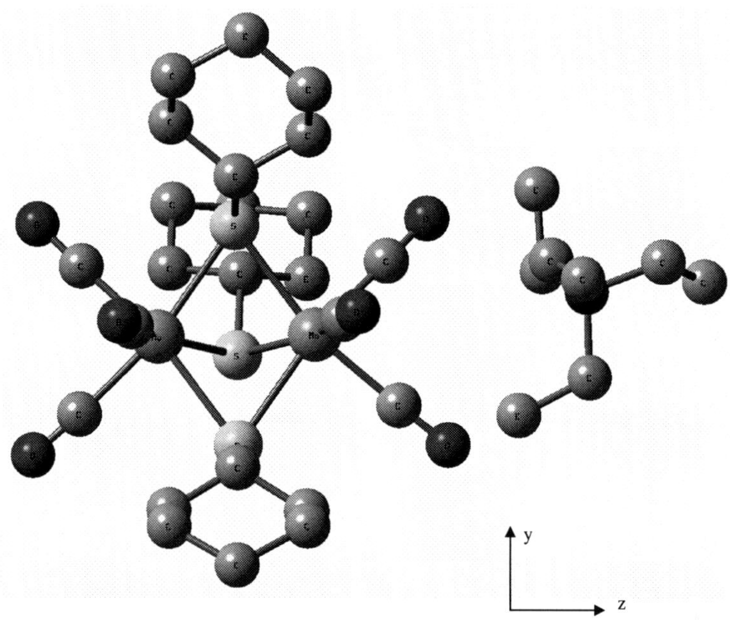

Fig. 1. The molecular structure of model **I** (hydrogen atoms are not shown).

The calculated results provided the valuable information to understand the NLO nature of the compound. The largest spatial average β value was in z direction, which was about three times larger than β_x and β_y. And in z direction we could find the connection of Mo_1-Mo_2, and the tetraethyl amine ionic group (Fig. 1). In our previous report [5], we stated that the charge transfer (CT) between the metal would be the origin the hyperpolarizability of the metal cluster. Once again, we found this situation in this subject cluster. In addition, the inter-group interaction between the metal cluster $[Mo_2S_3(C_6H_5)_3]^-$ and the positively charged organic group $[N(C_2H_3)_4]^+$ obviously play a role in the enhancement of the β_z value. The β_z value of model **II**, although kept to be the largest one among the three β_i values, reduced to about one fifth of that of model **I**. The electrostatic interaction between $[N(C_2H_3)_4]^+$ and $[Mo_2S_3(C_6H_5)_3]^-$ positively affect the CT inside the metal core. Along y direction, which is perpendicular to Mo_1-Mo_2 connection, β_y value of model **I** was about one sixth smaller than that of model **II**. This indicated that in y direction, the electrostatic interaction between ionic group and anionic group played a negative role that was unfavorable to the enhancement of β_y value. In x direction, this electrostatic interactions made almost no effect on β_x, which could be clearly expected from its isotropic structural orientation along x direction.

The diagonal components of the polarizability and second hyperpolarizability of model **I** were listed in Table 2. The anisotropic of α and γ values were impressive. Being similar to β_i, both the largest component of α and γ occurred in z direction. For example, γ_{zzzz} was about one-order larger in magnitude than other two diagonal components. γ_{zzzz} had a quite large value of -150×10^{-36} esu,

Table 2
The calculated diagonal components of polarizability ($\times 10^{-24}$ cm^3) and second hyperpolarizability ($\times 10^{-36}$ esu) of model **I**

α_{zz}	α_{zz}	α_{zz}	γ_{xxxx}	γ_{xxxx}	γ_{xxxx}
62.1	64.2	99.8	12	11	−150

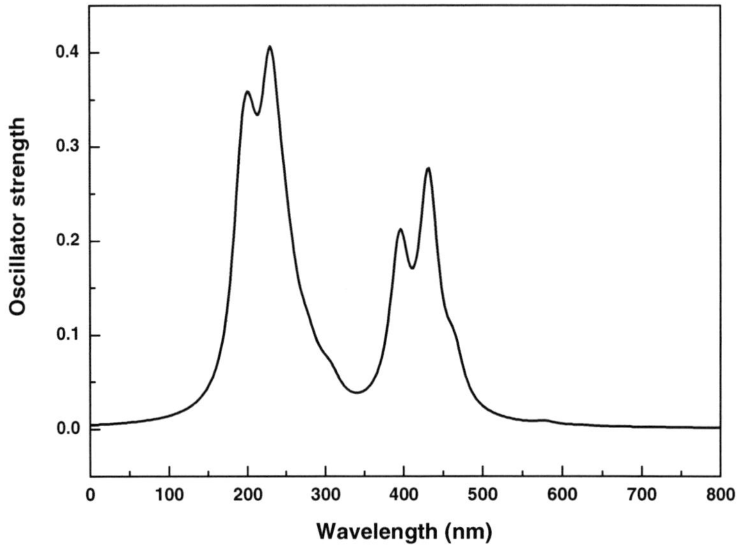

Fig. 2. The calculated electronic absorption spectrum of model **I**.

which means that, in addition to the practical potential in SHG application, the third-order NLO property of this compound, such as self-defocusing (γ_{zzzz} is negative) [9] is a very interesting subject for further studies.

The electronic excitation properties provided another way to help us to understand the origin of NLO effect [10]. We calculated the electronic excitation spectrum of model **I** by using the time-dependent HF (TDHF) method [11] at Lanl2dz level. The results in Fig. 2 showed that there were two absorption bands in the region of 200 nm–500 nm. The detail analysis of the excitation energy indicated that the low-energy (LE) absorption band centered at 431 nm was contributed by the excitation transition from HOMO-2 (the second orbital below the highest occupied molecular orbital) to LUMO (the lowest unoccupied molecular orbital). On the other hand, the high-energy (HE) absorption band centered at 230 nm was dominated by the excitation from the inner occupied orbitals HOMO-7 to the LUMO. Figure 3 showed the orbital contours surfaces [12] of HOMO-7, HOM-2 and LUMO, respectively. It clearly depicted that LE band was resulted from the CT inside the metal cluster, while the HE was resulted from the CT from hexamethyl thioalcohol ligands to the metal core. TDHF has the well-know systematic overestimation in predicting excitation energies, but the results qualitatively agreed with the above analysis we obtained from hyperpolarizability calculations. In addition, we found the important role that the ligands $(C_6H_{11})_3$ played in CT. The result agreed with our previous conclusion that in transition metal cluster the joint effect of the ligands and metal core cluster could sufficiently enhance the β value [13].

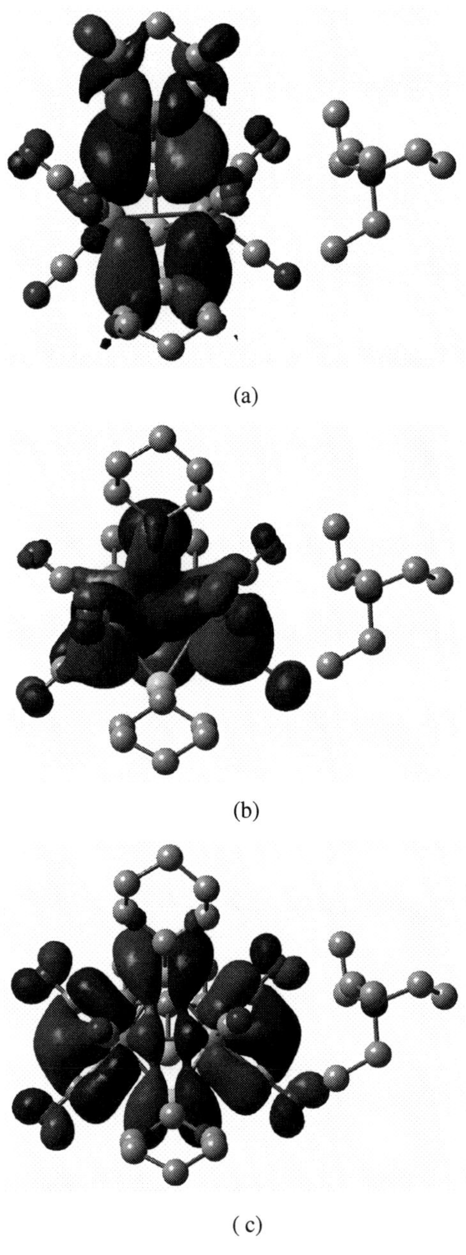

Fig. 3. The calculated molecular orbital surfaces of model **I**, (a) HOMO-7; (b) HOMO-2; (c) LUMO.

4. Conclusions

We have applied F-F approach within the first-principle theoretical framework to a new synthesized transition-metal cluster compound, $[Mo_2S_3(C_6H_{11})_3]\cdot N(C_2H_3)_4$. A large first hyperpolarizability along z direction was found to have the value of 13.8×10^{-30} esu. The practical value of the theoretical result lies in that the cluster component belongs to non-centric symmetric space group and the X-ray crystallographic experiments has showed no opposite orientations of the constituent clusters in the unit cell. It would be a promising candidate for the second-order NLO effect in IR region.

As have been mentioned, one of promising applications of transition metal clusters as NLO materials is IR SHG process due to the remarkable electronic structure characteristic and notable anisotropic structural configurations. Meanwhile, these special electronic and geometric structural properties make them very difficult for the first-principle calculation. Our study in this paper may provide an example of using *ab initio*/F-F method to effectively deal with the NLO properties of transition metal cluster. Another merit of F-F numerical method lies in the possibility of using higher-level calculation for polarizability and hyperpolarizability, such as MPn and CCSD. But in this paper, we did not describe the higher-level calculation because of the tremendous long computing time.

The theoretical results of NLO polarizability and electronic absorption spectrum presented the relationship between the structure of transition metal cluster and the NLO properties. The dominant source of NLO response came from the CT inside the metal core group $[Mo_2S_3(C_6H_{11})_3]$, which is clearly different to the D-π-A (D, electron donor, A, electron acceptor, π, conjugate π-bridge) mode in organic molecular system and is not identical to the MLCT (metal to ligand CT) or LMCT (ligand to metal CT) mode we often see in organometallic complexes as well. The metal core group may possess all the CT bridge (stereo d-p π system has been proposed), charge donor and accepter in one. All the necessary characters that play roles in NLO response may integrate in a metal cluster. In addition, they could have two-dimension CT or even three-dimension CT inside the metal core, which may make very high efficiency in CT process resulting in very large hyperpolarizability.

In our previous studies, the metal cluster has been focused with the neglecting of other structural unit. But since the subject metal cluster in this paper is charged, the organic ironic group $[N(C_2H_3)_4]^+$, came to play a role in the enhancement of β_z. The question still remains open that how the weak inter-group interaction between metal core and ironic group contribute to the NLO polarizability.

Although the F-F approach is very effective in transition-metal cluster NLO properties study, some restrictions exit. For instance, only static (hyper)polarizability are provided. They could not be able to directly compare with the experiments, which are usually measured at certain laser frequency. In spite of this, the F-F approach is one of the effective methods in NLO polarizability study, especially for those complex system containing transition-metal elements.

Acknowledgement

We thank National Science Foundation of China (69978021, 20173064, 90203017) for the financial supports.

References

[1] S.P. Karna, *J. Phys. Chem.* **104** (2000), 46571.
[2] A.D. Buckingham, *J. Chem. Phys.* **30** (1959), 1580.
[3] X. Chen, C. Lin, K. Wu, B. Zhuang, R. Sa, S. Peng and Z. Zhou, *Chin. J. Struct. Chem.* **20** (2001), 369.
[4] A.D. Buckingham, *Adv. Chem. Phys.* **12** (1967), 107.
[5] X. Chen, K. Wu and C. Lin, *Inorg. Chem.* **42** (20030, 532.
[6] P.J. Hay and W.R. Wadt, *J. Chem. Phys.* **82** (1985), 270.
[7] Gaussian 98, Revision A.11.3, M.J. Frisch, G.W. Trucks, H.B. Schlegel, G.E. Scuseria, M.A. Robb, J.R. Cheeseman, V.G. Zakrzewski, J.A. Montgomery, Jr., R.E. Stratmann, J.C. Burant, S. Dapprich, J.M. Millam, A.D. Daniels, K.N. Kudin, M.C. Strain, O. Farkas, J. Tomasi, V. Barone, M. Cossi, R. Cammi, B. Mennucci, C. Pomelli, C. Adamo, S. Clifford, J. Ochterski, G.A. Petersson, P.Y. Ayala, Q. Cui, K. Morokuma, N. Rega, P. Salvador, J.J. Dannenberg, D.K. Malick, A.D.

Rabuck, K. Raghavachari, J.B. Foresman, J. Cioslowski, J.V. Ortiz, A.G. Baboul, B.B. Stefanov, G. Liu, A. Liashenko, P. Piskorz, I. Komaromi, R. Gomperts, R.L. Martin, D.J. Fox, T. Keith, M.A. Al-Laham, C.Y. Peng, A. Nanayakkara, M. Challacombe, P.M.W. Gill, B. Johnson, W. Chen, M.W. Wong, J.L. Andres, C. Gonzalez, M. Head-Gordon, E.S. Replogle and J.A. Pople, Gaussian, Inc., Pittsburgh PA, 2002.

[8] C.G. Teng and A.F. Garito, *Phys. Rev. Lett.* **50** (1983), 350.

[9] Y.R. Shen, *The Principles of Nonlinear Optics*, John Wiley & Sons, Inc., 1984, p. 330.

[10] G. Ricciardi, A. Rosa, S.J.A. van Gisbergen and E.J. Baerends, *J. Phys. Chem. A* **104** (2000), 635.

[11] R. Bayermschmitt and R. Ahlrichs, *Chem. Phys. Lett* **256** (1996), 454.

[12] GaussView 2.1, Gaussian, Inc., Pittsburgh PA, 2000.

[13] K. Wu, P. Lin, X. Wu and L. Chen, *Laser Chem* **18** (2000), 193.

Journal of Computational Methods in Sciences and Engineering 4 (2004) 461–479
IOS Press

Real-space computational methods for linear and nonlinear polarizabilities

Jun-Ichi Iwata[a,*], Kazuhiro Yabana[b] and George F. Bertsch[c]
[a]*Nanotechnology Research Institute, National Institute of Advanced Industrial Science and Technology (AIST), Tsukuba 305-8568, Japan*
[b]*Institute of Physics, University of Tsukuba, Tsukuba 305-8571, Japan*
[c]*Department of Physics and National Institute for Nuclear Theory, University of Washington, Seattle, Washington 98195, USA*

Abstract. We present a real-space method for computing the linear and nonlinear polarizabilities in the time-dependent density-functional theory. In our method, the three-dimensional Cartesian coordinate is discretized on a uniform grid and the wave functions are represented on the grid points. The dynamic polarizabilities may be calculated perturbatively by expanding the dipole moment in a power series of the external field strength or directly by integrating the real-time equations of motion under the time-varying external field. We describe the equations and computational techniques for the both methods. Besides calculating second-order and third-order hyperpolarizabilities, the method can be applied to photon-absorption cross sections, and we show examples of this application as well. We also present here for the first time calculations of the nonlinear polarizabilities using the real-time method.

Keywords: Time-dependent density-functional theory, real-space method, real-time method, polarizability, hyperpolarizability, response function method

PACS: 42.65.An, 42.65.Ky, 42.50.Hz, 31.15.Ew, 31.15.Fx, 02.70.Bf

1. Introduction

Optical properties of atoms and molecules are characterized by their linear and nonlinear polarizabilities. They reflect time-dependent dynamics of electrons under periodic external fields. Among numerous theoretical approaches developed to calculate polarizabilities, the time-dependent density-functional theory (TDDFT) has attracted broad interest and is expected to be quite practical to describe optical responses of large systems [1–8].

In the TDDFT, many-electron dynamics is described by the time-dependent Kohn-Sham equation, an effective single-particle Schrödinger equation in which electron-electron interactions are treated through a time-varying potential [9]. We have been developing and applying the real-space and real-time approaches for the dynamic polarizabilities based on the TDDFT [3,4,10–14]. In this real-space method, the kinetic operator in the Kohn-Sham is represented by a finite different approximation on a uniform Cartesian grid. Since the potential is usually local in the coordinate representation, the real-space method offers the economy of sparse matrix operations.

*Corresponding author. E-mail: iwata@cm.ph.tsukuba.ac.jp.

We have developed two distinct approaches to calculate linear and nonlinear polarizabilities in the uniform grid representation. One is the real-time method in which the time-dependent Kohn-Sham equation is solved explicitly in time domain [3]. This is especially useful for the calculations of linear polarizability. The other is the perturbative formalism [4,14]. In this method, the response of the system is calculated for a perturbation of a given frequency.

In this paper, we first give an overview of the real-space and real-time method. We then present two new results. One is the real-time computation of the nonlinear polarizabilities. Our previous application of the real-time method was limited to the calculation of the linear polarizability [3]. We here report for the first time the calculation of the nonlinear polarizabilities in the real-space and real-time method, taking high order harmonic generation of Ar atom as example. The other is the calculation of the two-photon absorption cross section. The two-photon absorption is related to the imaginary part of a certain third-order nonlinear polarizability. In Ref. [4], we have reported a real-space computational method of the molecular hyperpolarizabilities. It was, however, limited to the computation of the real part. We here report the calculation of the imaginary part of the hyperpolarizability. Results are presented for Ar atom and ethylene molecule.

2. Formalism

2.1. Definitions of polarizabilities

In this section, we summarize the fundamental quantities, the polarizabilities, which characterize linear and nonlinear optical responses of atoms and molecules [15]. We consider an external field of a superposition of monochromatic waves

$$\mathbf{E}(t) = \frac{1}{2} \sum_{\omega_i \geqslant 0} \left\{ \mathbf{E}_{\omega_i} e^{-i\omega_i t} + \mathbf{E}_{-\omega_i} e^{i\omega_i t} \right\}. \tag{1}$$

When the strength of the fields are moderate, the induced dipole moment can be expanded as the following power series.

$$\mathbf{p}(t) = \mathbf{p}^{(0)} + \mathbf{p}^{(1)}(t) + \mathbf{p}^{(2)}(t) + \mathbf{p}^{(3)}(t) + \cdots. \tag{2}$$

Each $\mathbf{p}^{(k)}(t)$ can be expressed as

$$\mathbf{p}^{(k)}(t) = \frac{1}{2} \sum_{\omega_\sigma \geqslant 0} \left\{ \mathbf{p}^{(k)}_{\omega_\sigma} e^{-i\omega_\sigma t} + \mathbf{p}^{(k)}_{-\omega_\sigma} e^{i\omega_\sigma t} \right\}, \tag{3}$$

where the Cartesian μ-component of $\mathbf{p}^{(k)}_{\omega_\sigma}$ is given by

$$\begin{aligned} \left(\mathbf{p}^{(k)}_{\omega_\sigma}\right)_\mu = \frac{1}{k!} \sum_{\alpha_1 \cdots \alpha_k} \sum_{\{\omega_1 \cdots \omega_k\}} & K(-\omega_\sigma; \omega_1, \cdots, \omega_k) \chi^{(k)}_{\mu\alpha_1\cdots\alpha_k}(-\omega_\sigma; \omega_1, \cdots, \omega_k) \\ & (\mathbf{E}_{\omega_1})_{\alpha_1} \cdots (\mathbf{E}_{\omega_k})_{\alpha_k}. \end{aligned} \tag{4}$$

The $\chi^{(k)}_{\mu\alpha_1\cdots\alpha_k}(-\omega_\sigma; \omega_1, \cdots, \omega_k)$ is the k-th order nonlinear polarizability tensor. The sum $\sum_{\{\omega_1 \cdots \omega_k\}}$ is taken over all distinct sets of $\{\omega_1, \cdots, \omega_k\}$ whose sum is equal to ω_σ. $K(-\omega_\sigma; \omega_1, \cdots, \omega_k)$ is a numerical factor given by $2^{l+m-k}p$ where p is the number of distinct permutations of $\{\omega_1, \cdots, \omega_k\}$, k is the order of nonlinearity, and m is the number of frequencies which are equal to zero. $l = 1$ if $\omega_\sigma = 0$ and $l = 0$ otherwise.

At present, a few of the lowest order polarizabilities (up to third order in usual) are the subjects of many theoretical and experimental studies, because they are related to the important physical phenomena, such as harmonic generations, Kerr effects, photoabsorptions, wave-mixings, and so on [15]. The nonlinear polarizabilities up to third order are often described by the special symbols, $\alpha(-\omega_1;\omega_1)$ for linear polarizability ($k = 1$), $\beta(-\omega_\sigma;\omega_1,\omega_2)$ for $k = 2$ and $\gamma(-\omega_\sigma;\omega_1,\omega_2,\omega_3)$ for $k = 3$ nonlinear polarizabilities. These symbols may also be used in this paper.

2.2. *Time-dependent density-functional theory*

The density-functional theory has been remarkably successful in describing electronic ground states and now plays a central role in the first-principle calculations of material properties. The TDDFT is an extension of the density-functional theory to describe the electronic dynamics induced by a time-varying external field [9,11]. The strategy to describe linear and nonlinear response properties in TDDFT is given below.

We consider a quantum system under the external potential $v^{\text{ext}}(\mathbf{r}, t)$. In the TDDFT, the time evolution of the density distribution $n(\mathbf{r}, t)$ is described by the time-dependent Kohn-Sham equation,

$$i\hbar\frac{\partial}{\partial t}\psi_i(\mathbf{r}, t) = \left\{h[n(t)] + v^{\text{ext}}(\mathbf{r}, t)\right\}\psi_i(\mathbf{r}, t), \tag{5}$$

$$h[n(t)] = -\frac{\hbar^2}{2m}\nabla^2 + V_{\text{ion}}(\mathbf{r}) + e^2\int d\mathbf{r}'\frac{n(\mathbf{r}', t)}{|\mathbf{r} - \mathbf{r}'|} + \mu_{xc}[n(t)], \tag{6}$$

$$n(\mathbf{r}, t) = \sum_{i\in occ}|\psi_i(\mathbf{r}, t)|^2, \tag{7}$$

where $\mu[n(t)]$ expresses the exchange-correlation potential.

The dipole field $\mathbf{E}(t)$ of Eq. (1) is expressed as the external potential,

$$v^{\text{ext}}(\mathbf{r}, t) = \frac{1}{2}\sum_{\omega_i\geqslant 0}\left\{v^{\text{ext}}_{\omega_i}(\mathbf{r})e^{-i\omega_i t} + v^{\text{ext}}_{-\omega_i}(\mathbf{r})e^{i\omega_i t}\right\}, \tag{8}$$

with

$$v^{\text{ext}}_{\omega_i}(\mathbf{r}) = e\mathbf{r}\mathbf{E}_{\omega_i}. \tag{9}$$

When the external field is not so strong, one can expand the density $n(\mathbf{r}, t)$ in a power series as,

$$n(\mathbf{r}, t) = n^{(0)}(\mathbf{r}) + n^{(1)}(\mathbf{r}, t) + n^{(2)}(\mathbf{r}, t) + n^{(3)}(\mathbf{r}, t) + \cdots. \tag{10}$$

Each $n^{(k)}(\mathbf{r}, t)$ can be expressed in a similar way to Eq. (3) where the general form of the dipole moment $\mathbf{p}^{(k)}(t)$ is given,

$$n^{(k)}(\mathbf{r}, t) = \frac{1}{2}\sum_{\omega_\sigma\geqslant 0}\left\{n^{(k)}_{\omega_\sigma}(\mathbf{r})e^{-i\omega_\sigma t} + n^{(k)}_{-\omega_\sigma}(\mathbf{r})e^{i\omega_\sigma t}\right\}, \tag{11}$$

with

$$n^{(k)}_{\omega_\sigma}(\mathbf{r}) = \frac{1}{k!} \sum_{\alpha_1 \cdots \alpha_k} \sum_{\{\omega_1 \cdots \omega_k\}} K(-\omega_\sigma; \omega_1, \cdots, \omega_k) n^{(k)}_{\alpha_1 \cdots \alpha_k}(\mathbf{r}| - \omega_\sigma; \omega_1, \cdots, \omega_k)$$
$$(\mathbf{E}_{\omega_1})_{\alpha_1} \cdots (\mathbf{E}_{\omega_k})_{\alpha_k}. \tag{12}$$

The k-th order nonlinear polarizability tensor can then be expressed in terms of the k-th order density as

$$\chi^{(k)}_{\mu\alpha_1 \cdots \alpha_n}(-\omega_\sigma; \omega_1, \cdots, \omega_k) = \int d\mathbf{r}(-er_\mu) n^{(k)}_{\alpha_1, \cdots, \alpha_k}(\mathbf{r}| - \omega_\sigma; \omega_1, \cdots, \omega_k). \tag{13}$$

In short, a k-th order polarizability can be obtained through the perturbative expansion of the time-dependent density Eq. (10), and pulling out a component of the density or its dipole moment which has the same order and the same Fourier component of the k-th order polarizability.

2.3. Perturbative formalism

We will consider two distinct approaches to calculate linear and nonlinear polarizabilities in TDDFT. One is the real-time method where the polarizabilities are calculated directly from the real-time evolution of the wave functions. The method is non-perturbative, namely the effects of the all orders of perturbation series are involved. However, one can pull out a few lowest order responses to the applied external potential of moderate intensity by sorting the Fourier components of the induced dipole moment. In this method, the polarizabilities can be obtained directly from the definitions given in the previous subsection and no further development of the formalism is required, in other words the working equation of the method is nothing but Eqs (5)–(7). The details of the method will be discussed in Section 4.

The other approach is the perturbative method which is more traditional one to calculate polarizabilities [1,2], however the working equations are quite different from the original time-dependent Kohn-Sham equation. The perturbative method for nonlinear response problems has been developed for atoms [2] and molecules [4]. We here briefly summarize the basic formalism and the working equations suitable for practical numerical calculations.

The basic quantities to be calculated in the perturbative formalism is the individual k-th order density of definite Fourier component $n^{(k)}_{\mu\alpha_1, \cdots, \alpha_n}(\mathbf{r}| - \omega_\sigma; \omega_1, \cdots, \omega_k)$. A perturbation expansion gives a hierarchy of equations which determine these densities order by order. The first-order equation is the linear response equation [1],

$$n^{(1)}_{\alpha_1}(\mathbf{r}| - \omega_\sigma; \omega_\sigma) = \int d\mathbf{r}_1 \chi^{(1)}_0(\mathbf{r}, \mathbf{r}_1; \omega_\sigma) h^{(1)}_{\alpha_1}(\mathbf{r}_1| - \omega_\sigma; \omega_\sigma), \tag{14}$$

$$h^{(1)}_{\alpha_1}(\mathbf{r}| - \omega_\sigma; \omega_\sigma) = \int d\mathbf{r}_1 \left\{ \frac{e^2}{|\mathbf{r} - \mathbf{r}_1|} + f_{xc}(\mathbf{r}, \mathbf{r}_1; \omega_\sigma) \right\} n^{(1)}_{\alpha_1}(\mathbf{r}_1| - \omega_\sigma; \omega_\sigma) + er_{\alpha_1}, \tag{15}$$

where ω_σ is one of the frequency of the external field $\{\pm\omega_i\}$. The (independent-particle version) frequency-domain density-density response function $\chi^{(1)}_0(\mathbf{r}, \mathbf{r}'; \omega_\sigma)$ and exchange-correlation kernel

$f_{xc}(\mathbf{r},\mathbf{r}';\omega_\sigma)$ which is a functional derivative of the exchange-correlation potential with respect to the density, are introduced. The explicit expression of $\chi_0^{(1)}(\mathbf{r},\mathbf{r}';\omega_\sigma)$ is given by [1,2],

$$\chi_0^{(1)}(\mathbf{r},\mathbf{r}';\omega_\sigma) = \sum_i^{occ.} \Big[\phi_i^*(\mathbf{r})G^{(+)}(\mathbf{r},\mathbf{r}';\varepsilon_i+\hbar\omega_\sigma)\phi_i(\mathbf{r}') + \phi_i^*(\mathbf{r}')G^{(-)}(\mathbf{r}',\mathbf{r};\varepsilon_i-\hbar\omega_\sigma)\phi_i(\mathbf{r}) \Big], \tag{16}$$

where $G^{(\pm)}(\mathbf{r},\mathbf{r}';\omega_\sigma)$ is the (frequency-domain) single-particle Green function [1,14],

$$G^{(\pm)}(\mathbf{r},\mathbf{r}';E) = \sum_\mu \frac{\phi_\mu(\mathbf{r})\phi_\mu^*(\mathbf{r}')}{E-\varepsilon_\mu \pm i\eta}, \tag{17}$$

where η is a positive infinitesimal. Equations (14)–(16) have been solved directly with explicit construction of the response function for spherical systems such as atoms [1] or jellium clusters [16,17]. For systems without spherical symmetry, however, the integral Eq. (14) should be recasted into the equivalent differential equation,

$$n_{\alpha_1}^{(1)}(\mathbf{r}|-\omega_\sigma;\omega_\sigma) = \sum_{i\in occ.} \left\{ \phi_i^*(\mathbf{r})\psi_{i,\alpha_1}^{(1)}(\mathbf{r}|-\omega_\sigma;\omega_\sigma) + \phi_i(\mathbf{r})\psi_{i,\alpha_1}^{(1)*}(\mathbf{r}|\omega_\sigma;-\omega_\sigma) \right\}, \tag{18}$$

where $\psi_{i,\alpha_1}^{(1)}(\mathbf{r}|-\omega_\sigma;\omega_\sigma)$, $\psi_{i,\alpha_1}^{(1)*}(\mathbf{r}|\omega_\sigma;-\omega_\sigma)$ satisfy the following differential equations,

$$(\varepsilon_i+\hbar\omega_\sigma-h[n])\psi_{i,\alpha_1}^{(1)}(\mathbf{r}|-\omega_\sigma;\omega_\sigma) = h_{\alpha_1}^{(1)}(\mathbf{r}|-\omega_\sigma;\omega_\sigma)\phi_i(\mathbf{r}), \tag{19}$$

$$(\varepsilon_i-\hbar\omega_\sigma-h[n])\psi_{i,\alpha_1}^{(1)}(\mathbf{r}|\omega_\sigma;-\omega_\sigma) = h_{\alpha_1}^{(1)*}(\mathbf{r}|-\omega_\sigma;\omega_\sigma)\phi_i(\mathbf{r}). \tag{20}$$

These are known as the Sternheimer equations. We call the present differential form of the linear response problem as the modified Sternheimer method [18].

The second- and the third-order equations can be developed systematically. The second-order equations are

$$\begin{aligned} n_{\alpha_1\alpha_2}^{(2)}(\mathbf{r}|-\omega_\sigma;\omega_1,\omega_2) = \sum_{i\in occ.} \Big\{ &\phi_i^*(\mathbf{r})\psi_{i,\alpha_1\alpha_2}^{(2)}(\mathbf{r}|-\omega_\sigma;\omega_1,\omega_2) + \phi_i(\mathbf{r})\psi_{i,\alpha_1\alpha_2}^{(2)*}(\mathbf{r}|\omega_\sigma;-\omega_1,-\omega_2) \\ &+\hat{\mathbf{S}}_{12}\left[\psi_{i,\alpha_1}^{(1)*}(\mathbf{r}|\omega_1;-\omega_1)\psi_{i,\alpha_2}^{(1)}(\mathbf{r}|-\omega_2;\omega_2)\right]\Big\}, \end{aligned} \tag{21}$$

$$\begin{aligned} (\varepsilon_i+\hbar\omega_\sigma-h[n])\psi_{i,\alpha_1\alpha_2}^{(2)}(\mathbf{r}|-\omega_\sigma;\omega_1,\omega_2) = {}& h_{\alpha_1\alpha_2}^{(2)}(\mathbf{r}|-\omega_\sigma;\omega_1,\omega_2)\phi_i(\mathbf{r}) \\ &+\hat{\mathbf{S}}_{12}\left[h_{\alpha_1}^{(1)}(\mathbf{r}|-\omega_1;\omega_1)\psi_{i,\alpha_2}^{(1)}(\mathbf{r}|-\omega_2;\omega_2)\right], \end{aligned} \tag{22}$$

$$\begin{aligned} (\varepsilon_i-\hbar\omega_\sigma-h[n])\psi_{i,\alpha_1\alpha_2}^{(2)}(\mathbf{r}|\omega_\sigma;-\omega_1,-\omega_2) = {}& h_{\alpha_1\alpha_2}^{(2)}(\mathbf{r}|\omega_\sigma;-\omega_1,-\omega_2)\phi_i^*(\mathbf{r}) \\ &+\hat{\mathbf{S}}_{12}\left[h_{\alpha_1}^{(1)}(\mathbf{r}|\omega_1;-\omega_1)\psi_{i,\alpha_2}^{(1)*}(\mathbf{r}|\omega_2;-\omega_2)\right], \end{aligned} \tag{23}$$

where $\hat{\mathbf{S}}_{12}$ indicates that the summation should be taken over all the possible permutations of the indices $\{1,2\}$. $h^{(2)}_{\alpha_1\alpha_2}(\mathbf{r}|-\omega_\sigma;\omega_1,\omega_2)$ include the second- and first-order densities as

$$\begin{aligned} h^{(2)}_{\alpha_1\alpha_2}(\mathbf{r}|-\omega_\sigma;\omega_1,\omega_2) &= \int d\mathbf{r}_1 \left\{ \frac{e^2}{|\mathbf{r}-\mathbf{r}_1|} + f_{xc}(\mathbf{r},\mathbf{r}_1;\omega_\sigma) \right\} n^{(2)}_{\alpha_1\alpha_2}(\mathbf{r}_1|-\omega_\sigma;\omega_1,\omega_2) \\ &+ \int d\mathbf{r}_1 \int d\mathbf{r}_2\ g_{xc}(\mathbf{r},\mathbf{r}_1,\mathbf{r}_2;\omega_1,\omega_2) n^{(1)}_{\alpha_1}(\mathbf{r}_1|-\omega_1;\omega_1) n^{(1)}_{\alpha_2}(\mathbf{r}_2|-\omega_2;\omega_2), \end{aligned} \tag{24}$$

where $g_{xc}(\mathbf{r},\mathbf{r}_1,\mathbf{r}_2;\omega_1,\omega_2)$ is the second-order exchange-correlation kernel which is a second functional derivative of the exchange-correlation potential with respect to the density.

The third-order equations are

$$\begin{aligned} n^{(3)}_{\alpha_1\alpha_2\alpha_3}(\mathbf{r}|-\omega_\sigma;\omega_1,\omega_2,\omega_3) &= \sum_{i\in occ.} \Big\{ \phi_i^*(\mathbf{r})\psi^{(3)}_{i,\alpha_1\alpha_2\alpha_3}(\mathbf{r}|-\omega_\sigma;\omega_1,\omega_2,\omega_3) + \phi_i(\mathbf{r})\psi^{(3)*}_{i,\alpha_1\alpha_2\alpha_3} \\ &(\mathbf{r}|\omega_\sigma;-\omega_1,-\omega_2,-\omega_3) + \hat{\mathbf{S}}_{123}\left[\psi^{(1)*}_{i,\alpha_1}(\mathbf{r}|\omega_1;-\omega_1)\psi^{(2)}_{i,\alpha_2\alpha_3}(\mathbf{r}|-\omega_2-\omega_3;\omega_2,\omega_3)\right] \\ &+\hat{\mathbf{S}}_{123}\left[\psi^{(1)}_{i,\alpha_1}(\mathbf{r}|-\omega_1;\omega_1)\psi^{(2)*}_{i,\alpha_2\alpha_3}(\mathbf{r}|\omega_2+\omega_3;-\omega_2,-\omega_3)\right]\Big\}, \end{aligned} \tag{25}$$

$$\begin{aligned} (\varepsilon_i + \hbar\omega_\sigma - h[n])\psi^{(3)}_{i,\alpha_1\alpha_2\alpha_3}(\mathbf{r}|-\omega_\sigma;\omega_1,\omega_2,\omega_3) &= h^{(3)}_{\alpha_1\alpha_2\alpha_3}(\mathbf{r}|-\omega_\sigma;\omega_1,\omega_2,\omega_3)\phi_i(\mathbf{r}) \\ &+\hat{\mathbf{S}}_{123}\left[h^{(1)}_{\alpha_1}(\mathbf{r}|-\omega_1;\omega_1)\psi^{(2)}_{i,\alpha_2\alpha_3}(\mathbf{r}|-\omega_2-\omega_3;\omega_2,\omega_3)\right] \\ &+\hat{\mathbf{S}}_{123}\left[h^{(2)}_{\alpha_2\alpha_3}(\mathbf{r}|-\omega_2-\omega_3;\omega_2,\omega_3)\psi^{(1)}_{i,\alpha_1}(\mathbf{r}|-\omega_1;\omega_1)\right], \end{aligned} \tag{26}$$

$$\begin{aligned} (\varepsilon_i - \hbar\omega_\sigma - h[n])\psi^{(3)}_{i,\alpha_1\alpha_2\alpha_3}(\mathbf{r}|\omega_\sigma;-\omega_1,-\omega_2,-\omega_3) &= h^{(3)}_{\alpha_1\alpha_2\alpha_3}(\mathbf{r}|\omega_\sigma;-\omega_1,-\omega_2,-\omega_3)\phi_i^*(\mathbf{r}) \\ &+\hat{\mathbf{S}}_{123}\left[h^{(1)}_{\alpha_1}(\mathbf{r}|-\omega_1;\omega_1)\psi^{(2)*}_{i,\alpha_2\alpha_3}(\mathbf{r}|\omega_2+\omega_3;-\omega_2,-\omega_3)\right] \\ &+\hat{\mathbf{S}}_{123}\left[h^{(2)}_{\alpha_2\alpha_3}(\mathbf{r}|-\omega_2-\omega_3;\omega_2,\omega_3)\psi^{(1)*}_{i,\alpha_1}(\mathbf{r}|\omega_1;-\omega_1)\right], \end{aligned} \tag{27}$$

where $\hat{\mathbf{S}}_{123}$ indicates that the summation should be taken over all the possible permutations of the indices $\{1,2,3\}$. $h^{(3)}_{\alpha_1\alpha_2\alpha_3}(\mathbf{r}|-\omega_\sigma;\omega_1,\omega_2,\omega_3)$ includes the first-, second- and third-order densities as

$$\begin{aligned} h^{(3)}_{\alpha_1\alpha_2\alpha_3}(\mathbf{r}|-\omega_\sigma;\omega_1,\omega_2,\omega_3) &= \int d\mathbf{r}_1 \left\{ \frac{e^2}{|\mathbf{r}-\mathbf{r}_1|} + f_{xc}(\mathbf{r},\mathbf{r}_1;\omega_\sigma) \right\} n^{(3)}_{\alpha_1\alpha_2\alpha_3}(\mathbf{r}_1|-\omega_\sigma;\omega_1,\omega_2,\omega_3) \\ &+ \int d\mathbf{r}_1 \int d\mathbf{r}_2\ g_{xc}(\mathbf{r},\mathbf{r}_1,\mathbf{r}_2;\omega_1,\omega_2+\omega_3) n^{(1)}_{\alpha_1}(\mathbf{r}_1|-\omega_1;\omega_1) n^{(2)}_{\alpha_2\alpha_3}(\mathbf{r}_2|-\omega_2-\omega_3;\omega_2,\omega_3) \\ &+ \int d\mathbf{r}_1 \int d\mathbf{r}_2\ g_{xc}(\mathbf{r},\mathbf{r}_1,\mathbf{r}_2;\omega_2,\omega_3+\omega_1) n^{(1)}_{\alpha_2}(\mathbf{r}_1|-\omega_2;\omega_2) n^{(2)}_{\alpha_3\alpha_1}(\mathbf{r}_2|-\omega_3-\omega_1;\omega_3,\omega_1) \\ &+ \int d\mathbf{r}_1 \int d\mathbf{r}_2\ g_{xc}(\mathbf{r},\mathbf{r}_1,\mathbf{r}_2;\omega_3,\omega_1+\omega_2) n^{(1)}_{\alpha_3}(\mathbf{r}_1|-\omega_3;\omega_3) n^{(2)}_{\alpha_1\alpha_2}(\mathbf{r}_2|-\omega_1-\omega_2;\omega_1,\omega_2) \end{aligned} \tag{28}$$

$$+\int d\mathbf{r}_1 \int d\mathbf{r}_2 \int d\mathbf{r}_3\ h_{xc}(\mathbf{r},\mathbf{r}_1,\mathbf{r}_2,\mathbf{r}_3;\omega_1,\omega_2,\omega_3)$$
$$\times n^{(1)}_{\alpha_1}(\mathbf{r}_1|-\omega_1;\omega_1)n^{(1)}_{\alpha_2}(\mathbf{r}_2|-\omega_2;\omega_2)n^{(1)}_{\alpha_3}(\mathbf{r}_3|-\omega_3;\omega_3).$$

where $h_{xc}(\mathbf{r},\mathbf{r}_1,\mathbf{r}_2,\mathbf{r}_3;\omega_1,\omega_2,\omega_3)$ is the third-order exchange-correlation kernel which is a third functional derivative of the exchange-correlation potential with respect to the density.

In the TDDFT, exchange and correlation effects are included in the exchange-correlation potential μ_{xc} and kernels f_{xc}, g_{xc}, and h_{xc}. In most calculations, the exchange-correlation potential which is local in time (adiabatic approximation) and in space (local-density approximation) is employed. This is sometimes called the adiabatic local-density approximation (ALDA). In the ALDA, the exchange-correlation potential and kernels are given as follows,

$$f_{xc}(\mathbf{r},\mathbf{r}_1;\omega) = \frac{d\mu_{xc}}{dn}\delta(\mathbf{r}-\mathbf{r}_1), \tag{29}$$

$$g_{xc}(\mathbf{r},\mathbf{r}_1,\mathbf{r}_2;\omega_1,\omega_2) = \frac{d^2\mu_{xc}}{dn^2}\delta(\mathbf{r}-\mathbf{r}_1)\delta(\mathbf{r}_1-\mathbf{r}_2), \tag{30}$$

$$h_{xc}(\mathbf{r},\mathbf{r}_1,\mathbf{r}_2,\mathbf{r}_3;\omega_1,\omega_2,\omega_3) = \frac{d^3\mu_{xc}}{dn^3}\delta(\mathbf{r}-\mathbf{r}_1)\delta(\mathbf{r}_1-\mathbf{r}_2)\delta(\mathbf{r}_2-\mathbf{r}_3). \tag{31}$$

3. Real-space computational method

3.1. Hamiltonian

Our method utilizes the uniform grid representation in the three-dimensional Cartesian coordinates. The uniform grid representation is convenient in the Kohn-Sham theory where the self-consistent potential is almost local in the coordinate representation. It is also advantageous to describe isolated molecules since the boundary condition can be imposed naturally. The uniform grid representation is also considered to be suitable for the parallel computation, although our present calculations are achieved with a single processor.

In the uniform grid representation, one cannot explicitly treat core electrons, since the description of core electrons requires very fine grid spacing. Instead, one employs pseudopotential which describes interaction between valence electrons and ion. Among existing methods to construct pseudopotentials, we employ the norm-conserving pseudpotential with a prescription proposed by Troullier and Martins [19]. The norm-conserving pseudopotential is nonlocal, dependent on the relative angular momentum between the electron and ion. Denoting the potential for l-th partial wave as $V_l^{\mathrm{ion}}(r)$, the pseudopotential is expressed as

$$V^{\mathrm{ion}}(\mathbf{r},\mathbf{r}') = \sum_{\mathrm{lm}} V_l^{\mathrm{ion}}(r)\frac{\delta(r-r')}{rr'}Y^*_{\mathrm{lm}}(\hat{r})Y_{\mathrm{lm}}(\hat{r}') \tag{32}$$

This form of potential, local in the radial distance r and nonlocal in the angular variable $\hat{r}$, is not convenient in the uniform grid representation. We instead employ separable approximation to the potential which is known as the Kleinman-Bylander procedure [20],

$$V_{KB}^{\rm ion}(\mathbf{r},\mathbf{r}') = \sum_{\rm lm} \frac{V_l^{\rm ion}(r)\phi_{\rm lm}(\mathbf{r})V_l^{\rm ion}(r')\phi_{\rm lm}^*(\mathbf{r}')}{\langle\phi_{\rm lm}|V_l^{\rm ion}|\phi_{\rm lm}\rangle}, \tag{33}$$

where $\phi_{\rm lm}(\mathbf{r})$ is the valence orbital wave functions of an atom with angular momentum lm, constructed with the pseudopotential $V_l^{\rm ion}(r)$.

In the uniform grid representation, it is known that, for a fixed grid spacing, one can obtain more accurate results by employing higher-order finite-difference formula to deal with the kinetic energy operator [21]. We will employ nine-point finite-difference formula.

3.2. Ground state construction

For response calculations, one first need to construct Kohn-Sham ground state. This step is achieved as follows. We first prepare an initial guess for the Kohn-Sham orbitals ϕ_i and the density. For a given density $n(\mathbf{r})$, the Kohn-Sham Hamiltonian $h[n]$ is constructed. With this Hamiltonian, we improve the orbital wave functions employing a conjugate gradient method. At this step, the number of iterations of the conjugate gradient procedure is set to be rather small, usually 3 or 4. The orbital wave functions are then orthonormalized by the Schmidt procedure. The density and the Kohn-Sham Hamiltonian is then updated. This iterative procedure is continued until convergence. Although this procedure to obtain the ground state is not very sophisticated, it works robustly.

In constructing the Kohn-Sham Hamiltonian from the density, one need to calculate the Hartree potential. We calculate it by solving the Poisson equation employing the conjugate gradient method. To start the procedure, the boundary condition for the Hartree potential is required. We make use of the multipole expansion of the electron density around each atom to prepare the boundary condition.

3.3. Time evolution

The real-time evolution of the wave function is achieved by the Taylor expansion of the time evolution operator [3,22]. For a short period Δt, we ignore the change of the Hamiltonian $h[n(t)]$ in time. The wave function at $t+\Delta t$, $\psi(t+\Delta t)$, is obtained from the wave function at t, $\psi(t)$, by

$$\psi_i(t+\Delta t) \simeq \exp\left[-\frac{i}{\hbar}h[n(t)]\Delta t\right]\psi_i(t) \simeq \sum_k^{N_{max}} \frac{1}{k!}\left(\frac{-i\Delta t}{\hbar}h[n(t)]\right)^n \psi_i(t) \tag{34}$$

To achieve stable time evolution, one should choose N_{max} and Δt so that no eigenvalue of the approximated operator has a modulus greater than unity [11]. The maximum eigenvalue ε_{max} of $h[n(t)]$ is essentially determined by the kinetic energy operator; it may be estimated from the grid spacing Δx as $\varepsilon_{max} \sim 3(\hbar^2/2m)(\pi/\Delta x)^2$. We use the choice $N_{max} = 4$, for which the stability condition is

$$\Delta t \leqslant \sqrt{8}\frac{\hbar}{\varepsilon_{max}}. \tag{35}$$

In the Taylor expansion method, the time evolution is not exactly unitary. In practice, however, we found that the violation of the unitarity has negligible effect on the observables calculated here. We also find that the orthogonality of the wave functions ψ_i is preserved in the time evolution without requiring orthonormalization in the time evolution.

3.4. Perturbative calculation

In the perturbative method described in Section 2.3, the k-th order density $n^{(k)}(\mathbf{r})$ can be obtained by solving the Sternheimer equations order by order. We here explain this procedure in more detail. Looking carefully at the structure of the equations Eqs (18)–(29), the k-th order density $n^{(k)}_{\alpha_1\cdots\alpha_k}(\mathbf{r}|-\omega_\sigma;\omega_1\cdots\omega_k)$ which we simply denote $n^{(k)}(\mathbf{r})$ below satisfies the equation which has the same structure as the linear response Eq. (14).

$$n^{(k)}(\mathbf{r}) = \int d\mathbf{r}_1 \chi_0^{(1)}(\mathbf{r},\mathbf{r}_1;\omega_\sigma) \int d\mathbf{r}_2 \left\{ \frac{e^2}{|\mathbf{r}_1-\mathbf{r}_2|} + f_{xc}(\mathbf{r}_1)\delta(\mathbf{r}_1-\mathbf{r}_2) \right\} n^{(k)}(\mathbf{r}_2) + n_0^{(k)}(\mathbf{r}), \quad (36)$$

where $n_0^{(k)}(\mathbf{r})$ indicates a term which includes quantities of lower orders ($\leqslant k-1$). This equation for an unknown function $n^{(k)}(\mathbf{r})$ can be regarded as a linear algebraic equation in the uniform grid representation. Since the coefficient matrix of the equation is not Hermitian, we employ the generalized-conjugate-residual method which is an extension of the well-known conjugate-gradient method for non-Hermitian problems [4,23].

In practice, integrals in Eq. (36) are treated by converting them into equivalent differential equations, going back to the original differential form given by Eqs (18)–(29). In calculating $h^{(k)}_{\alpha_1\cdots\alpha_k}(\mathbf{r})$, the integral involving Coulomb potential is converted into the Poisson equation, and it is solved by the conjugate gradient method. The Sternheimer equations are the linear algebraic equation with a symmetric coefficient matrix after discretization. This is solved again by the conjugate gradient method for real frequencies. For complex frequencies, the coefficient matrix is not Hermitian. We found the bi-conjugate-gradient method works well for this case, even when the frequency is near the resonance energy of the system.

In solving the Sternheimer equations, it is important to treat the component of occupied orbitals separately to accelerate the convergence. For example, Eq. (19) is solved first in the space orthogonal to the occupied orbitals. Namely, we solve

$$(\varepsilon_i + \hbar\omega_\sigma - h[n])\tilde{\psi}_i^{(1)}(\mathbf{r}) = h^{(1)}(\mathbf{r})\phi_i(\mathbf{r}) - \sum_{j\in occ.} \phi_j(\mathbf{r})\langle\phi_j|h^{(1)}|\phi_i\rangle. \quad (37)$$

The occupied orbital components are then added.

$$\psi_i^{(1)}(\mathbf{r}) = \tilde{\psi}_i^{(1)}(\mathbf{r}) + \sum_{j\in occ.} \frac{1}{\varepsilon_i - \varepsilon_j + \hbar\omega_\sigma} \phi_j(\mathbf{r})\langle\phi_j|h^{(1)}|\phi_i\rangle. \quad (38)$$

This procedure is always applied when we solve the Sternheimer equations.

4. Real-time method

In this section, we discuss a real-time computational method for polarizabilities. The real-time method works most efficiently for the calculations of linear polarizabilities, and has been applied to calculating photoabsorption of various systems [3,10]. We first briefly explain it.

4.1. Instantaneous external field

We consider a system under a weak external electric field $E(t)$ of any time profile. We introduce the linear polarizability function in time domain, $\alpha(t)$. This function relates the external electric field $E(t)$ with the induced dipole moment $p(t)$ by

$$p(t) = \int_{-\infty}^{t} dt' \alpha(t-t') E(t'). \tag{39}$$

The frequency-dependent polarizability is obtained from $\alpha(t)$ by the Fourier transformation,

$$\alpha(\omega) = \int_{0}^{+\infty} dt e^{i\omega t} \alpha(t). \tag{40}$$

The lower limit of the integral reflects the causality condition.

If Eq. (39) can be inverted, one may employ any external field $E(t)$ to obtain $\alpha(t)$. The simplest choice for $E(t)$ is the instantaneous external field, $eE(t) = k\delta(t)$, where k is a small parameter to ensure the linear response. Right after this instantaneous dipole field, all electrons in the system receive impulsive force. Taking the direction of this field as z-axis, all the orbital wave functions are multiplied by a plane wave, $\phi_i(\mathbf{r}) \to e^{ikz}\phi_i(\mathbf{r})$. The time-dependent Kohn-Sham equation is then solved with this initial wave function and without any external potential. The dipole polarizability in time is obtainable from the dipole moment,

$$\alpha(t) = -\frac{e^2}{k} \int d\mathbf{r} z n(\mathbf{r}, t). \tag{41}$$

In this real-time method with the instantanous external field, one can obtain frequency-dependent polarizability $\alpha(\omega)$ of full spectral region from a single time evolution calculation. A finite time period of the time evolution limits the resolution of the obtained frequency-dependent polarizability, which we denote ΔE. It is related to the length of the time evolution, T, by the uncertainty relation, $\Delta E = 2\pi/T$.

This real-time method for the linear polarizability has been successfully applied for various molecules and clusters [3,10–13]. The usefulness of the instantaneous external field is, however, limited to the calculation of the linear polarizability.

4.2. Sinusoidal external field

We here propose a new real-time approach which can be applied to the nonlinear polarizabilities as well as the linear one. This is an extension of the finite field approach for frequency-independent responses [24]. In this approach, we apply an external field consisting, in general, of several sinusoidal time-dependences. For simplicity, we consider a case of calculating high harmonic generation where a monochromatic external field is applied to a system. We apply the dipole external field of the following form,

$$V_{\text{ext}}(t) = f(t) k z \cos \omega t, \tag{42}$$

where the function $f(t)$ describes the time profile of the external field. We employ a function $f(t)$ that increases smoothly from zero to unity to simulate the adiabatic switching. In the adiabatic limit, we expect the electrons to respond with the frequency equal to the external one and its harmonics. In practicical calculations, we will employ the function

$$f(t) = \begin{cases} 0 & t < 0 \\ \frac{1}{2}\left(1 - \cos \pi \frac{t}{\tau}\right) & 0 < t < \tau \\ 1 & \tau < t \end{cases}, \quad (43)$$

where the parameter τ controlls the adiabaticity.

If the adiabaticity is well satisfied, the electrons in the system oscillates periodically with the period of the external field, $T = 2\pi/\omega$ at $t > \tau$. The dipole moment $d(t) = \int d\mathbf{r} z n(\mathbf{r}, t)$ should be periodic as well, $d(t + T) = d(t)$. If we ignore the nonlinear polarizabilities higher than third order, the dipole moment can be expressed

$$d(t) = (d^{(0)} + d_0^{(2)}) + (d_\omega^{(1)} + d_\omega^{(3)}) \cos \omega t + d_{2\omega}^{(2)} \cos 2\omega t + d_{3\omega}^{(3)} \cos 3\omega t. \quad (44)$$

We introduce the Fourier components of the dipole moments as

$$d_n = \frac{2}{T} \int_{mT}^{(m+1)T} dt e^{in\omega t} d(t), \quad (45)$$

which should be independent of which cycle m is chosen, provided it is in the domain where $f = 1$. The linear and nonlinear polarizabilities can be expressed with these Fourier components as

$$\alpha_{zz}(-\omega; \omega) = -\frac{e^2}{k} d_1, \quad (46)$$

$$\beta_{zzz}(-2\omega; \omega\omega) = -\frac{4e^3}{k^2} d_2, \quad (47)$$

$$\gamma_{zzzz}(-3\omega; \omega\omega\omega) = -\frac{24e^4}{k^3} d_3. \quad (48)$$

These relations allow us to calculate frequency-dependent linear and nonlinear polarizabilities of high harmonic generation from the real-time calculation. We should note that these relationships hold on condition that the parameter k, which controlls the strength of the external field, is small enough so that the higher order effects are negligible. On the other hand, the parameter k should be taken sufficiently large to avoid rounding error.

4.3. Example: Ar atom

As an example of the real-time method, we calculate here the linear and nonlinear polarizabilities of Ar atom. We apply an external dipole field of time profile given by Eqs (42) and (43) with the parameters, $\hbar\omega = 1.175$ eV and $\tau = 2T$. The strength of the external field k is varied in the range $0.5 \sim 1.5$ eV/Å. We take a grid points inside a sphere of 6 Å with grid spacing 0.3 Å. We employ the exchange-correlation potential and kernels of Perdew and Zunger [25] in the Kohn-Sham operator.

The linear and nonlinear polarizabilities of Ar atom have been reported in Ref. [4] employing the modified Sternheimer method which was explained in Section 2.3. The linear polarizability, $\alpha(\omega) = 12.2$ a.u., and the hyperpolarizability of third harmonic generation, $\gamma(-3\omega; \omega, \omega, \omega) = 2270$ a.u.

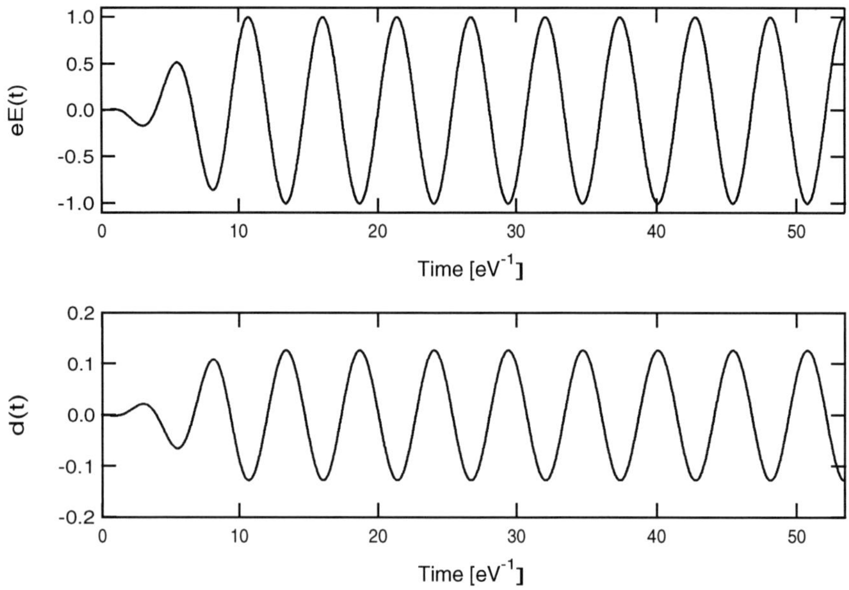

Fig. 1. Time profile of the external field $eE(t)$ and dipole moment $d(t)$. The intensity parameter k is set at 1 [eV/Å].

We show in Fig. 1 the time-dependence of the external field $eE(t)$ (upper panel) in eV/Å, and the induced dipole moment $d(t)$ (lower panel) in Å for the case of $k = 1.0$ eV/Å. It is apparent from the figure that the dipole moment closely follows the instantaneous electric field. Taking the Fourier transform of the computed $d(t)$ in the interval $[mT, (m+1)T]$ with $m = 2$, we can express $d(t)$ as the Fourier series

$$d(t) = -0.127 \cos \omega t - 0.000395 \cos 3\omega t + \cdots \tag{49}$$

The neglected terms in the series, $\cos n\omega t$ with even n, the higher harmonics $n \geqslant 5$, and terms proportional to $\sin n\omega t$, are small.

From the coefficient of $\cos \omega t$ and using Eq. (46), we can extract a value of 1.83 Å^3 for ordinary polarizability. This value is close to the polarizability $\alpha(\omega) = 12.2$ a.u. $\simeq 1.81$ Å^3 obtained by the modified Sternheimer method. The coefficient of $\cos 3\omega t$ represents the third harmonic generation. We found the coefficient of $\cos 2\omega t$ is very small, less than 10^{-6}; in principle it vanishes by symmetry and its numerical size gives an indication of the numerical precision of the calculation.

We next calculate the dipole polarizability $\alpha(\omega)$ as a function of field strength, using Eq. (46). We take the same adiabatic turn-on function f as in the previous paragraph. The calculated dipole polarizabilities do not depend much on the time period specified by m, but depend on k value. We take an average over m values in the range $2 \leqslant m \leqslant 10$ and show results in Fig. 2 as a function of the square of the strength parameter, k^2. The imaginary part of the calculated dipole polarizability is negligible.

We should note that, from Eq. (44), the coefficient of $\cos \omega t$ includes a contribution from the third order as well as the linear polarizability. This effect is seen in Fig. 3 as a linear dependence of the linear polarizability as a function of k^2. The polarizability extrapolated to $k \to 0$ coincides well with the calculated value $\alpha(\omega) = 12.2$ a.u. in the modified Sternheimer method.

We next turn to the hyperpolarizability of third harmonic generation, $\gamma(-3\omega; \omega, \omega, \omega)$. The hyperpolarizability is calculated according to Eq. (48) and the average is taken over the period $[2T, 10T]$, as in

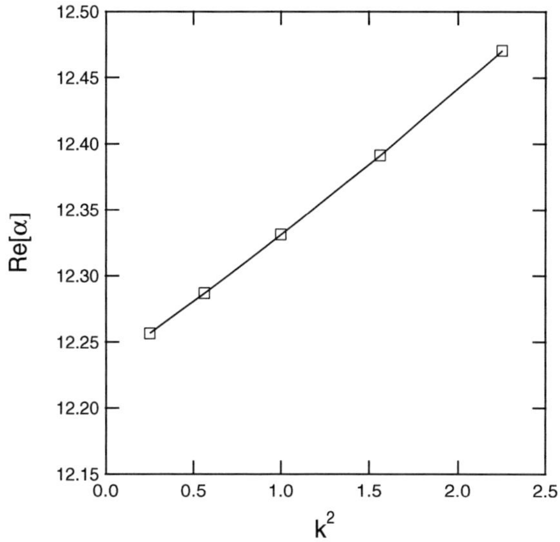

Fig. 2. Linear polarizability $\alpha(\omega)$ in a.u. is plotted against the square of the external field strength parameter, k^2.

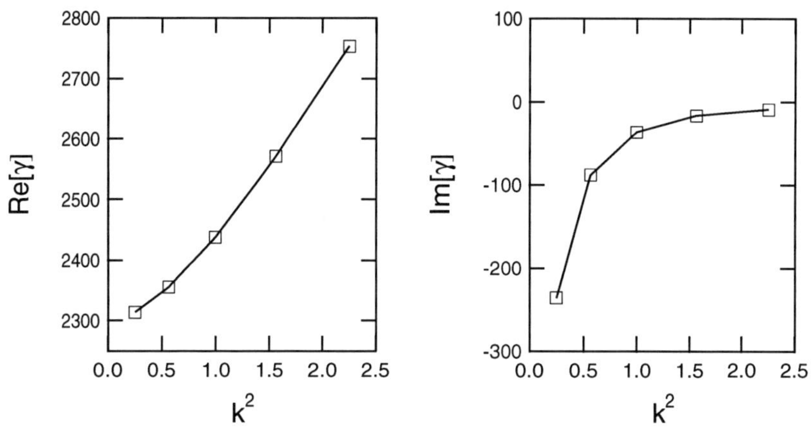

Fig. 3. Second hyperpolarizability of high order harmonics $\gamma(-3\omega;\omega,\omega,\omega)$ in a.u. is plotted against the square of the external field strength parameter, k^2. Real part (left-hand panel) and imaginary part (right-hand panel) are shown.

the case of the linear polarizability. The real and the imaginary part of the hyperpolarizability are plotted in Fig. 3 against k^2.

The hyperpolarizability calculated with the modified Sternheimer method is 2270 a.u. This is close to the value in the present real-time method taking a small k limit. There are, however, two problems which are not yet resolved. First the Re$[\gamma]$ does not show quadratic dependence on k, although fifth order effect is expected to appear as the quadratic dependence in the present finite field calculation. Second, there appears substantial imaginary part when we make calculations with small k value. The hyperpolarizability should be real since $3\hbar\omega = 3.525$ eV is well below the ionization threshold. We cannot explain the origin of these problems, although, of course, we imagine the rounding errors at small k calculation could be a candidate of the possible origins.

Summarizing this section, we have proposed a new computational method for the frequency-dependent linear and nonlinear polarizabilities. The most attractive feature of the present method is that it does not require any perturbative expansion. The polarizabilities are obtainable directly from the real-time calculation which requires rather small addition in the computational programming. Since the perturbative method requires development of lengthy formula and coding as shown in Section 2.3, this is a significant advantage of the present real-time method. Of course, the present real-time calculation can readily be applicable to the electron dynamics under the intense short pulse laser. In fact, the TDDFT has been extensively applied to the phenomena related to the strong laser field such as the high harmonic generation and ionization processes [9,26].

5. Two-photon absorption cross sections

In this section, we present results of the TDDFT calculation for hyperpolarizability in the complex energy plane. The imaginary part of the second hyperpolarizability $\gamma(-\omega;-\omega,\omega,\omega)$ provides two-photon absorption (TPA) cross section of the system. To our knowledge, this is a first attempt to calculate TPA in the TDDFT. When the frequency of the external field coincides with half of the excitation energies, the TPA takes place. In addition, the real-part of the hyperpolarizability shows divergent behavior at the frequencies. This enhanced hyperpolarizabilities at resonant frequency region are of technological interests. We will show that, employing the modified Sternheimer method explained in Section 2.3, the TPA and the resonant enhancement in the hyperpolarizability can be calculated stably. We show results taking Ar atom and ethylene molecule as examples.

5.1. Ar atom

We first show the TPA cross section of Ar atom. In order to get the information on the imaginary part of the hyperpolarizability, we introduce small imaginary component in the frequency and replace all the frequency arguments in the Sternheimer equations as

$$\hbar\omega \longrightarrow \hbar\omega + i\eta \tag{50}$$

The perturbative calculaitons are done order by order according to the procedure explained in Sections 2.3 and 3.4. The small positive parameter η is chosen as $\eta = 0.1$ eV in the practical calculations. The uniform grid points inside a sphere of 5 Å with the grid spacing of 0.3 Å is employed. We use the LB94 exchange-correlation potential [27] in the ground state construction. The LB94 potential includes gradient correction which describes asymptotic $1/r$ behavior of the potential correctly. The adiabatic local-density approximation is assumed for the exchange-correlation kernels in the response calculations.

In Fig. 4, we show results of two calculations for the imaginary part of the hyperpolarizability $\gamma(-\omega;-\omega,\omega,\omega)$ varying $\hbar\omega$. The TDDFT calculation is shown by the solid curve and the response calculation in the independent-particle approximation (IPA) is shown by the dashed curve. The IPA implies that the induced field is ignored, namely the k-th order density is calculated by the k-th order density-density response function as

$$n^{(k)IPA}_{\alpha_1,\cdots,\alpha_k}(\mathbf{r}|-\omega_\sigma;\omega_1,\cdots,\omega_k)=\int d\mathbf{r}_1\cdots d\mathbf{r}_k\chi_0^{(k)}(\mathbf{r},\mathbf{r}_1,\cdots,\mathbf{r}_k;\omega_1,\cdots,\omega_k)(e\mathbf{r}_1)_{\alpha_1}\cdots(e\mathbf{r}_k)_{\alpha_k} \tag{51}$$

In the IPA, the TPA appear when 2ω coincides with the poles of $\chi_0^{(k)}$, which are the orbital energy differences in the static Kohn-Sham calculation.

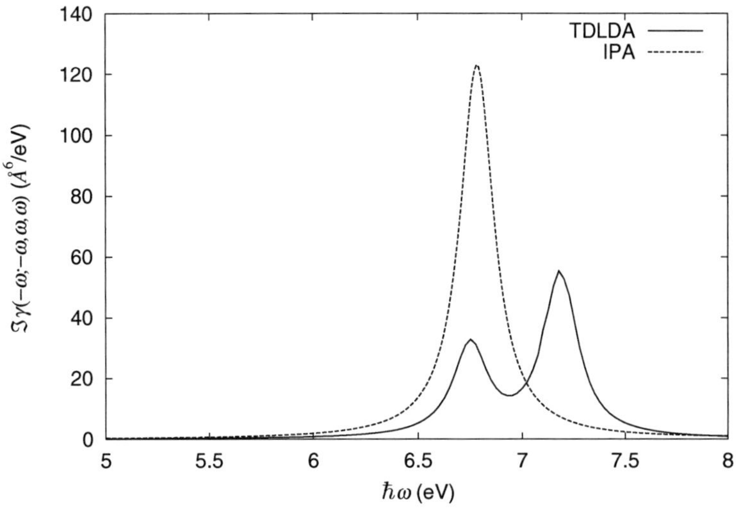

Fig. 4. Imaginary part of the hyperpolarizability $\gamma(-\omega; -\omega, \omega, \omega)$ of Ar atom. The solid line is the present result, and dashed line is a result of calculation with independent particle approximation (IPA) (see text). The prominent peaks correspond to two-photon absorption transitions from the ground $^1S^e$ state to excited even-parity states.

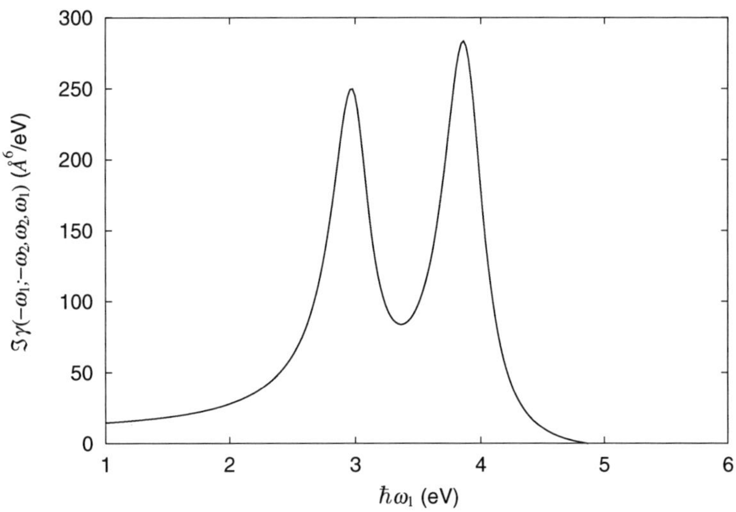

Fig. 5. Imaginary part of the hyperpolarizability $\gamma(-\omega_1; -\omega_2, \omega_2, \omega_1)$ of Ar atom. $\hbar\omega_2$ is fixed at 10.49 eV.

In Fig. 4, one prominent peak is seen in the IPA calculation, and two peaks in the TDDFT. The energy of the IPA peak locates at 6.78 eV which is half of the energy difference between $3p$ and $4p$ orbitals in the static Kohn-Sham calculation. In the TDDFT calculation, two peaks appear at 6.75 and 7.2 eV. We show in Table 1 the excitation energies of $^1S^e$ and $^1D^e$ states of Ar atom in the TDDFT calculation as well as measured values. The peak positions in the response calculation shown in Fig. 4 coincide with half of the excitation energies in the TDDFT with LB94. The dynamical screening effect included in the TDDFT calculation thus splits the strength of the TPA of $3p \rightarrow 4p$ transition into two multiplets of $^1S^e$ and $^1D^e$ states. The integrated strength of the two peaks in TDDFT is somewhat smaller than that of IPA.

Table 1
Excitation energies of Ar atom obtained by the TDDFT calculation and measurements [28]. In the TDDFT, two kinds of exchange-correlation potentials, simple LDA and the LB94 functional [27], are employed. The unit is in eV

state	LB94	LDA	EXP.
$^1D_2^e$	13.50	12.35	13.17
$^1S_0^e$	14.37	13.03	13.48

Table 2
Two-photon transition rate constants in unit of cm^{4}s between the ground state and $^1D_2^e$ and $^1S_0^e$ states of Ar atom

state	present	CCSD^a	MCSCF^b	EXP.^{c}
$\hbar\omega_1 = \hbar\omega_2$				
$^1D_2^e$	0.55×10^{-51}	0.412×10^{-51}	0.426×10^{-51}	
$^1S_0^e$	0.99×10^{-51}	3.04×10^{-51}		
$\hbar\omega_2 = 10.49$(eV)				
$^1D_2^e$	11×10^{-51}	1.83×10^{-51}	1.91×10^{-51}	2.05×10^{-51}
$^1S_0^e$	14×10^{-51}	9.52×10^{-51}		

[a] from Ref. [29], [b] from Ref. [30].
[c] from Ref. [31].

We also carried out calculations of the TPA for two-color laser field, $\gamma(-\omega_1; -\omega_2, \omega_2, \omega_1)$ with $\hbar\omega_1 \neq \hbar\omega_2$. The measurements and results of other calculations are available for the case of $\hbar\omega_2 = 10.49$ eV. In Fig. 5, we show the TDDFT results where $\hbar\omega_1$ is varied. There appear peaks at $\hbar\omega_1 = 2.98$ eV and 3.86 eV. The summed energies, $\hbar\omega_1 + \hbar\omega_2$, coincide with the excitation energies of $^1S^e$ and $^1D^e$ states. The shape of the spectra is seen to be quite different from the degenerate type TPA spectra of Fig. 4. Especially, the peak intensities are substantially enhanced, about 5 times larger than the degenerate case.

In Table 2, we show the TPA strength obtained by the present calculation and compare it with measurements and other calculations. The TPA strength is defined by

$$K = \frac{2\pi^2\hbar\omega^2}{c^2}\frac{1}{2R_y}\int_\Gamma d(\hbar\omega)\mathrm{Im}[\chi^{(3)}(-\omega; -\omega, \omega, \omega)] \tag{52}$$

and

$$K = \frac{4\pi^2\hbar\omega_1\omega_2}{c^2}\frac{1}{2R_y}\int_\Gamma d(\hbar\omega_1)\mathrm{Im}[\chi^{(3)}(-\omega_1; -\omega_2, \omega_2, \omega_1)] \tag{53}$$

where R_y is the Rydberg constant, and the integration is performed in the region around a resonance peak.

We find common features between the TDDFT and CCSD calculations: The TPA strength of $^1S^e$ state is larger than that of $^1D^e$. The TPA strength with two-color laser ($\hbar\omega_1 \neq \hbar\omega_2$) is much larger than the strength with the monochromatic laser ($\hbar\omega_1 = \hbar\omega_2$) in both calculations. However, the difference of strengthes between the $^1S^e$ and $^1D^e$ transitions is much smaller in the TDDFT calculation than those in the CCSD calculation. The experimental value is reported for $^1D^e$ transition in two-color measurement, which agrees well with CCSD and MCSCF calculations and is five times smaller than the TDDFT result.

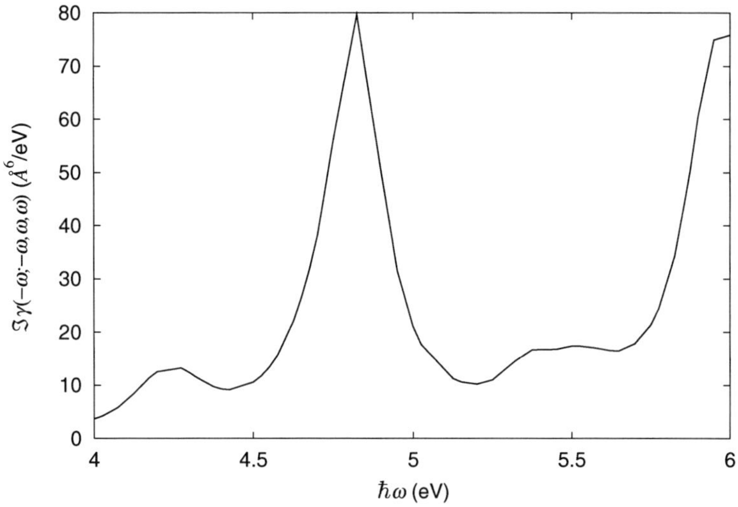

Fig. 6. Imaginary part of the second-hyperpolarizability of ethylene molecule.

5.2. Ethylene molecule

We next show the TPA spectra of ethylene molecule in the TDDFT. This molecule has many electronic excited states below the ionization threshold. The $\pi \rightarrow \pi^*$ excited states appear at 4.36 eV (triplet) and 7.6 eV (singlet), and their symmetries are assigned as B_{1u}. Since the ground state of the ethylene molecule has ${}^1A_{1g}$ symmetry, vertical electronic transitions to the $\pi \rightarrow \pi^*$ excitations and other ungerade states are forbidden in the TPA. The TPA transitions to the gerade states, $\pi \rightarrow p$ and $\pi \rightarrow f$ Rydberg states, are allowed.

The linear optical absorption in the ethylene molecule in the TDDFT was reported in Ref. [14], in which the measured spectra are excellently reproduced by the calculation. In view of this success in the one-photon absorption, we think the two-photon absorption in the ethylene molecule will give a good opportunity to examine accuracy of the TDDFT for the TPA. Unfortunately, we could not find any experimental report.

The TDDFT calculation is achieved employing a grid points inside a sphere of 6 Å radius with 0.3 Å grid spacing. The LB94 exchange-correlation potential is utilized. In Figs 6 and 7, the imaginary and the real parts of the hyperpolarizability, $\gamma(-\omega; -\omega, \omega, \omega)$, are shown. The average over orientation of the molecule is taken into account. Several structures are seen in the TPA, namely, the imaginary part of $\gamma(-\omega; -\omega, \omega, \omega)$, below half of the ionization potential (11.7 eV in the calcualtion and 11.0 eV in the measurements). In the real part, an enhancement is seen at the frequency where substantial TPA is observed. A substantial peak appears at 4.83 eV in the imaginary part. At twice of this energy 9.6 eV, we found $2A_g$ state appear in the TDDFT excitation energy calculation [32,33]. This state is known to appear at 8.29 eV in the measurement. The calculated excitation energy of this state is too high by 1.3 eV. This state is a π to Rydberg p_x transition [34,35]. Weak features are seen around 4.3 and 5.5 eV. At twice of these energies, several excited states can be found in the TDDFT excitation energy calculation, and they are gerade states which are forbidden by one photon absorption transitions.

Summarizing this section, the hyperpolarizability in the complex energy plane can be calculated in the uniform grid representation in the perturbative formalism with the modified Sternheimer method. Since measurements of the TPA is rather limited, it is at this stage difficult to make an assessment on the accuracy of the TDDFT for the TPA.

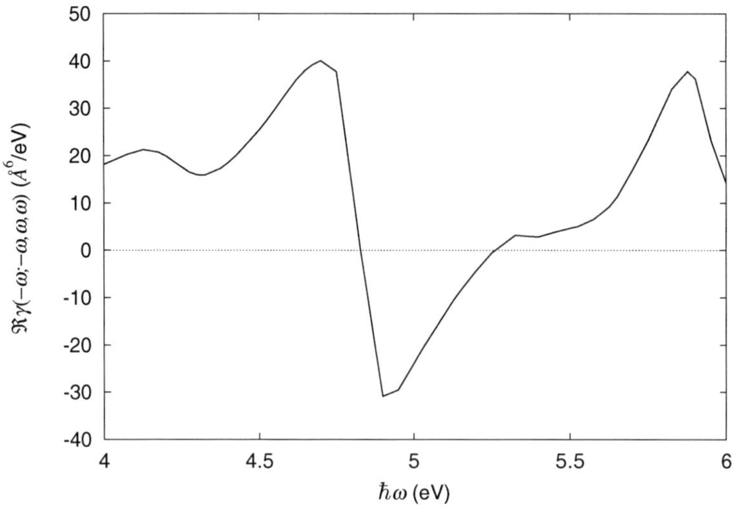

Fig. 7. Real part of the second-hyperpolarizability of ethylene molecule.

6. Summary

We present real-space computational methods for linear and nonlinear polarizabilities based on the time-dependent density-functional theory. The real-space method implies usage of the uniform grid representation in the three-dimensional Cartesian coordinates. The wave functions are expressed with their values on the grid points. Two distinct methods are discussed, the real-time method and the perturbative formalism with the modified Sternheimer method.

The real-time, real-space method has been most successful for calculations of linear polarizability. In this article, we have shown that the real-time method is also useful for the calculations of nonlinear polarizabilities. Since the method does not require any perturbative expansion, this approach is much simpler in both formalism and programming than the perturbative method.

The uniform grid representation also provides efficient representation to solve the response equations, in which the integral equations are converted into equivalent differential equations. We have shown that the nonlinear polarizabilites in the complex frequency plane can be efficiently calculated. To demonstrate it, the two-photon absorption cross sections which is related to the imaginary part of the hyperpolarizability are analyzed.

Acknowledgments

K.Y. acknowledges support by the Ministry of Education, Culture, Sports, Science and Technology of Japan, Contract No. 14540369. G.B. also acknowledges support by the US Department of Energy under Grant DE-FG03-00-ER41132 and the Computational Materials Science Network. The authors acknowledge the Institute for Solid State Physics, Univ. of Tokyo, the Institute of Physical and Chemical Research (RIKEN), for the use of supercomputers.

References

[1] A. Zangwill and P. Soven, *Phys. Rev.* **A21** (1980), 1561.

[2] G. Senatore and K.R. Subbaswamy, *Phys. Rev.* **A35** (1987), 2440.
[3] K. Yabana and G.F. Bertsch, *Phys. Rev.* **B54** (1996), 4484.
[4] J.-I. Iwata, K. Yabana and G.F. Bertsch, *J. Chem. Phys.* **115** (2001), 8773.
[5] S.J.A. van Gisbergen, J.G. Snijders and E.J. Baerends, Phys. Rev. Lett. **78**, 3097 (1997).
[6] S.J.A. van Gisbergen, J.G. Snijders and E.J. Baerends, *J. Chem. Phys.* **109** (1998), 10644; (erratum) **111** (1999), 6652.
[7] H.H. Henze, F. Della and A. Görling, *J. Chem. Phys.* **116** (2002), 9624.
[8] P. Salek, O. Vahtras, T. Helgaker and H. Ågren, *J. Chem. Phys.* **117** (2002), 9630.
[9] E.K.U. Gross, J.F. Dobson and M. Petersilka, *Density Functional Theory*, Springer Series Topics in Current Chemistry, edited by R.F. Nalewajski (Springer, Heidelberg, 1996), p. 81.
[10] K. Yabana and G.F. Bertsch, *Int. J. Quantum Chem.* **75** (1999), 55.
[11] G.F. Bertsch and K. Yabana, Density functional theory, in: *Introduction to Modern~@Methods of quantum many-body theory and their applications*, A. Fabrocini, S. Fantoni and E. Krotscheck, eds, (World Scientific, 2002), pp. 1–48.
[12] K. Yabana and G.F. Bertsch, *Phys. Rev.* **A60** (1999), 1271.
[13] G.F. Bertsch, J.-I. Iwata, A. Rubio and K. Yabana, *Phys. Rev.* **B62** (2000), 7998.
[14] T. Nakatsukasa and K. Yabana, *J. Chem. Phys.* **114** (2001), 2550.
[15] P.N. Butcher and D. Cotter, *The Elements of Nonlinear Optics*, Cambridge University Press, Cambridge, 1990.
[16] W. Ekardt, *Phys. Rev.* **B29** (1984), 1558.
[17] G.F. Bertsch, *Comp. Phys. Comm.* **60** (1990), 247.
[18] G.D. Mahan, *Phys. Rev.* **A22** (1980), 1780.
[19] N. Troullier and J. Martins, *Phys. Rev.* **B44** (1991), 1993.
[20] L. Kleinman and D. Bylander, *Phys. Rev. Lett.* **48** (1982), 1425.
[21] J.R. Chelikowsky, N. Troullier, K. We and Y. Saad, *Phys. Rev.* **B50** (1994), 11355.
[22] H. Flocard, S.E. Koonin and M.S. Weiss, *Phys. Rev.* **C17** (1978), 1682.
[23] R. Brett et al., *Templates for the Solution of Linear Systems: Building Blocks for Iterative Methods*, Society for Industrial and Applied Mathmatics, Philadelphia, 1994.
[24] J. Guan, P. Duffy, J.T. Carter, D.P. Chong, K.C. Casida, M.E. Casida, M. Wrinn, *J. Chem. Phys.* **98** (1993), 4753.
[25] J.P. Perdew and A. Zunger, *Phys. Rev.* **B23** (1981), 5048.
[26] F. Calvayrac, P.G. Reinhard, E. Suradud, C.A. Ullrich, Phys. Rep. **337**, 493 (2000).
[27] R. van Leeuwen and E.J. Baerends, *Phys. Rev.* **A49** (1994), 2421.
[28] C.E. Moore, *Natl, Bur. Stand.* (*U.S.*) *Circ.* **467** (1949), 212.
[29] C. Hätting, O. Christiansen and P. Jorgensen, *J. Chem. Phys.* **108** (1998), 8355.
[30] D. Sundholm, A. Rizzo and P. Jorgensen, *J. Chem. Phys.* **101** (1994), 4931.
[31] M.P. McCann, C.H. Chen and M.G. Payne, *J. Chem. Phys.* **89** (1988), 5429.
[32] M.E. Casida, in: *Recent Advances in Density Functional Methods, Part I*, D.P. Chong, ed., Singapore, World Scientific, 1995, p. 155.
[33] A. Muta, J.-I. Iwata, Y. Hashimoto and K. Yabana, *Prog. Theor. Phys.* **108** (2002), 1065.
[34] K.B. Wiberg et al., *J. Phys. Chem.* **96** (1992), 10756.
[35] P.R. Schipper et al., *J. Chem. Phys.* **112** (2000), 1344.

Journal of Computational Methods in Sciences and Engineering 4 (2004) 481–491
IOS Press

Rational Design of atomic gaussian basis sets for Ab initio calculations of the dipole polarizabilities and hyperpolarizabilities. I. Optimized polarization sets for the first-row atoms from B to F

Mikhail B. Zuev* and Sergey E. Nefediev
A.E. Arbuzov Institute of Organic and Physical Chemistry of Kazan Scientific Centre, Russian Academy of Sciences, Arbuzov str., 8, 420088, Kazan, Russia

Abstract. The rational design rules for atomic Gaussian basis sets, tailored for the calculations of molecular polarizabilities α_{ij} and hyperpolarizabilities β_{ijk} at the SCF level, are elaborated. The sequence of the optimized polarization sets (2d), (3d), (4d), (3d1f), (4d3f) for the first-row atoms from B to F is constructed, the optimization technique is supposed to provide a rapid convergence of both α_{ij} and β_{ijk} with the size of the polarization set. Adding of these polarization sets to the [6s4p] and [6s5p] substrates, derived from the energy-optimized (10s6p) primitive set of Huzinaga, gives the [6s4pNdMf] and [6s5pNdMf] bases families. They are combined with the corresponding [4sNpMd] and [5sNpMd] bases for Hydrogen built up in the same manner. The convergence of SCF results with the size of the basis set has been thoroughly investigated by the examples of H_2CO, CH_3CN, and HCN molecules. It has been established that both α_{ij} and β_{ijk} values converge to their saturation limits at the [6s4p3d/4s3p] level. This basis set provides the results, which are very close to near HF limit values, estimated with the most sophisticated polarized Gaussian basis sets used up to now.

Keywords: Gaussian basis sets, convergence, SCF calculations of electric properties, polarizability and hyperpolarizability

Mathematics Subject Classification: 81-08 – Quantum Theory, Computational Methods

1. Introduction

The dipole electric properties of an individual molecule are commonly defined using the Taylor series expansion for the dipole polarization $\mathbf{p}$ in the presence of an applied dc electric field $\mathbf{F}$

$$p_i = \mu_i + \alpha_{ij} F_j + 1/2! \beta_{ijk} F_j F_k + 1/3! \gamma_{ijkl} F_j F_k F_l + \dots \quad (1)$$

The permanent dipole moment (μ_i), the electronic polarizability (α_{ij}), the first (β_{ijk}) and second (γ_{ijkl}) electronic hyperpolarizabilities are evaluated at the equilibrium geometry of a free molecule. The subscripts denote Cartesian components, and a repeated subscript implies summation over x, y, and z.

*Corresponding author. E-mail: mikhail@iopc.knc.ru.

We are interested in determining the reliable values of β_{ijk} for conjugated organic molecules which are of primary importance in molecular design of electro-optic polymer systems [1]. The precise calculation of the linear and nonlinear polarizabilities is a very demanding task since the major part of the response to an applied electric field comes from the outer part of the atomic and/or molecular electron density [2]. It has long been realized [3] that the dipole polarization of atoms and molecules can not be taken into account with traditional atomic Gaussian basis sets, such as triple zeta valence one plus two sets of polarization functions located in the tight region of the atomic valence shell. The polarized triple zeta valence bases generally give excellent SCF results for energies and energy-related properties such as spectroscopic constants of free molecules. Unfortunately, these basis sets can not provide an adequate description of the polarizability and hyperpolarizability of atoms and molecules since they have not been adjusted in the outer parts of the atomic valence shells. Extension of energy-optimized primitive Gaussian basis sets with both diffuse valence and diffuse polarization functions is required to eliminate the trouble [3].

At present accurate ab initio calculations of electric properties of molecules are performed with very large one-particle basis sets including typically about 80 basis functions on the first row atoms. The design rules for these almost saturated atomic Gaussian basis sets have been formulated and thoroughly tested by Maroulis et al. [4–9] and Rice et al. [10–12] in the course of the investigations of small organic molecules. The developed computational methodologies are extremelly important for fundamental understanding of the basis set requirements for the linear and nonlinear polarizabilities estimation at both SCF and correlated level of theory. They can be used for benchmark calculations at least within the SCF HF approximation. Unfortunately, application of the rigorous computational techniques in order to aid the design of novel conjugated NLO chromophores is impossible because of a large size of the favoured systems; the simplest of them are p-nitroaniline and 4-nitro-4'-aminoazobenzene containing 10 and 18 first-row atoms, respectively.

In this paper we have made an attempt to rationalize the design rules for atomic Gaussian basis sets, tailored for the calculations of α_{ij} and β_{ijk} in organic π-electron systems. The rationalization is based on intuitive physical ideas in contrast to the basis set polarization method of Sadlej [13], derived from quite formal mathematical consideration. The main goal is to search for an optimal radial positioning of polarization functions producing a rapid convergence of SCF results with the size of the polarization set at a given sp substrate.

2. Basis set design

The performance of one-particle Gaussian basis set in electron structure calculations depends mainly on (i) the completeness of the atomic primitive Gaussian sets for each of l values (s, p, d, etc), which are thought to be essential for properties under consideration, and (ii) the flexibility of the contraction of s, p, and, perhaps, d primitive orbitals located in the valence region of atomic electron shells. The radial location of a Gaussian-type orbital may be characterized by its effective radius r that depends on the orbital exponent α and the l value

$$r^2 = \frac{l+1}{2}\alpha^{-1}. \tag{2}$$

By definition, r is the distance from the nuclei at which the radial distribution of the orbital density reaches its maximum. For the first-row atoms, one can accept that the orbitals with r larger than $0.4a_0$ fall into the valence part of the primitive space.

Since each primitive set can be contracted in various ways, it is extremely desirable to optimize the exponents of primitive Gaussian functions without regard to any special contraction scheme in use. It is also desirable that precepts and rules, employed in basis set construction, could be explained physically. We follow here these guidelines as far as possible.

At the SCF level the atomic primitive set requirements are rather simple. First, the basis set has to be well balanced with respect to the SCF AO energies and total atomic SCF energies as well. In other words, an accuracy of the SCF energies above (estimated with respect to the corresponding Hartree-Fock limits) has to be approximately the same for all the atoms in a molecule. Second, an atomic primitive Gaussian set has to provide a steady approximation of valence SCF AO at various distances from the nucleus. This requirement is well fulfilled if the exponents of the valence s and p primitives are generated as even-tempered sequences

$$\alpha_i = \alpha_1^{(m-i)/(m-1)} \cdot \alpha_m^{(i-1)/(m-1)} \quad \text{for } i = 1, 2, \ldots, m \tag{3}$$

defined for each l value.

First-row atoms. We start from the energy-optimized ($10s6p$) primitive Gaussian set of Huzinaga [14] that approximately satisfies both requirements above. This primitive set is hereafter referred to as a parent set.

An arrangement of the parent set for calculations of the polarizabilities requires first adding of the diffuse functions to both $10s$ and $6p$ subsets. Construction of diffuse sp sets is not a difficult task. The exponents of the diffuse primitives with a given l can be obtained as an even-tempered sequence with the two last exponents of the original l subset. Doing so, we come to the rule for adding the diffuse s and p functions to the ($10s6p$) set:

$$\alpha_{m+i}^l = \alpha_m^l(\alpha_m^l/\alpha_{m-1}^l)^i \quad \text{for} \quad i = 1, 2, \ldots \tag{4}$$

where l stands for s or p, and m is the number of the outermost exponent of the original primitive l subset (i.e. $m = 10$ for s subset and $m = 6$ for p subset). We have used this approach to construct the ($11s7p$) and ($12s8p$) primitive Gaussian sets for the atoms from B to F. For input all the diffuse exponents are truncated to four decimal places. Notice that the outermost p functions of the ($12s8p$) set appear to be inappropriately diffuse, their effective radii (ranging from 7 to $11a_0$) being far beyond the valence region of the first-row atoms. Therefore, there is rather high probability that these functions have little if any effect on the polarizabilities. The ($11s7p$) primitive set seems to be saturated in both s and p subsets.

The next step is the determination of polarization d and f subsets for ($11s7p$) primitive set referred to as a sp substrate below. From a physical viewpoint, it is obvious that a minimal polarization set, that could work for the polarizabilities, must include two d subsets located in the inner and outer parts of valence region, respectively. This requirement follows from the point that a polarization set is now used to describe two different effects: (i) polarization of the electron density in covalent bonds, and (ii) polarization of the electron density induced by a dc electric field applied to a free atom or molecule. We begin with optimizing the electric response of free atoms at the ($11s7p1d$) level. The aim can be achieved via minimization of the atomic SCF energy with respect to the d exponent in the presence of an external electric field. This procedure is clearly equivalent to maximization of the dipole polarization of an atom because the polarization functions do not contribute to atomic SCF energies when an applied field is removed.

The trouble with this approach is possible dependence of the optimized d exponent on the electric field strength F inducing the polarization. With this in mind, we have calculated the optimal d exponents

Table 1
Exponents of polarization functions for the first-row atoms and hydrogen

Polarization set		B	C	N	O	F	Polarization set		H
(2*d*)	α_1	0.50	0.72	0.98	1.28	1.62	(2*p*)	α_1	1.0
	α_2	0.068	0.115	0.144	0.225	0.298		α_2	0.125
(3*d*)	α_1	0.50	0.72	0.98	1.28	1.62	(3*p*)	α_1	1.0
	α_2	0.112	0.182	0.233	0.347	0.455		α_2	0.210
	α_3	0.041	0.073	0.089	0.146	0.195		α_3	0.074
(4*d*)	α_1	0.824	1.139	1.583	1.977	2.474	(4*p*)	α_1	1.682
	α_2	0.304	0.455	0.607	0.829	1.061		α_2	0.595
	α_3	0.112	0.182	0.233	0.347	0.455		α_3	0.210
	α_4	0.041	0.073	0.089	0.146	0.195		α_4	0.074
(1*f*)	α_1	0.091	0.153	0.192	0.300	0.397	(1*d*)	α_1	0.188
(3*f*)	α_1	0.667	0.960	1.307	1.707	2.160	(3*d*)	α_1	1.5
	α_2	0.245	0.384	0.501	0.716	0.927		α_2	0.530
	α_3	0.091	0.153	0.192	0.300	0.397		α_3	0.188

for some first-row atoms with various F values ranging from 0.003 to 0.01 a.u.; that is just the range of the field strengths used for estimating the static electronic hyperpolarizabilities by numerous finite field techniques. It has been established that the optimal d exponents are not affected by the field strength F falling in this range. Nevertheless, the choice of a standard electric field is necessary to impart the completeness to the d exponent optimization procedure. The field strength of 0.01 a.u. is accepted here as the standard one.

We employ this procedure to define the diffuse subset of a (2*d*) polarization set under consideration. The d exponents minimizing the atomic SCF energy at the standard electric field, are calculated with totally uncontracted $(11s7p1d)$ primitive set by the ROHF technique; the five cubic harmonical d functions (i.e. d_{z^2}, $d_{x^2-y^2}$, d_{xz}, d_{yz}, and d_{xy}) are always used. These exponents are denoted as d_2^{2d} below. Let us notice that reoptimization of α_2^{2d} at the $(12s8p1d)$ level does not affect the results.

The exponent α_1^{2d}, defining the tight subset of the (2*d*) polarization set is determined by the empirical formula of Ahlrichs and Taylor [15]

$$\alpha_1^{2d} = 0.02Z^2 \quad \text{for } B - Ne, \tag{5}$$

obtained by fitting the results of correlated calculations for closed-shell hydrides. These d exponents together with a suggested p exponent of 1.0 for hydrogen are the best for traditional double zeta plus polarization basis sets, used in the SCF calculations of the energies and energy-related properties of molecules. The values of α_1^{2d} and α_2^{2d} for the first-row atoms from B to F are given in Table 1.

The radial positioning of the (2*d*) set in the valence shell of an atom can be represented by the effective radii of the tight and diffuse d orbitals which are denoted as r_1^{2d} and r_2^{2d} below. By definition, they mark out the radial domains, which are mostly affected by the polarization effects of interest. With this in mind, we use the exponents α_1^{2d} and α_2^{2d} in the design of more saturated polarization sets, providing a better description of the electron density in this region.

The (4*d*) primitive set is first constructed. It is built up with an often-used splitting technique: each exponent of the (2*d*) set is splitted into two ones, the splitting factor is chosen to obtain the four-membered geometrical progression. This gives the even-tempered (4*d*) set defined by Eq. (2) with

$$\alpha_1^{4d} = \alpha_1^{2d}(\alpha_1^{2d}/\alpha_2^{2d})^{1/4} \quad \text{and} \quad \alpha_4^{4d} = \alpha_4^{4d} = \alpha_2^{2d}(\alpha_1^{2d}/\alpha_2^{2d})^{-1/4}. \tag{6}$$

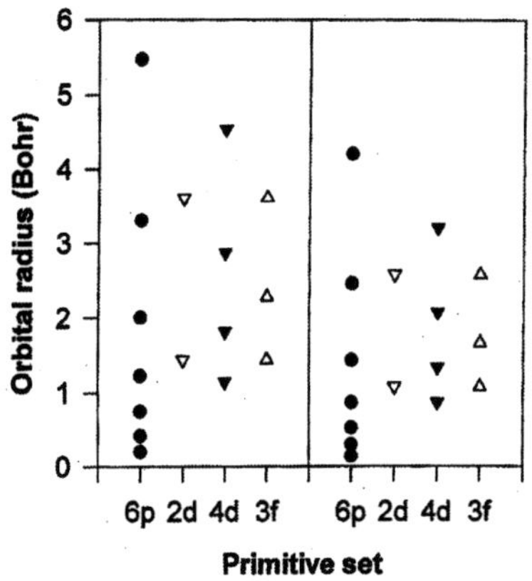

Fig. 1. Radial positioning of primitive Gaussian orbitals for Carbon (left) and Oxyden (right).

It is interesting to note that the denominator of the obtained progression (that is a ratio between r_1^{2d} and r_2^{2d}) ranges from 2.7 for B to 2.3 for F. This is very close to the ratio 2.5 for the successive exponents of ($6d$) and ($4f$) primitive sets supposed by Almlöf and Taylor [16] for the first-row atoms.

The radial positioning of the ($4d$) set (expressed in terms of the orbital radii r_i^{4d} for $i = 1, \ldots, 4$) is shown in Fig. 1 by the examples of carbon and oxygen. It can be seen that the orbitals of the ($4d$) set are located between successive valence p orbitals of the ($7p$) primitive set. Closer inspection shows that the effective radius of d orbital is approximately equal to the geometrical mean of the radii of adjacent p orbitals, the relationship

$$r_i^{4d} \approx \left(r_{2+i}^{7p} \cdot r_{3+i}^{7p}\right)^{1/2} \quad \text{for} \quad i = 1, \ldots, 4 \tag{7}$$

being valid within the limits of $0.2a_0$. This observation may be converted into a general rule for generating the exponents of the $(n-1)$-membered polarization $l+1$ set from the exponents of a given n-membered primitive l set

$$\alpha_i^{(n-1)(l+1)} = \frac{l+2}{l+1}\left(\alpha_i^{nl}\alpha_{i+1}^{nl}\right)^{1/2} \quad \text{for} \quad i = 1, \ldots, n-1. \tag{8}$$

We use this rule to construct the even-tempered ($3f$) primitive set defined by Eq. (2) with

$$\alpha_1^{3f} = 4/3\,\alpha_1^{2d} \quad \text{and} \quad \alpha_2^{3f} = 4/3\,\alpha_2^{2d}. \tag{9}$$

The radial positioning of the ($3f$) set is illustrated in Fig. 1. Note that a scaling factor of $4/3$ between d and f exponents is very close to that recommended by Ahlrichs and Taylor [15].

As mentioned above, the even-tempered ($4d$) set has been built up by splitting both α_1^{2d} and α_2^{2d} exponents. Since the electric response of an atom comes from the outer part of the valence shell, the splitting of the larger exponent α_1^{2d} might not be indispensable at the SCF level. Keeping this in mind, we also introduce the ($3d$) polarization set with the exponents

$$\alpha_1^{3d} = \alpha_1^{2d}, \quad \alpha_2^{3d} = \alpha_3^{4d}, \quad \alpha_3^{3d} = \alpha_4^{4d}. \tag{10}$$

The ($3d$) set may be augmented by one primitive f set with the exponent equal to the outermost exponent of the ($3f$) set, i.e.

$$\alpha^{1f} = 4/3\,\alpha_2^{2d}. \tag{11}$$

Table 2
Dipole polarization (in a.u.) of the first-row atoms calculated at the SCF level in the presence of the electric field of 0.01 a.u.

Basis set	B	C	N	O	F
(11s7p)	−0.2297	−0.1016	−0.0533	−0.0248	−0.0135
(11s7p2d)	−0.2700	−0.1290	−0.0729	−0.0440	−0.0291
(11s7p3d)	−0.2718	−0.1306	−0.0736	−0.0457	−0.0306
(11s7p4d)	−0.2720	−0.1307	−0.0736	−0.0457	−0.0307
(11s7p3d1f)	−0.2720	−0.1306	−0.0736	−0.0457	−0.0307
(11s7p4d3f)	−0.2723	−0.1307	−0.0737	−0.0459	−0.0309

The polarization exponents for the first-row atoms from B to F are presented in Table 1. Using the data from Table 1, one can determine a consecution of the atomic polarization sets ($2d$), ($3d$), ($4d$), ($3d1f$), ($4d3f$) which are ordered with their size and expected performance. By adding these sets to ($11s7p$) substrate we obtain a family of the polarized Gaussian primitive sets, tailored for the electric response calculations. This family, below referred to as ($11s7p$NdMf), was used in ROHF calculations of atomic dipole polarization in the presence of the standard electric field of 0.01 a.u. The aim is to study the convergence rate of SCF results with the improvement of a polarization set. The obtained SCF data for the first-row atoms are given in Table 2. It can be seen that within the ($11s7p$NdMf) bases family the convergence of calculated atomic response is reached at the ($11s7p3d$) level. Further refinement of the primitive basis set has negligible effect on the SCF dipole polarization of atoms. The calculation errors at the ($11s7p2d$) level are within 1% for B, C, and N and increase to 5% for O and F. It is shown that the ($11s7p2d$) set is not perfectly balanced with respect to the electric dipole polarization of the first-row atoms. Significance of the second diffuse sp set has been also evaluated. It has been established that extension of ($11s7p$NdMf) bases up to ($12s8p$NdMf) ones does not affect the data presented in Table 2. This result is in accordance with the qualitative physical arguments considered above.

The arrangement of ($11s7p$NdMf) bases for calculations of molecular polarizabilities requires the contraction of the primitive Gaussian sets in segmented or general fashion [16]. Generally speaking, it can be done in various ways. We employ here both segmented contractions developed by Dunning [17] for the ($10s6p$) parent set. The first is triple zeta valence contraction that reduces ($10s6p$) to [$5s3p$]. Application of this contraction to ($11s7p$NdMf) bases produces the [$6s4p$NdMf] family of contracted Gaussian basis sets for the first-row atoms. The second Dunning's contraction reduces ($10s6p$) to more flexible [$5s4p$] set. Being applied to ($11s7p$NdMf), it produces the [$6s5p$NdMf] basis set family.

Hydrogen. We start from ($5s$) and ($6s$) parent sets developed by Huzinaga [14]. The extension of these sets to primitive sets of ($6s$NpMd) and ($7s$NpMd) families is performed in the same manner as has been employed for the first-row atoms. Let us illustrate it by the example of the ($5s$) parent set. The diffuse s primitives are first obtained using Eq. (3) with $m = 5$; the obtained ($6s$) set is then used to optimize the outermost exponent of the polarization ($2p$) set; the ($3p$), ($4p$), ($1d$) and ($3d$) polarization sets are finally determined in perfect analogy to the ($3d$), ($4d$), ($1f$) and ($3f$) sets for the first-row atoms except that the scaling factor of 3/2 between p and d exponents predicted by Eq. (7) has been used. It is important to note that the reoptimization of diffuse exponents of the ($2p$) set with ($7s1p$) basis, derived from Huzinaga's ($6s$) parent set, does not affect the results. The obtained Hydrogen polarization exponents are presented in Table 1.

The primitive bases of ($6s$NpMd) family are contracted to [$3s$NpMd] with Dunning triple-zeta valence scheme [17]; the contracted [$3s$NpMd] bases are combined with the corresponding basis sets of [$6s4p$NdMf] family. Similarly, ($7s$NpMd) primitive sets are contracted to [$5s$NpMd] and combined with the corresponding [$6s5p$NdMf] bases. Contraction coefficients used here together with the corresponding values of s-orbital exponents are as follows:

Exponents: 68.1600, 10.2465, 2.34648, 0.67332, 0.22466, 0.082217, 0.0301

Coefficients: 0.024379, 0.185228, 0.887204, 1.0, 1.0, 1.0, 1.0

The following consideration is focused on a convergence of the molecular first hyperpolarizability with the size of the polarization set at a given sp substrate. The main purpose is the elucidation of basis set requirements to the estimation of longitudinal and transverse components of the β_{ijk} tensor at the SCF level. The appropriate benchmark tests are required to achieve this purpose.

3. Benchmarks and computational details

To date there are just three acentric molecules with π-bonds that have been studied with the sophisticated basis sets, producing the SCF results for electronic polarizabilities and hyperpolarizabilities, expected to be very close to the HF limit. They are H_2CO [11], CH_3CN [12] and HCN [5]. The SCF data for these molecules, obtained in [5,11,12], were chosen as the benchmark tests in our consideration.

H_2CO. The contracted Gaussian basis set, applied to the study of this molecule in [11], is [$7s6p4d3f/6s4p3d$], derived from the primitive set ($15s10p7d5f/10s7p5d$) using the general contraction scheme, based on the Atomic Natural Orbitals (ANO) approach of Almlöf and Taylor [15]. The five-membered d sets and seven-membered f sets of cubic harmonic functions are used with this basis set, that are referred to as $5d/7f$ sets below.

CH_3CN. As the size of this molecule is larger than that of H_2CO, the authors of [12] were forced to reduce the size of the basis set compared to the previous case and to use ($15s10p7d1f/10s6p$), contracted to [$7s6p4d1f/6s3p$] with the ANO approach. It is worth mentioning that this basis is somewhat unbalanced for the first-row atoms and Hydrogen, this fact may cause the distortion of the results for dipole polarizabilities and hyperpolarizabilities.

HCN. The Gaussian basis set, used for this molecule in [5], is ($15s10p5d3f/9s5p2d$) contracted to [$10s7p5d3f/9s5p2d$] with segmented fashion. This basis is the largest among those we cite here for the benchmark calculations.

The electric properties of the molecules are defined here in the coordinate systems, sketched below with the z-axis coinciding with the longitudinal axis of a molecule defined by the permanent dipole moment vector:

The symmetry-produced relationships between the non-zero components of static α_{ij} and β_{ijk} tensors in these coordinate systems are listed below for all the molecules under consideration.

$$H_2CO : \alpha_{xx}, \alpha_{yy}, \alpha_{zz}; \beta_{zxx}, \beta_{zyy}, \beta_{zzz}.$$

$$CH_3CN : \alpha_{xx} = \alpha_{yy}, \alpha_{zz}; \beta_{zxx} = -\beta_{xyy}, \beta_{zxx} = \beta_{zyy}, \beta_{zzz}.$$

$$HCN : \alpha_{xx} = \alpha_{yy}, \alpha_{zz}; \beta_{zxx} = \beta_{zyy}, \beta_{zzz}.$$

The calculations were performed at the same geometries which were used in benchmark calculations [5, 11,12]. They are $R_{CO} = 1.203$ Å, $R_{CH} = 1.099$ Å and $\angle HCH = 116.5°$ for H_2CO, $R_{CN} = 1.153$ Å, $R_{CH} = 1.066$ Å for HCN, and $R_{CN} = 1.166$ Å, $R_{CH} = 1.083$ Å, $R_{CC} = 1.459$ Å, $\angle HCC = 108.9°$, $\angle HCH = 110.1°$ for CH_3CN.

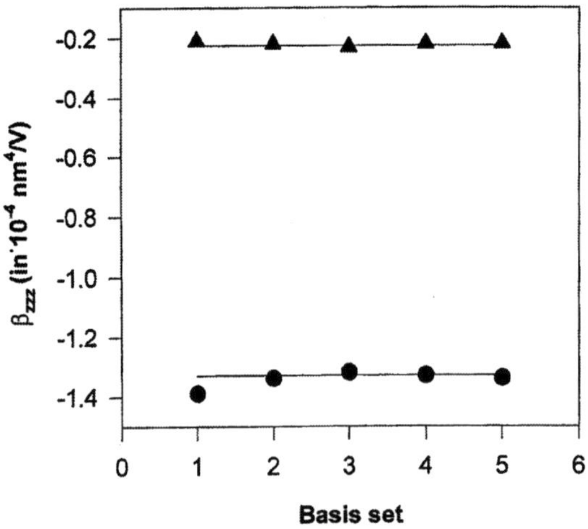

Fig. 2. Convergence of longitudinal hyperpolarizabilities of H2CO (circules), CH_3CN (triangles) and HCN (squadres) for the [6s4pNdMf/4sNpMd] bases families. The basis sets are ordered according to N and M numbers given as (N,M) pairs: (2,0) - **1**, (3,0) - **2**, (4,0) - **3**, (3,1) - **4**, (4,3) –**5**.

The calculations of the static linear and nonlinear polarizabilities are performed by coupled perturbed Hartree-Fock method [18] with Gaussian98 program package [19]. The pure cubic harmonical $5d/7f$ functions are always used during the SCF calculations.

The SI units are used throughout the paper. They are 10^{-2} nm^3 for α and 10^{-4}nm^4/V for β. The conversion factors are the following: 1 a.u. $= 0.1862 \cdot 10^{-2}$ nm^3 for α and 1 a.u. $= 0.0362 \cdot 10^{-4}$ nm^4/V for β. The permanent dipole moment is expressed in Debye (1 a.u. $= 2.542$ D).

4. Results and discussions

The convergence of the SCF molecular energy with the size of the basis set is trivial: the larger is the basis set, the higher is the accuracy of the result. The consideration of the convergence issues for the properties other than energy is much more complicated. Usually one deals with a family of basis sets organized in the finite sequence according to their size and/or expected efficiency. Typically, the extension of the basis set within a given family is meaningful only up to some definite size, at which the apparent convergence is reached. Further improvement of the basis set results in small random oscillations of calculated property in the neighbourhood of the saturation limit corresponding to a bases family under consideration, the amplitude of these oscillations bounding the calculation accuracy.

Such behavior is illustrated in the Fig. 2 by the example of longitudinal hyperpolarizabilities β_{zzz}, producing the dominant contribution to the nonlinear response of the molecules under study. We consider here the [$6s4p$NdMf/$4s$NpMd] bases family, ordered according to the quality of polarization set given by N and M values. It can be seen from Fig. 2, that the extension of ($2d/2p$) set to ($3d/3p$) results in the pronounced improvement of the calculated β_{zzz} value. The addition of the following d and f polarization functions slightly affects the results, causing only small oscillations near the limiting value. To resume one may say, that the convergence of β_{zzz} is reached at the ($3d/3p$) level. The ($4d/4p$), ($3d1f/3p1d$) and ($4d3f/4p3d$) polarization sets, added to [$6s4p$] substrate, fall in the convergence region and produce the results with the same uncertainty as the ($3d/3p$) one. This uncertainty may be estimated as half of a maximal difference between the data falling in the convergenc domain. Its value depends on the quality

Table 3
SCF results for permanent dipole moment μ (in D), dipole polarizability α (in 10^{-2} nm^3), and dipole hyperpolarizability β (in 10^{-4} nm^4/V) of H_2CO

Basis set	μ	α_{xx}	α_{yy}	α_{zz}	β_{zxx}	β_{zyy}	β_{zzz}
[6s4p2d/4s2p]	2.85	2.27	3.00	3.94	−0.34	−1.16	−1.39
[6s4p3d/4s3p]	2.84	2.32	3.03	3.95	−0.27	−1.12	−1.34
[6s4p4d/4s4p]	2.84	2.32	3.03	3.95	−0.27	−1.11	−1.32
[6s4p3d1f/4s3p1d]	2.84	2.32	3.03	3.95	−0.26	−1.11	−1.33
[6s4p4d3f/4s4p3d]	2.84	2.32	3.03	3.95	−0.27	−1.11	−1.34
[6s5p2d/5s2p]	2.85	2.27	3.00	3.94	−0.34	−1.16	−1.39
[6s5p3d/5s3p]	2.84	2.32	3.03	3.95	−0.28	−1.13	−1.33
[6s5p4d/5s4p]	2.84	2.32	3.03	3.95	−0.28	−1.12	−1.32
[6s5p3d1f/5s3p1d]	2.84	2.32	3.03	3.95	−0.27	−1.12	−1.33
[6s5p4d3f/5s4p3d]	2.84	2.32	3.03	3.95	−0.27	−1.12	−1.33
[7s6p4d3f/6s4p3d]*	2.84	2.32	3.03	3.95	−0.24	−1.08	−1.31

*Benchmark test from [11].

Table 4
SCF results for permanent dipole moment μ (in D), dipole polarizability α (in 10^{-2} nm^3), and dipole hyperpolarizability β (in 10^{-4} nm^4/V) of CH_3CN

Basis set	μ	α_{xx}	α_{zz}	β_{xxx}	β_{zxx}	β_{zzz}
[6s4p2d/4s2p]	4.29	4.36	7.07	−0.19	−0.11	0.48
[6s4p3d/4s3p]	4.29	4.39	7.11	−0.20	−0.08	0.51
[6s4p4d/4s4p]	4.29	4.39	7.11	−0.21	−0.08	0.50
[6s4p3d1f/4s3p1d]	4.29	4.39	7.11	−0.20	−0.08	0.52
[6s4p4d3f/4s4p3d]	4.28	4.39	7.12	−0.21	−0.08	0.51
[6s5p2d/5s2p]	4.29	4.36	7.07	−0.19	−0.11	0.48
[6s5p3d/5s3p]	4.29	4.39	7.11	−0.21	−0.08	0.51
[6s5p4d/5s4p]	4.29	4.39	7.11	−0.21	−0.08	0.50
[6s5p3d1f/5s3p1d]	4.29	4.39	7.11	−0.20	−0.08	0.52
[6s5p4d3f/5s4p3d]	4.28	4.39	7.12	−0.21	−0.08	0.51
[7s6p4d1f/6s3p]*	4.29	4.39	7.07	−0.20	−0.07	0.51

*Benchmark test from [12].

of the sp substrate, and defines the number of β_{zzz} significant digits, obtained reliably with bases of the $[6s4p\text{N}d\text{M}f/4s\text{N}p\text{M}d]$ family.

The calculated electronic polarizabilities and hyperpolarizabilities of H_2CO, CH_3CN and HCN molecules are presented in Tables 3–5. Let us begin the consideration with the SCF results obtained with the basis sets of $[6s4p\text{N}d\text{M}f/4s\text{N}p\text{M}d]$ family. According to the data of Tables 3–5, the convergence of both α_{ij} and β_{ijk} is reached at $[6s4p3d/4s3p]$ level for all three molecules under study. The estimated uncertainties do not exceed $0.005 \cdot 10^{-2}$ nm^3 for α_{ij} and $0.01 \cdot 10^{-4}$ nm^4/V for β_{ijk}. For H_2CO and HCN the values of α_{ij}, calculated at the $[6s4p3d/4s3p]$ level, coincide with the near-HF values obtained in the benchmark calculations [5,11]. In the case of CH3CN molecule the α tensor components, calculated with $[6s4p3d/4s3p]$ basis, are somewhat overestimated, however the errors are smaller than 1%.

Similar situation takes place for the dominant longitudinal β_{zzz}. For all three molecules the β_{zzz} values, calculated with the $[6s4p3d/4s3p]$ basis, coincide with the benchmark data within their uncertainty. As for the small transverse β components, maximal discrepancy with the benchmark data does not exceed $0.03 \cdot 10^{-4}$ nm^4/V. It allows us to recommend this basis for the benchmark SCF calculations of conjugated organic molecules, which are too large to be studied with more sophisticated Gaussian basis sets.

One may see from Tables 3–5 that the use of more flexible contraction of sp substrate is not necessary

Table 5
SCF results for permanent dipole moment μ (in D), dipole polarizability α (in 10^{-2} nm^3), and dipole hyperpolarizability β (in 10^{-4} nm^4/V) of HCN

Basis set	μ	α_{xx}	α_{zz}	β_{zxx}	β_{zzz}
[6s4p2d/4s2p]	3.30	2.55	4.16	−0.11	−0.21
[6s4p3d/4s3p]	3.30	2.59	4.18	−0.11	−0.22
[6s4p4d/4s4p]	3.30	2.59	4.18	−0.12	−0.23
[6s4p3d1f/4s3p1d]	3.30	2.59	4.18	−0.11	−0.22
[6s4p4d3f/4s4p3d]	3.29	2.59	4.18	−0.11	−0.22
[6s5p2d/5s2p]	3.30	2.55	4.16	−0.11	−0.23
[6s5p3d/5s3p]	3.30	2.59	4.18	−0.11	−0.22
[6s5p4d/5s4p]	3.30	2.59	4.18	−0.12	−0.23
[6s5p3d1f/5s3p1d]	3.30	2.59	4.18	−0.11	−0.22
[6s5p4d3f/5s4p3d]	3.29	2.59	4.18	−0.11	−0.22
[10s7p5d3f/9s5p2d]*	3.30	2.59	4.18	−0.11	−0.23

*Benchmark test from [5].

at the SCF level: the results, obtained with the [$6s5p$NdMf/$4s$NpMd] and [$6s5p$NdMf/$5s$NpMd] bases, are perfectly equivalent. The adding of the second diffuse sp set to the [$6s5p$NdMf/$4s$NpMd] bases has also a negligible effect on the results of the calculation of both α_{ij} and β_{ijk}.

It is interesting to note that the [$6s4p2d$/$4s2p$] basis provides the results which are very close to those obtained with [$6s4p3d$/$4s3p$] one. This basis may be recommended for the estimation of dominant components of α and β tensors with the errors not exceeding 5%.

It is worth emphasizing that all the conclusions above are valid only for the SCF level of theory. It is well known [3], that the electric properties of small molecules converge with the size increasing and the composition perfecting of the basis set much more slowly when the dynamic electron correlation is taken into account. The capacity of basis sets, developed here for the correlated calculations, needs special investigation.

As for the SCF HF approximation, the most significant result of this study is that the accuracy of the calculation of hyperpolarizability is controlled mostly by the quality of the polarization set, and depends slightly on the segmented contraction of sp substrate used. This observation gives grounds to hope that the elaborated approach may be applied to the ($9s5p$) parent set of Huzinaga [14] with various contraction schemes, suggested by Dunning for this set [20]. It would allow to further reduce the size of the polarized Gaussian basis sets, tailored for the *ab initio* calculations of hyperpolarizabilities of promising organic NLO chromophores. This problem is investigated thoroughly in the next paper of this series.

Acknowlegments

The authors are grateful to Dr. M.Yu. Balakina for fruitful discussion. This work was supported by Russian Foundation for Basic Research (grant N 02-03-32897).

References

[1] L.R. Dalton, W.H. Steier, B.H. Robinson, Ch. Zhang, A. Ren, S. Garner, A.Chen, T. Londergan, L. Irwin, B. Carlson, L. Fifield, G. Phelan, C. Kincaid, J. Amend and A. Jen, From molecules to opto-chips: organic electro-optic materials, *J. Mater. Chem.* **9** (1999), 1905–1920.

[2] D.R. Kanis, M.A. Ratner and T.J. Marks, Design and construction of molecular assemblies with large second-order optical nonlinearities. Quantum chemical aspects, *Chem. Rev.* **94** (1994), 195–242.

[3] D.P. Shelton and J.E. Rice, Measurements and calculations of the hyperpolarizabilities of atoms and small molecules in the gas phase, *Chem. Rev.* **94** (1994), 3–29.

[4] G. Maroulis, A systematic study of basis set, electron correlation, and geometry effects on the electric multipole moments, polarizability and hyperpolarizability of HCl, *J. Chem. Phys.* **108** (1998), 5432–5448.

[5] G. Maroulis and C. Pouchan, Molecules in static electric fields: linear and nonlinear polarizability of HC≡N and HC≡P, *Phys. Rev. A.* **57** (1998), 2440–2447.

[6] G. Maroulis, Hyperpolarizability of H_2O revisited: accurate estimate of the basis set limit and the size of electron correlation effects, *Chem. Phys. Lett.* **289** (1998), 403–411.

[7] G. Maroulis, On the accurate theoretical determination of the static hyperpolarizability of trans-butadiene, *J. Chem. Phys.* **111** (1999), 583–591.

[8] D. Xenides and G. Maroulis, Basis set and electron correlation effects on the first and second static hyperpolarizability of SO_2, *Chem. Phys. Lett.* **319** (2000), 618–624.

[9] G. Maroulis, Accurate electric multipole moment, static polarizability and hyperpolarizability derivatives for N_2, *J. Chem. Phys.* **118** (2003), 2673–2687.

[10] P.R. Taylor, T.J. Lee, J.E. Rice and J. Almlöf, The polarizabilities of Ne, *Chem. Phys. Lett.* **163** (1989), 359–365.

[11] J.E. Rice, R.D. Amos, S.M. Colwell, N.C. Handy and J. Sanz, Frequency dependent hyperpolarizabilities with application to formaldehyde and methyl fluoride, *J. Chem. Phys.* **111** (1999), 583–591.

[12] M. Stähelin, C.R. Moylan, D.M. Burland, A. Willetts, J.E. Rice, D.P. Shelton and E.A. Donley, A comparison of calculated and experimental hyperpolarizabilities for acetonitrile in gas and liquid phases, *J. Chem. Phys.* **98** (1993), 5595–5603.

[13] A.J. Sadlej, Medium-size polarized basis sets for high-level correlated calculations of molecular electric properties, *Collection Czechoslovak Chem. Commun* **53** (1988), 1995–2016.

[14] S. Huzinaga, Gaussian-type functions for polyatomic systems. I, *J. Chem. Phys.* **32** (1960), 1293–1303.

[15] R. Ahlrichs and P.R. Taylor, The choice of Gaussian basis sets for molecular electronic structure calculations, *J. Chim. Phys.* **78** (1981), 315–321.

[16] J. Almlöf and P.R. Taylor, General contraction of Gaussian basis sets. I. Atomic natural orbitals for first- and second-row atoms, *J. Chem. Phys.* **86** (1987), 4070–4077.

[17] T.H. Dunning, Gaussian basis functions for use in molecular calculations. III. Contraction of (10s6p) atomic basis sets for the first-row atoms, *J. Chem. Phys.* **55** (1971), 716–723.

[18] S.P. Karna and M. Dupuis, Frequency dependent nonlinear optical properties of molecules: formulation and implementation in HONDO program, *J. Comp. Chem.* **12** (1991), 487–504.

[19] M.J. Frisch et al., Gaussian 98, Revision A.6, Gaussian, Inc., Pittsburgh PA, 1998.

[20] T.H. Dunning, Gaussian basis functions for use in molecular calculations. I. Contraction of (9s5p) atomic basis sets for the first-row atoms, *J. Chem. Phys.* **53** (1970), 2823–2833.

Journal of Computational Methods in Sciences and Engineering 4 (2004) 493–500
IOS Press

Rational Design of atomic Gaussian basis sets for *Ab initio* Calculations of the dipole polarizabilities and hyperpolarizabilities II. Moderate Size Basis Sets for the First-Row Atoms from B to F

Mikhail B. Zuev*, Marina Yu. Balakina and Sergey E. Nefediev
A.E. Arbuzov Institute of Organic and Physical Chemistry of Kazan Scientific Centre, Russian Academy of Sciences, Arbuzov str., 8, 420088, Kazan, Russia

Abstract. The design rules, elaborated earlier in the first paper of this series for the construction of atomic Gaussian basis sets, tailored for the calculations of molecular polarizabilities α_{ij} and hyperpolarizabilities β_{ijk} at the SCF level, are tested here by the example of the basis set families, generated from the energy-optimized (9s5p) parent set. Adding the polarization sets to contracted *sp* substrates of various flexibility, we generate the [4s3pNdMf], [5s3pNdMf] and [6s4pNdMf] bases families, which are combined with the corresponding [3sNpMd] and [4sNpMd] bases for Hydrogen. The SCF results for H_2CO, CH_3CN, and HCN molecules, obtained with (11s7p3d/6s3p)/[6s4p3d/4s3p], are used as a benchmark here. The contraction flexibility is shown to affect the convergence of both α_{ij} and β_{ijk} values only insignificantly, slightly increasing the uncertainties of the calculated data. The smallest [4s3p3d/3s3p] basis is recommended for the calculation of (hyper)polarizabilities of large molecules within the SCF approximation. For the conjugated organic molecules with dominant π-electron polarization rather reliable SCF results are obtained at the [4s3p2d/3s] level. So, for *p*-nitroaniline this basis set reproduces the results obtained with [6s4p3d/4s3p] basis with the error 1% for α_{ij} and 3% for dominant longitudinal hyperpolarizability β_{zzz}.

Keywords: Gaussian basis sets, convergence, SCF calculations of electric properties, polarizability and hyperpolarizability

Mathematics Subject Classification: 78-0 – Optics, Computational Methods

1. Introduction

In the previous paper [1] we have elaborated the strategy for the construction of atomic Gaussian basis sets, tailored for the calculations of electronic polarizabilities α_{ij} and first hyperpolarizabilities β_{ijk} of organic molecules at the SCF level. The strategy starts with the energy-optimized Huzinaga (10s6p) basis set for the first row atoms, combined with the (5s) set for Hydrogen, and finally produces the [6s4p3d/4s] Gaussian basis set, which provides the results close to the HF limit, estimated for H_2CO, CH_3CN, and HCN molecules with the most sophisticated polarized Gaussian basis sets, used up to now [2–4]. The

*Corresponding author. E-mail: mikhail@iopc.knc.ru.

contraction of the used sp substrate has been shown to slightly affect the accuracy of the calculation of (hyper)polarizabilities; the uncertainty of the calculated values are controlled mostly by the quality of the polarization set, added to the contracted sp substrate. In the present paper we examine the validity of the elaborated strategy for the construction of the polarized Gaussian basis set, derived from (9s5p) primitive set of Huzinaga [5] in various Dunning's contractions [6]. The main goal here is to reduce substantially the size of the one-particle basis set without significant loss of the calculation accuracy for both α_{ij} and β_{ijk} at the SCF level of theory. The impetus for this study is given by the aspiration to obtain the computational methodology capable to aid in molecular design of nonlinear optical (NLO) chromophores, and the elucidation of the relationship between their electron structure and NLO activity of conjugated organic molecules.

2. Basis set construction

The detailed consideration of the main physical concepts, underlying the construction of the polarized Gaussian basis sets for the calculation of α_{ij} and β_{ijk}, is presented in the first paper of this series [1]. Here we formulate briefly the precepts for the construction of such bases for the first-row atoms and Hydrogen.

First-row atoms. The starting point here is the (9s5p) primitive set developed by Huzinaga [5]; it is referred to as a parent one throughout the paper. The strategy to arrange this basis for the calculation of the electric properties is the following:

Step 1. The augment of the parent set with the single set of diffuse s and p Gaussian primitives with the exponents, forming the even-tempered sequence with the two last exponents of the original s and p subsets:

$$\alpha_{10}^{s} = \alpha_{9}^{s}(\alpha_{9}^{s}/\alpha_{8}^{s}), \quad \alpha_{6}^{p} = \alpha_{5}^{p}(\alpha_{5}^{p}/\alpha_{4}^{p}). \tag{1}$$

The obtained (10s6p) set is referred to as sp primitive substrate below.

Step 2. The determination of the (2d) polarization set, consisting of two d subsets located in the inner and outer parts of the valence region, respectively. The exponent of the tight subset is given by the formula of Ahlrichs and Taylor [7]:

$$\alpha_{1}^{2d} = 0.02Z^{2}. \tag{2}$$

The diffuse exponent α_{2}^{2d} of (2d) set is obtained by means of maximization of the atomic dipole polarization in the presence of the standard external electric field of 0.01 a.u. by the ROHF technique at the (10s6p1d) level.

Step 3. The construction of the sequence of polarization sets (2d), (3d), (4d), (3d1f), (4d3f) is fulfilled with the help of the splitting and scaling procedure, described in the previous paper [1]. The polarization exponents for these sets are given by the formulae:

(3d) set:

$$\alpha_{1}^{3d} = \alpha_{1}^{2d}; \quad \alpha_{2}^{3d} = \alpha_{3}^{4d}; \quad \alpha_{3}^{3d} = \alpha_{4}^{4d}. \tag{3}$$

(4d) set:

$$\alpha_{m}^{4d} = \alpha_{m-1}^{4d}\left(\alpha_{2}^{2d}/\alpha_{1}^{2d}\right)^{1/2}, \quad \text{for } m = 2,3,4 \ \text{ with } \ \alpha_{1}^{4d} = \alpha_{1}^{2d}(\alpha_{1}^{2d}/\alpha_{2}^{2d})^{1/4}. \tag{4}$$

Table 1
Exponents of polarization functions for the first-row atoms and Hydrogen

Polarization set		B	C	N	O	F	Polarization set		H
(2*d*)	α_1	0.50	0.72	0.98	1.28	1.62	(2*p*)	α_1	1.0
	α_2	0.068	0.115	0.144	0.225	0.298		α_2	0.125
(3*d*)	α_1	0.50	0.72	0.98	1.28	1.62	(3*p*)	α_1	1.0
	α_2	0.112	0.182	0.233	0.347	0.455		α_2	0.210
	α_3	0.041	0.073	0.089	0.146	0.195		α_3	0.074
(4*d*)	α_1	0.824	1.139	1.583	1.977	2.474	(4*p*)	α_1	1.682
	α_2	0.304	0.455	0.607	0.829	1.061		α_2	0.595
	α_3	0.112	0.182	0.233	0.347	0.455		α_3	0.210
	α_4	0.041	0.073	0.089	0.146	0.195		α_4	0.074
(1*f*)	α_1	0.091	0.153	0.192	0.300	0.397	(1*d*)	α_1	0.188
(3*f*)	α_1	0.667	0.960	1.307	1.707	2.160	(3*d*)	α_1	1.5
	α_2	0.245	0.384	0.501	0.716	0.927		α_2	0.530
	α_3	0.091	0.153	0.192	0.300	0.397		α_3	0.188

(1f) set:

$$\alpha^{1f} = 4/3\ \alpha_2^{2d}, \tag{5}$$

(3f) set:

$$\alpha_m^{3f} = \alpha_{m-1}^{3f}\left(\alpha_2^{2d}/\alpha_1^{2d}\right)^{1/2}, \quad \text{for } m = 2,3 \text{ with } \alpha_1^{3f} = 4/3\ \alpha_1^{2d}. \tag{6}$$

The estimated exponent values are rounded off to three decimal places. This technique produces completely the same polarization exponents as those obtained earlier in [1] for the polarization subsets added to (11s7p) primitive set. For convenience we present them here in Table 1. Adding these (NdMf) polarization sets to the (10s6p) primitive substrate, we generate the (10s6pNdMf) primitive bases family.

Step 4. The contraction of the obtained primitive sets is carried out by means of the Dunning's schemes [6], developed for the parent (9s5p) set, resulting in three families of polarized contracted Gaussian basis sets, namely [4s3pNdMf], [5s3pNdMf] and [6s4pNdMf] bases families.

Hydrogen. We start here from (4s) and (5s) primitive bases of Huzinaga [5], forming the (5sNpMd) and (6sNpMd) families of primitive sets, constructed in perfectly the same manner as described above. The only difference is that the scaling factor of 3/2 between p and d exponents has been used instead of the factor 4/3, coupling d and f exponents (see Eq. (5) and (6)). As in the case of the first-row atoms, the obtained polarization exponents are the same as those obtained earlier in [1]; they are given in Table 1.

The (5sNpMd) sets are contracted to [3sNpMd] and [4sNpMd] [6], the (6sNpMd) ones only to [4sNpMd] with the corresponding Dunning's contractions [8]. The (6sNpMd)/[4sNpMd] basis sets family has been constructed to study the effect of Hydrogen primitive set size on the SCF results for electronic (hyper)polarizabilities.

Further we analyze the results of the probation of the constructed basis sets by the example of H_2CO, CH_3CN and HCN molecules. As the benchmarks we use here the SCF results, obtained for these molecules with [6s4p3d/4s3p] basis set built up in the previous paper [1]. This set provides the results, which are very close to the near HF limit values, estimated with almost saturated polarized Gaussian basis sets [2,3] and [4].

The NLO response of the above-mentioned molecules comprises almost equal contributions from the dipole polarization of σ- and π-electron systems. As for the conjugated π-electron molecules,

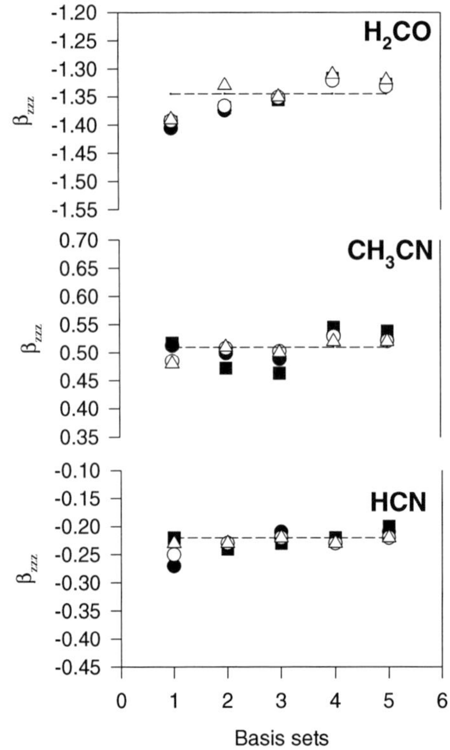

Fig. 1. Convergence of longitudinal hyperpolarizabilities of H_2CO, CH_3CN and HCN for the [4s3pNdMf/3sNpMd] (black circles), [5s3pNdMf/3sNpMd] (black squares), [6s4pNdMf/4sNpMd] (derived from (10s6pNdMf/5sNpMd) – empty circles, derived from (10s6pNdMf/5sNpMd) - empty triangles) bases families. The basis sets are ordered according to N and M numbers given as (N,M) pairs: (2,0) – **1**, (3,0) – **2**, (4,0) – **3**, (3,1) – **4**, (3,3) –**5**.

intuitively it is clear that the π-electron polarization is dominant. It is reasonable to assume, that Hydrogen polarization functions might not be indispensible for the description of the electric properties of conjugated organic molecules. If it is so, it would result in a significant reduction of the size of the developed bases applied to the case of conjugated π-electron systems. This possibility is studied in detail here by the example of p-nitroaniline (PNA), the simplest of the dipolar NLO chromophores used at present in the design of NLO organic materials.

3. Computational details

The calculations of the static linear and nonlinear polarizabilities are performed by coupled perturbed Hartree-Fock method [9] with Gaussian98 program package [10]. The pure cubic harmonic $5d/7f$ functions are always used in the SCF calculations.

The calculations were performed at the following geometries:

H_2CO: $R_{CO} = 1.203$ Å, $R_{CH} = 1.099$ Å and $\angle HCH = 116.5°$;

CH_3CN: $R_{CN} = 1.166$ Å, $R_{CH} = 1.083$ Å, $R_{CC} = 1.459$ Å, $\angle HCC = 108.9°$, $\angle HCH = 110.1°$;

$HCN : R_{CN} = 1.153$ Å, $R_{CH} = 1.066$ Å.

We have to note that the same geometries of H_2CO, CH_3CN and HCN molecules are used in the calculations presented in the previous paper [1].

Table 2
SCF results for permanent dipole moment μ (in D), dipole polarizability α (in 10^{-2} nm^3), and dipole hyperpolarizability β (in 10^{-4} nm^4/V) of H_2CO

Primitive set	Contracted set	μ_z	α_{xx}	α_{yy}	α_{zz}	β_{zxx}	β_{zyy}	β_{zzz}
(10s6p2d/5s2p)	[4s3p2d/3s2p]	2.85	2.24	2.97	3.94	−0.33	−1.21	−1.40
(10s6p3d/5s3p)	[4s3p3d/3s3p]	2.85	2.30	3.02	3.96	−0.27	−1.12	−1.37
(10s6p4d/5s4p)	[4s3p4d/3s4p]	2.85	2.30	3.03	3.96	−0.26	−1.11	−1.35
(10s6p3d1f/5s3p1d)	[4s3p3d1f/3s3p1d]	2.85	2.32	3.03	3.96	−0.26	−1.10	−1.32
(10s6p4d3f/5s4p3d)	[4s3p3d3f/3s3p3d]	2.85	2.33	3.03	3.96	−0.28	−1.09	−1.33
(10s6p2d/5s2p)	[5s3p2d/3s2p]	2.85	2.25	2.97	3.94	−0.33	−1.18	−1.39
(10s6p3d/5s3p)	[5s3p3d/3s3p]	2.85	2.31	3.03	3.96	−0.26	−1.12	−1.37
(10s6p4d/5s4p)	[5s3p4d/3s4p]	2.85	2.31	3.03	3.96	−0.26	−1.10	−1.35
(10s6p3d1f/5s3p1d)	[5s3p3d1f/3s3p1d]	2.85	2.32	3.03	3.96	−0.26	−1.10	−1.32
(10s6p4d3f/5s4p3d)	[5s3p3d3f/3s3p3d]	2.85	2.33	3.03	3.96	−0.28	−1.09	−1.33
(10s6p2d/5s2p)	[6s4p2d/4s2p]	2.85	2.27	2.99	3.94	−0.34	−1.16	−1.39
(10s6p3d/5s3p)	[6s4p3d/4s3p]	2.85	2.31	3.03	3.96	−0.26	−1.12	−1.37
(10s6p4d/5s4p)	[6s4p4d/4s4p]	2.85	2.32	3.03	3.96	−0.27	−1.09	−1.35
(10s6p3d1f/5s3p1d)	[6s4p3d1f/4s3p1d]	2.85	2.32	3.03	3.96	−0.27	−1.10	−1.32
(10s6p4d3f/5s4p3d)	[6s4p3d3f/4s3p3d]	2.85	2.33	3.03	3.96	−0.27	−1.10	−1.33
(10s6p2d/6s2p)	[6s4p2d/4s2p]	2.85	2.26	2.99	3.94	−0.33	−1.18	−1.39
(10s6p3d/6s3p)	[6s4p3d/4s3p]	2.85	2.32	3.03	3.95	−0.28	−1.13	−1.33
(10s6p4d/6s4p)	[6s4p4d/4s4p]	2.85	2.32	3.03	3.95	−0.26	−1.11	−1.35
(10s6p3d1f/6s3p1d)	[6s4p3d1f/4s3p1d]	2.85	2.32	3.03	3.95	−0.27	−1.11	−1.31
(10s6p4d3f/6s4p3d)	[6s4p3d3f/4s3p3d]	2.85	2.33	3.03	3.95	−0.28	−1.10	−1.32
(11s7p3d/6s3p)*	[6s4p3d/3s3p]	2.84	2.32	3.03	3.95	−0.27	−1.12	−1.34

*) Benchmark test from [1].

The PNA molecule has C_{2v} symmetry, its geometry have been optimized with the 6-31G(d) basis set. The values of the main geometrical parameters are presented on the scheme below:

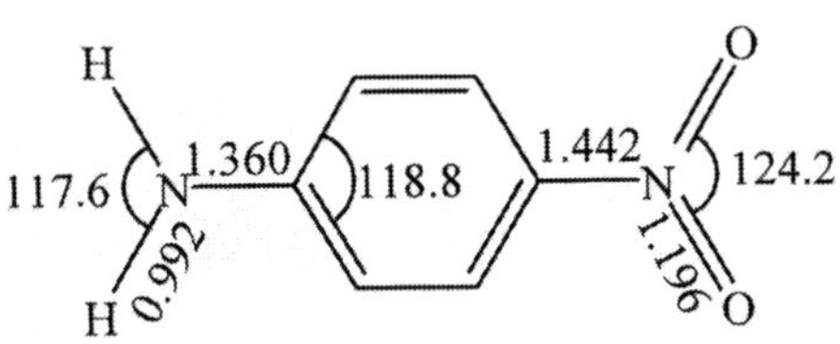

The molecule is in (x,z) coordinate plane with the z-axis coinciding with the longitudinal axis of a molecule, defined by the permanent dipole moment vector.

The SI units are used throughout the paper. They are 10^{-2} nm^3 for α and 10^{-4} nm^4/V for β. The conversion factors are the following: 1 a.u. $= 0.1862 \cdot 10^{-2}$ nm^3 for α and 1 a.u. $= 0.0362 \cdot 10^{-4}$ nm^4/V for β. The permanent dipole moment is expressed in Debye (1 a.u. $= 2.542$ D).

4. Results and Discussions

Our first task is to examine the convergence of SCF results for α_{ij} and β_{ijk} with the size of the polarization set and to estimate the uncertainties, determining the number of significant digits of the calculated values. These uncertainties are due to the small random oscillation of calculated property in the neighborhood of the saturation limit, corresponding to a given bases family once the apparent convergence is reached. We may suppose, that the magnitude of these uncertainties is specified mainly

Table 3
SCF results for permanent dipole moment μ (in D), dipole polarizability α (in 10^{-2} nm^3), and dipole hyperpolarizability β (in 10^{-4} nm^4/V) of CH_3CN

Primitive set	Contracted set	μ_z	α_{zz}	α_{xx}	β_{xxx}	β_{zxx}	β_{zzz}
(10s6p2d/5s2p)	[4s3p2d/3s2p]	4.29	4.34	7.05	−0.22	−0.15	0.51
(10s6p3d/5s3p)	[4s3p3d/3s3p]	4.29	4.38	7.09	−0.22	−0.11	0.50
(10s6p4d/5s4p)	[4s3p4d/3s4p]	4.29	4.38	7.10	−0.22	−0.11	0.49
(10s6p3d1f/5s3p1d)	[4s3p3d1f/3s3p1d]	4.29	4.39	7.10	−0.20	−0.07	0.53
(10s6p4d3f/5s4p3d)	[4s3p3d3f/3s3p3d]	4.29	4.40	7.12	−0.20	−0.08	0.52
(10s6p2d/5s2p)	[5s3p2d/3s2p]	4.27	4.34	7.05	−0.21	−0.14	0.52
(10s6p3d/5s3p)	[5s3p3d/3s3p]	4.29	4.38	7.09	−0.22	−0.10	0.47
(10s6p4d/5s4p)	[5s3p4d/3s4p]	4.29	4.39	7.10	−0.22	−0.09	0.46
(10s6p3d1f/5s3p1d)	[5s3p3d1f/3s3p1d]	4.29	4.39	7.11	−0.20	−0.08	0.54
(10s6p4d3f/5s4p3d)	[5s3p3d3f/3s3p3d]	4.29	4.40	7.12	−0.20	−0.08	0.54
(10s6p2d/5s2p)	[6s4p2d/4s2p]	4.27	4.36	7.08	−0.20	−0.13	0.48
(10s6p3d/5s3p)	[6s4p3d/4s3p]	4.29	4.39	7.11	−0.20	−0.08	0.51
(10s6p4d/5s4p)	[6s4p4d/4s4p]	4.29	4.39	7.11	−0.21	−0.08	0.50
(10s6p3d1f/5s3p1d)	[6s4p3d1f/4s3p1d]	4.29	4.39	7.11	−0.20	−0.08	0.53
(10s6p4d3f/5s4p3d)	[6s4p3d3f/4s3p3d]	4.29	4.40	7.12	−0.20	−0.08	0.52
(10s6p2d/6s2p)	[6s4p2d/4s2p]	4.29	4.45	7.07	−0.19	−0.12	0.48
(10s6p3d/6s3p)	[6s4p3d/4s3p]	4.29	4.39	7.10	−0.21	−0.08	0.51
(10s6p4d/6s4p)	[6s4p4d/4s4p]	4.29	4.39	7.11	−0.21	−0.08	0.50
(10s6p3d1f/6s3p1d)	[6s4p3d1f/4s3p1d]	4.29	4.39	7.11	−0.20	−0.08	0.52
(10s6p4d3f/6s4p3d)	[6s4p3d3f/4s3p3d]	4.29	4.40	7.12	−0.21	−0.08	0.52
(11s7p3d/6s3p)*	[6s4p3d/3s3p]	4.29	4.39	7.11	−0.20	−0.08	0.52

*) Benchmark test from [1].

by the contraction flexibility of the sp substrate defining a bases family under consideration. We start with the examination of this assumption.

The convergence of the longitudinal hyperpolarizability β_{zzz}, calculated with the basis sets of [4s3pNdMf/3sNpMd], [5s3pNdMf/3sNpMd], [6s4pNdMf/4sNpMd] families, is shown on Fig. 1. It can be seen that the above assumption is valid, the oscillations of the calculated β_{zzz} in the convergence domain tend to increase with the flexibility of the contraction used. This effect is the most pronounced in the case of CH_3CN molecule. For more detailed consideration let us turn to Tables 2–4.

Let us begin the consideration with the SCF results obtained with the basis sets [6s4pNdMf/4sNpMd], derived from (10s6pNdMf/5sNpMd) and (10s6pNdMf/6sNpMd) primitive sets. It can be seen that the size of the primitive set on Hydrogen does not affect significantly the results of the calculations. We would like to note only, that the increasing of the size of the primitive set on Hydrogen may result in the small reduction of the uncertainty values estimated for the longitudinal β_{zzz}. Since this effect takes place only for the case of H_2CO molecule, we limit ourselves hereafter by the consideration of the results, obtained with the bases of the (10s6pNdMf/5sNpMd)/[6s4pNdMf/4sNpMd] family, having a smaller dimension of the primitive space.

According to the data of Tables 2–4, the convergence of both α_{ij} and β_{ijk} is reached at [6s4p3d/4s3p] level for all three molecules under study. The uncertainties (estimated as half a maximal difference between the data falling in the convergence domain) do not exceed $0.01 \cdot 10^{-2}$ nm^3 for α_{ij} and $0.02 \cdot 10^{-4}$ nm^4/V for β_{ijk}. The SCF results for all three molecules coincide within their uncertainties with the benchmarks obtained by us in [1].

The use of less flexible double-zeta and double-zeta valence Dunning's contractions (the [5s3pNdMf/3sNpMd] and [4s3pNdMf/3sNpMd] bases in Tables 2–4) does not affect essentially the convergence behavior, resulting only in the slight increase of the uncertainties of the calculated data.

Table 4
SCF results for permanent dipole moment μ (in D), dipole polarizability α (in 10^{-2} nm^3), and dipole hyperpolarizability β (in 10^{-4}nm^4/V) of HCN

Primitive set	Contracted set	μ_z	α_{xx}	α_{zz}	β_{zxx}	β_{zzz}
(10s6p2d/5s2p)	[4s3p2d/3s2p]	3.28	2.53	4.15	−0.12	−0.27
(10s6p3d/5s3p)	[4s3p3d/3s3p]	3.28	2.57	4.18	−0.13	−0.23
(10s6p4d/5s4p)	[4s3p4d/3s4p]	3.28	2.57	4.18	−0.13	−0.21
(10s6p3d1f/5s3p1d)	[4s3p3d1f/3s3p1d]	3.30	2.58	4.17	−0.11	−0.23
(10s6p4d3f/5s4p3d)	[4s3p3d3f/3s3p3d]	3.30	2.58	4.18	−0.12	−0.21
(10s6p2d/5s2p)	[5s3p2d/3s2p]	3.28	2.53	4.15	−0.11	−0.22
(10s6p3d/5s3p)	[5s3p3d/3s3p]	3.30	2.57	4.17	−0.11	−0.24
(10s6p4d/5s4p)	[5s3p4d/3s4p]	3.30	2.58	4.18	−0.11	−0.23
(10s6p3d1f/5s3p1d)	[5s3p3d1f/3s3p1d]	3.30	2.58	4.18	−0.11	−0.22
(10s6p4d3f/5s4p3d)	[5s3p3d3f/3s3p3d]	3.30	2.59	4.19	−0.12	−0.20
(10s6p2d/5s2p)	[6s4p2d/4s2p]	3.30	2.55	4.16	−0.11	−0.25
(10s6p3d/5s3p)	[6s4p3d/4s3p]	3.30	2.58	4.18	−0.11	−0.23
(10s6p4d/5s4p)	[6s4p4d/4s4p]	3.30	2.59	4.18	−0.12	−0.22
(10s6p3d1f/5s3p1d)	[6s4p3d1f/4s3p1d]	3.30	2.58	4.18	−0.12	−0.23
(10s6p4d3f/5s4p3d)	[6s4p3d3f/4s3p3d]	3.30	2.59	4.19	−0.11	−0.22
(10s6p2d/6s2p)	[6s4p2d/4s2p]	3.30	2.55	4.16	−0.11	−0.23
(10s6p3d/6s3p)	[6s4p3d/4s3p]	3.30	2.58	4.18	−0.11	−0.23
(10s6p4d/6s4p)	[6s4p4d/4s4p]	3.30	2.59	4.18	−0.12	−0.22
(10s6p3d1f/6s3p1d)	[6s4p3d1f/4s3p1d]	3.30	2.58	4.18	−0.11	−0.23
(10s6p4d3f/6s4p3d)	[6s4p3d3f/4s3p3d]	3.30	2.59	4.19	−0.11	−0.22
(11s7p3d/6s3p)*	[6s4p3d/3s3p]	3.30	2.59	4.18	−0.11	−0.22

*) Benchmark test from [1].

Table 5
SCF results for permanent dipole moment μ (in D), dipole polarizability α (in 10^{-2} nm^3), and dipole hyperpolarizability β (in 10^{-4} nm^4/V) of PNA

Primitive set	Contracted set	μ_z	α_{xx}	α_{yy}	α_{zz}	β_{zxx}	β_{zyy}	β_{zzz}
(10s6p2d/5s)	[4s3p2d/3s]	7.39	18.0	9.9	24.8	−5.83	−1.26	35.1
(10s6p2d/5s2p)	[4s3p2d/3s2p]	7.39	18.2	10.0	24.9	−6.01	−1.39	34.9
(10s6p3d/5s)	[4s3p3d/3s]	7.38	18.2	10.0	25.0	−5.90	−1.33	34.2
(10s6p3d/5s3p)	[4s3p3d/3s3p]	7.39	18.3	10.0	25.0	−6.04	−1.38	34.0
(10s6p2d/5s)	[5s3p2d/3s]	7.40	18.0	10.0	24.8	−5.81	−1.27	35.0
(10s6p2d/5s2p)	[5s3p2d/3s2p]	7.40	18.2	10.0	24.9	−6.00	−1.39	34.8
(10s6p3d/5s)	[5s3p3d/3s]	7.38	18.2	10.0	25.0	−5.90	−1.33	34.1
(10s6p3d/5s3p)	[5s3p3d/3s3p]	7.39	18.3	10.0	25.0	−6.04	−1.38	34.0
(10s6p2d/5s)	[6s4p2d/4s]	7.42	18.1	10.0	24.9	−5.86	−1.28	35.4
(10s6p2d/5s2p)	[6s4p2d/4s2p]	7.39	18.2	10.0	24.9	−6.01	−1.36	34.9
(10s6p3d/5s)	[6s4p3d/4s]	7.40	18.2	10.0	25.0	−5.94	−1.34	34.3
(10s6p3d/5s3p)	[6s4p3d/4s3p]	7.38	18.3	10.0	25.0	−6.05	−1.38	34.0

These uncertainties do not exceed $0.02 \cdot 10^{-2}$ nm^3 for α_{ij} and $0.04 \cdot 10^{-4}$ nm^4/V for β_{ijk} at both [4s3p3d/3s3p] and [5s3p3d/3s3p] levels. These are just the bases sets, starting from which the apparent convergence is reached. Thus, for the calculation of (hyper)polarizabilities of large molecules within the SCF approximation the smallest [4s3p3d/3s3p] basis can be recommended. It is worth stressing that the use of this basis provides the results close to the HF limit values (see [2,3] and [4]) at much lower computational costs.

We would like to draw attention to the fact, that even the [4s3p2d/3s2p] basis provides the results rather close to those obtained with optimal [6s4p3d/4s3p] basis, developed in the previous paper [1]. The dominant components of α and β tensors in H_2CO and CH_3CN can be estimated at the [4s3p2d/3s2p] level with the errors not exceeding 3% for α_{ij} and 5% for β_{ijk}. The estimation of β tensor components of the order of $0.3 \cdot 10^{-4}$ nm^4/V (such as β_{zzz} for HCN) should be carried out carefully, the absolute errors in these cases typically are of the order of $0.05 \cdot 10^{-4}$ nm^4/V, while the relative errors may be rather large, since the absolute value of β_{zzz} is small.

The last point of the discussion concerns the results for the PNA molecule presented in Table 5. These data show that at all levels of theory the polarization functions on Hydrogen affect the results insignificantly, thus we may exclude them from the basis set in the calculations of hyperpolarizabilities of conjugated organic molecules. For systems of this class we may hope to get rather reliable SCF results at the [4s3p2d/3s] level. So, for the PNA molecule this basis set reproduces the results obtained with [6s4p3d/4s3p] basis with the error 1% for α_{ij} and 3% for dominant longitudinal hyperpolarizability β_{zzz}.

We hope that the [4s3p2d/3s] basis set will be useful for solving the problems, related to the design of promising NLO organic chromophores and to the study of the effect of the surrounding dielectric medium on their NLO response.

Acknowlegment

This work was supported by Russian Foundation for Basic Research (grant N 02-03-32897).

References

[1] M.B. Zuev and S.E. Nefediev, Rational Design of Atomic Gaussian Basis Sets for Ab Initio Calculations of the Dipole Polarizabilities and Hyperpolarizabi. I. Optimized Polarization Sets for the First-Row Atoms from B to F, *JCMSE* **4** (2004), 481–491.
[2] J.E. Rice, R.D. Amos, S.M. Colwell, N.C. Handy and J. Sanz, Frequency dependent hyperpolarizabilities with application to formaldehyde and methyl fluoride, *J. Chem. Phys.* **93** (1990), 8828–8839.
[3] M. Stähelin, C.R.Moylan, D.M. Burland, A.Willetts, J.E. Rice, D.P. Shelton and E.A. Donley, A comparison of calculated and experimental hyperpolarizabilities for acetonitrile in gas and liquid phases, *J. Chem. Phys.* **98** (1993), 5595–5603.
[4] G. Maroulis and C. Pouchan, Molecules in static electric fields: linear and nonlinear polarizability of HC≡N and HC≡P, *Phys. Rev. A.* **57** (1998), 2440–2447.
[5] S. Huzinaga, Gaussian-type functions for polyatomic systems. I, *J. Chem. Phys.* **42** (1965), 1293–1302.
[6] T.H. Dunning, Gaussian basis functions for use in molecular calculations. I. Contraction of (9s5p) atomic basis sets for the first-row atoms, *J. Chem. Phys.* **53** (1970), 2823–2833.
[7] R. Ahlrichs and P.R. Taylor, The choice of Gaussian basis sets for molecular electronic structure calculations, *J. Chim. Phys.* **78** (1981), 315–321.
[8] T.H. Dunning, Gaussian basis functions for use in molecular calculations. III. Contraction of (10s6p) atomic basis sets for the first-row atoms, *J. Chem. Phys.* **55** (1971), 716–723.
[9] S.P. Carna, M.Dupuis, Frequency dependent nonlinear optical properties of molecules: formulation and implementation in HONDO program, *J. Comput. Chem.* **12** (1991), 487–504.
[10] M.J. Frisch et al., Gaussian 98, Revision A.6, Gaussian, Inc., Pittsburgh PA, 1998.

Journal of Computational Methods in Sciences and Engineering 4 (2004) 501–516
IOS Press

Analysis of accuracy in *ab initio* calculations of the static dipole polarizability components. Examples of the water molecule and hydroxide ion

Gaétan Weck[a], Anne Milet[b], Robert Moszynski[c] and Elise Kochanski[a,*]
[a]*Laboratoire de Chimie Théorique, UMR 7551 CNRS/ULP, Institut Le Bel, Université Louis Pasteur, BP 296, F-67008 Strasbourg Cedex, France*
[b]*Laboratoire d'Etudes Dynamiques et Structurales de la Sélectivité, LEDSS Chimie Théorique, UMR 5616 CNRS/UJF, 301 rue de la Chimie D.U., B.P. 53, 38041 Grenoble Cedex 9, France*
[c]*Department of Chemistry, University of Warsaw, Pasteura 1, 02-093 Warsaw, Poland*

Abstract. A comparison of *ab initio* calculations of the static dipole polarizability components for the water molecule and hydroxide anion is reported. The discussion is focused on the difference between the computational needs for an accurate treatment of these two systems. A particular attention is paid to the danger of extrapolating to the anion some approximate techniques that are valid for the water molecule. Four important points are considered in this analysis: the analytical/numerical calculation of the components, the effect of the electronic correlation, the characteristics of the basis set, and the vibrational effects. In the finite-field approach, the interdependence of some control parameters, such as the energy convergence threshold and the associated field strength, are recalled within the frame of parabolic or quartic polynomial fits. The correlation effect is examined within the coupled-cluster method restricted to single, double, and approximate triple excitations, CCSD(T), and the nth-order Møller-Plesset perturbation theory, MPn. The oscillating behaviour of the perturbation series is very pronounced for the anion case, and its convergence is doubtful. It is crucial to include much more diffuse functions (valence and polarization) in the basis set for the anion than for the water molecule. A comparison of many basis sets is presented. The role of the vibrational effect is briefly addressed. Because *ab initio* calculations on supersystems generally cannot be performed at the best level reachable for the monomer polarizabilities, lower (but reasonnable) accuracy of the polarizability components is also discussed.

Keywords: Static dipole polarizability, *ab initio* calculations, finite-field method, hydroxide ion, water molecule

1. Introduction

A huge amount of theoretical and experimental studies of the static dipole polarizability was published in the course of the last century. These studies were greatly motivated by the fundamental role of the polarizability in many fields, as various as the Raman spectroscopy, nonlinear optics, or the long range interactions between molecules [1]. The interest for this property was recently enhanced by emerging

*Corresponding author. E-mail: weck@phys.chem.ntnu.no.

applications in various fields such as nonlinear optics or interaction-induced spectroscopy, see, e.g., Ref. [2].

Accurate calculations are now available for some small neutral molecules, though the limit of the accuracy may not be yet well established, even for the static dipole polarizability [3]. In particular, an accurate ab initio calculation of the anisotropic part may be very delicate [4], the alternative approach proposed in Refs. [5,6] being used only for a very few diatomic molecules.

The study of the static dipole polarizability components of anions faces specific problems, which explains that studies of anions are rather scarce compared with those of neutral molecules. We recently reported some work on the hydroxide ion [7,8], which is particularly illustrative of the problems encountered in such treatments. Comparing calculations of the static dipole polarizability components of both the water molecule and the hydroxide anion allows to analyse the similarities and differences in the treatment of these two categories of compounds.

In the present paper, we recall some methods commonly used in general-purpose programs, based on *ab initio* wave functions at various correlated or uncorrelated levels. The use of such programs as "black-boxes" often leads to underestimate the complexity of the problem, which may even be unknown to the user. For this reason, we discuss in this paper a few points that are familiar to specialists of these questions, but often not to the "black-boxes" users. We must be aware that quantum chemistry is becoming a tool of easy use for researchers in other fields. This success requires, in parallel, to give them general analyses about the underlying limits of various quantum-chemical approaches.

We will not consider here the density functional approach (DFT), although it is becoming more and more popular. Because all functionals available in the literature depend on some fitted parameters, the accuracy of the computed polarizability components may show strong dependencies on these parameters. An analysis of the polarizability components obtained from DFT calculations is then out of the scope of this paper.

The choice of a computational approach to obtain the molecular static dipole polarizability by *ab initio* calculations implies, in fact, a triple choice:

1. the choice of the (approximate) wave function,
2. the choice of the basis set,
3. the choice of the method used (analytical or numerical) to obtain the polarizability components.

For accurate calculations a post-Hartree-Fock treatment is generally needed. The methods used in the literature are mostly based on the Møller-Plesset expansion (hereafter denoted by MPn) or the coupled-cluster (CC) techniques. In the course of the last years, the MP2 and the CCSD(T) levels became clearly favoured among the papers devoted to the study of this molecular property. Some problems in the convergence of the MPn series deserve a special attention and will be discussed in this paper for both the water and the anion cases.

The adequacy of the basis set to correctly describe the molecular polarizability of the water molecule is now rather well known. Until recently, it was not so for the hydroxide anion, though specific bases were already proposed fourteen years ago [9]. Rather surprisingly sophisticated *ab initio* studies on the $OH^-(H_2O)$ system appeared in the literature since 1987 using basis sets adequate for the description of the water polarizability, but not for the hydroxide part of the system [10–22]. Though some diffuse functions are included, we recently showed [8] that they are not optimized to give the proper polarizability components of the hydroxide anion. In fact, some cancellation of errors occur in the supermolecule treatment [8], leading to sensible values of ΔH even if the polarizability of OH^- is strongly underestimated. However, it is difficult to know *a priori* when advantageous cancellations will occur. Furthermore, the knowledge of the molecular static dipole polarizability components does require adequate basis sets. A large range of basis sets widely used in the literature will thus be examined in the present paper, leading to results of variable accuracies.

Analytical expressions are available for some categories of wave functions. Only minor problems to get accurate results are encountered in these cases. The question is more delicate when the finite-field method is used. We already discussed some of the problems that can arise, on the example of the hydroxide anion [7]. We will recall the main lines of our analysis in the present paper. Indeed, because the finite-field method is quite generally used when no analytical expressions are available (and this occurs at many levels of the theory), recalling some rules to obtain reliable results with this method is not a minor question.

Studies of the vibrational corrections to the polarizability are scarce. Indeed, only a very few papers treating this point for the water molecule are available, none for the hydroxide anion. This question will be addressed in the present analysis.

Our paper is organized as follows. In Section 2 we will briefly recall some of the significant works found in the literature on the water molecule and hydroxide anion static dipole polarizability. In Section 3 we describe the computational methods that will be considered in the present paper, and give the computational details in Section 4. The following section (Section 5) is devoted to numerical results, before concluding remarks reported in Section 6.

2. *ab initio* calculations of the static dipole polarizability of the water molecule and hydroxide anion in the literature

Many papers concerning *ab initio* calculations of the static dipole polarizability of the water molecule were reported in the literature. Pioneering calculations appeared in the seventies and early eighties [23–32]. The need for diffuse polarization functions to correctly describe this molecular property was recognized during this period and the choice of the exponents of the additional functions was rationalized [27]. The effect of the post-Hartree-Fock levels were explored. The interest for this property was enhanced in the late eighties, and did not decrease since this date [4,22,33–50]. The papers published in this last period generally report the role of both the basis sets and the correlation contribution. This last point is closely related to the development of the methods in quantum chemistry accounting for the electronic correlation. Polarization functions are now commonly added to the usual atomic basis sets used in molecular calculations. As we will show in Section 4, several categories of diffuse polarized basis sets were generated in this way and widely used in the literature.

The polarizability of the hydroxide anion did not receive much attention in the past. *Ab initio* calculations were reported in a few papers only published between 1986 and 1992 [9,51–54]. Both the basis sets and the accuracy of the wave function were examined, within the frame of the methods available at that time. Maybe the good quality of these calculations explains that further studies appeared only ten years later, when new quantum chemistry methods and computer facilities became available [7, 8,22]. However, this probably involved that many studies appeared on the hydroxide anion hydrates with basis sets that are adequate for the description of the water polarizability and not for OH^-. Unfortunately, we may suspect that the early conclusions on the basis sets requirements to treat OH^- remained purely ignored.

The zero-point and vibrational averaging correction was considered in a few studies on the water molecule [27,37,38,43,48]. The pioneering work appeared in 1976 [27], and reported numerical results at the SCF level. Ref. [27] gave an order of magnitude of these corrections for the geometry and for the polarizability components. The largest contribution (0.33 a.u.) occurs for the in-plan component orthogonal to the C_{2v} molecular axis. The smaller one (0.07 a.u.) is found for the out-of-plan component, an intermediate value (0.15 a.u.) being obtained along the C_{2v} molecular axis. These contributions were used to correct PNO-CI and CEPA calculations. More recent works lead to a similar behaviour for the three components, though the values are somewhat larger: around 0.43, 0.11–0.15, and 0.3 a.u., respectively, at the MP2 [38] and MCSCF [48] levels.

3. Methods of calculations

3.1. Wave functions and energies

Results obtained at the Hartree-Fock (HF), MP2, MP3, MP4 and CCSD(T) levels are available in the literature fot both the water molecule and the hydroxide anion. They will be used to illustrate the present analysis. A particular attention will be paid to the MP2 and CCSD(T) levels that are the most favoured in the literature for the study of the molecular polarizability. These two levels of the theory are representative of the problems encountered in the calculations of the dipole polarizability components at other levels, and thus allow a general analysis. The convergence of the MPn series will be discussed for both the water molecule and the hydroxide anion.

One of the difficulties encountered is related to the convergence of the wave functions and energies. In the MP2 treatment, built on the use of the SCF occupied and virtual orbitals, an excellent convergence can be obtained since the MP2 convergence depends solely on the convergence of the SCF iterations, and on the accuracy of the four-index transformation. The convergence of the CCSD calculations may be much more delicate. This question will be examined in Section 5.

3.2. Analytical expressions or finite-field method

Analytic formulas to compute the polarizability components are available for some of the post-Hartree-Fock correlation methods, in particular at the MP2 level. In other cases, the finite-field method is generally used [55–58]. A second-order perturbation expression, as proposed in early works [31,59], did not received much attention. We previously reported a detailed analysis performed for the hydroxide anion [7] taking advantage of the possibility to compute the polarizability components from analytical formulas and in the frame of the finite-field method. In the latter case, they can be obtained either as the second derivative of the energy or as the first derivative of the dipole moment. Since the main features of this analysis can be quite general, we recall here the global conclusions.

The polarizability components are given as the numerical second-order derivatives of the energy, according to the expression [56,58]:

$$\alpha_{ii} = \frac{2E(0) - E(F_i) - E(-F_i)}{F_i^2} + O(F_i^2), \quad i = x,\ y \text{ or } z \tag{1}$$

where $E(F_i)$ is the energy of the system computed in the presence of an electric field of strength F_i. An intermediate approach combines the analytical and numerical derivatives if the second derivative of the energy is obtained as the first (numerical) derivative of the (analytical) dipole moment:

$$\alpha_{ii} = \frac{\mu(F_i) - \mu(-F_i)}{2F_i} + O(F_i), \quad i = x,\ y \text{ or } z \tag{2}$$

Table 1
Implicit numerical error (in au)

Field (a.u.)	0.002	0.001	0.0002
δ_{conv} (a.u.)			
10^{-7}	≈0.05+0.08/0.17	≈0.2+0.02/0.04	≈5+0.0008/0.0017
	≈0.13/0.22	≈0.22/0.24	≈5/5
10^{-9}	≈0.0005+0.08/0.17	≈0.002+0.02/0.04	≈0.05+0.0008/0.0017
	≈0.08/0.17	≈0.022/0.045	≈0.05/0.05

A rough estimate of the implicit numerical error on the polarizability components computed from the expression (1) can be obtained using a power series expansion:

$$\Delta\alpha_{ii} = \frac{2\delta_{\text{conv}}}{F_i^2} - \frac{2}{4!}F_i^2\gamma_{iiii} + \cdots \tag{3}$$

This expression requires the knowledge of the second hyperpolarizability γ_{iiii} components. The value of such a property is very different for the water molecule and hydroxide anion. The largest component (γ_{yyyy}, perpendicular to the molecular plane) is roughly equal to 2650 a.u. for the water molecule [4]. For the anion, our knowledge of the second hyperpolarizability is poor. Early values at the SCF level were reported in Ref. [51]. More sophisticated calculations were recently published [7]. In the latter work, the values of the largest component (γ_{xxxx}, perpendicular to the molecular axis) given by the finite-field method are 238350 and 512000 a.u. at the MP2 and CCSD(T) levels, respectively [7]. These values are only qualitative since the basis set was not fully optimized for this property and the accuracy limits were not explored. However, for the purposes of the present analysis only the order of magnitude is significant and it clearly shows that the hyperpolarizability of the hydroxide anion is at least 100 times larger than that for the water molecule. A high accuracy in the finite-field calculations of the polarizability components will be less easily reached for the hydroxide anion than for the water molecule.

The values of the implicit numerical error reported in Ref. [7] for the delicate OH^- case are illustrative. They were computed for electric field strengths of 0.002, 0.001 and 0.0002 a.u., and convergence thresholds of 10^{-9} and 10^{-7} a.u., using the values of γ_{xxxx} mentioned above. Equation (3) leads to the values reported in Table 1 for $\Delta\alpha_{xx}$. The analysis given in Ref. [7] for the hydroxide anion can be generalized to the case of the water molecule, allowing a comparison.

For each threshold, the decomposition into the first and second terms of Eq. (3) is reported in the first line of Table 1. The first term depends only on the field strength and the convergence threshold, whatever the molecule. It is then valid for OH^- and the water molecule. Two values are given for the second term: MP2/CCSD(T). They are based on the largest second hyperpolarizability component of the hydroxide anion, γ_{xxxx}, perpendicular to the molecular axis. Because the γ_{zzzz} component (approximately 7500 a.u.,) of OH^- and the components of the water molecule (approximately 2650 a.u.) are much smaller, the values given in Table 1 may be considered as the largest estimates of the implicit errors on the static dipole polarizability of the two molecules.

It is clear from Table 1 that the part due to the convergence threshold (the first term in Eq. (3) is dominant for the smallest field, whatever the molecule. A tight convergence threshold is thus necessary to get an acceptable accuracy with such a small field. However, it was shown [7] that, in practice, a low convergence may not be so unfavorable (we will discuss this point later). For large fields, the situation is quite different for the water and the hydroxide anion. As previously discussed [7], the effect of higher-order terms (the second term in Eq. (3) is then either dominant or similar to the convergence effect in the case of OH^-. The electric field strength of 0.001 a.u. is a compromise between the effects mentioned above: for the weak convergence threshold, the convergence effect is dominant, while the opposite is found for the tight convergence threshold. Given the values of the water second hyperpolarizability components, the effect of the higher order term becomes considerably smaller (roughly by a factor of 100), involving that the implicit error on the static polarizability components will be dominated by the convergence effect. For the water molecule, rather large fields can be used, giving an acceptable accuracy even with a low convergence threshold.

This analysis shows the limits of a parabolic fit. With quartic polynomials fits the error introduced by higher-order terms are smaller. This was examplified in the calculations performed on OH^- [7] with both field strengths of 0.001 and 0.002 a.u. (see Section 5.1). The accuracy should even be better with a fifth-order polynomial fit, as used in Ref. [22], with an electric field strength ranging from −0.004 to 0.004 a.u.

4. Basis sets and geometries

As mentioned in the Introduction, the characteristics of adequate basis sets to correctly describe the static polarizability components of the water molecule are rather well known but the situation was more confuse for the hydroxide anion, despite some early good quality calculations [9,51–54]. We give here the main characteristics of some of the basis sets that will be discussed in the present paper. More information may be found in Refs. [4,8,22] and in the papers cited therein. We will use either their specific acronyms or the standard (GTO)[CGTO] notation for sets of GTO primitive and CGTO contracted Gaussian functions. The total bases for both the oxygen and hydrogen atoms will be denoted by (sO.pO.dO.../sH.pH...)[sO.pO.dO.../sH.pH...]. We can group them into several categories, because they are often built from initial sets by adding polarization and diffuse functions.

The R-ANO basis is built from a (14.9.4.3/8.4.3)[6.5.3.2/4.3.2] ANO basis [60], extended with three diffuse s-, two diffuse p- and one diffuse d-type functions on the oxygen atom. This leads to a [9.7.4.2/4.3.2] contracted basis. The R-ANO basis set was successfully used to describe the electric properties of the first Rydberg excited states of water [34,40].

The polarized medium-size [5.3.2/3.2] basis, referred to as basis “S” or “Pol1” in the literature, is a relatively small sequential basis proposed by Sadlej that gives good performance in calculations of the electric properties for neutral molecules in their ground state [33]. Extension with three s- and one p-type diffuse functions on the oxygen atom leads to a (13.7.4/6.4)[8.4.2/3.2] basis, denoted by S+ [41]. The addition of more diffuse and polarization functions generated bases denoted by S++; S++,+; S++1f,+1d; S++2f,+2d; S++f′1f,+d′1d and S++f′2f,+d′2d [8]. Results obtained with some of these bases will be reported.

A third category of bases is represented by the cc-pVXZ, aug-cc-pVXZ and x-aug-cc-pVXZ series proposed by Dunning et al. The polarized cc-pVXZ bases, derived to account for the correlation effects [61], were extended with some diffuse functions to properly describe electron affinities, leading to the aug-cc-pVXZ series [62]. Aug-cc-pVXZ bases are often used to treat polarizabilities and intermolecular interactions. However, still more diffuse functions were further added to improve the description of the electric properties (x-aug-cc-pVXZ series [63]). Studies using x-aug-cc-pVXZ bases are scarce [63, 42,45,64,65,8] and comparison of the aug-cc-pVXZ and x-aug-cc-pVXZ bases performances quite rare. Recent work shows that the improvement due to the use of x-aug-cc-pVXZ bases is small for diatomic radicals [65] and can be quite important for anions [63]. A comparison of aug-cc-pVXZ and d-aug-cc-pVXZ bases was recently given for both the water molecule and the hydroxide anion [8] (see section 5.2-3).

The 6-311++G and 6-311++G** sets [66] are illustrative of the fourth category of bases. They are often used in the literature to treat dimers (or heterodimers) of molecules (see Ref. [67] and references therein), in particular for the $OH^-(H_2O)$ system [17,21].

The KT1 basis is a very large diffuse and polarized [9s.6p.6d.3f/6s.4p.2d.1f] basis, recently used in the study of the water molecule [4].

The orbital basis used in the R12 calculations [68] is composed of (15s.9p.7d.5f/9s.7p.5d) uncontracted functions. It was used in a recent work reporting polarizability components of the water molecule and of the hydroxide anion [22].

These bases lead to static dipole polarizability components of variable accuracy. Some of them should not be used for such a purpose. However, because they are used in the literature in studies of the hydroxide anion hydrates (see our comments on this question in the Introduction), they will be discussed as well. In particular, beyond the aim to obtain a good accuracy, we will also consider the possibility to get an "acceptable" accuracy for the monomers when very large basis sets cannot be employed in the study of hetrodimers or larger clusters.

In most calculations the experimental geometry was used for both the hydroxide ion ($r(OH^-)$=0.964 Å [69]) and the water molecule (r(OH)=0.9572 Å and ∠HOH=104.5° [70]). The molecular axis of the hydroxide anion lies along the z axis. The water molecule was in the yz plane, with the C_{2v} axis along the z axis. We will mention when other geometries are used.

5. Numerical results and discussion

As recalled in the Introduction, *ab initio* calculations of the static dipole polarizability components imply a triple choice: accuracy of the wave function, size and quality of the basis set, and method used (analytical or numerical) to obtain the components. These points will be discussed in the present section, illustrated with results taken from Refs. [4,7,8,22].

5.1. Accuracy of finite-field calculations

This point was examined in details in the case of OH^- [7] and summarized in a more general paper [8]. We just give here the main lines and conclusions, the details being available in these papers. The calculations were performed with the Gaussian 98 package [71], at the MP2 and CCSD(T) levels. They are representative of the general problems of accuracy encountered at other levels. The finite-field method had to be employed at the CCSD(T) level because no analytical expression is available. It can be used as a "black-box" by non-specialists of this domain, the needed parameters being automatically selected in this package when required. However, the counterpart of this great facilities is to hid some difficulties and it is worth to recall again the limits in the accuracy.

The static dipole polarizability components were computed from an analytical expression at the MP2 level, and with the finite-field technique using the two expressions Eqs (1) and (2) at the MP2 and CCSD(T) levels. This allows a comparison of the three different approaches. A special attention was paid to the use of the strength of the electric field 0.001 a.u. because this value is automatically selected when the finite-field method is employed in the Gaussian 98 package. The values 0.002, 0.0005 and 0.0002 a.u. were also considered.

One of the difficulties is related to the convergence of the wave functions and energies. A convergence threshold of 10^{-12} a.u. was easily reached at the MP2 level, while 10^{-9} a.u. was the best convergence that could be achieved for the CCSD energies. The default criterion for the CCSD treatment in the Gaussian 98 package being 10^{-7} a.u., these three thresholds were then especially considered.

Following the analysis detailed in Ref. [7], the "unconverged" decimals were retained in all calculations, whatever was the convergence threshold, as done in general-purpose quantum chemistry codes. The theoretical justification to keep the "unconverged" decimals is quite unclear. However, the results were systematically improved when all decimals were retained, suggesting [7] that the unconverged decimals are not purely random numbers but somewhat consistent in the converging process.

5.1.1. MP2 calculations

The MP2 analytical calculations of the static dipole polarizability components were quite stable with an SCF convergence threshold varying from 10^{-7} to 10^{-12} a.u.. With the finite-field method and a parabolic fit, the results were dependent on both the field strength and the convergence of the wave function. The error in the largest polarizability component (α_{xx}, perpendicular to the molecular axis) with respect to the MP2 analytical value was 0.02, 0.02 and 0.03 a.u. for the convergence thresholds of 10^{-12}, 10^{-10} and 10^{-7} a.u., respectively, and a field strength of 0.001 a.u. (the analytical value being 36.149 a.u. with the basis S+). An excellent agreement with the analytical results could be obtained with an optimized field strength of 0.0002 a.u. and a tight convergence threshold, but the error reached 0.2 a.u. with this field and a convergence threshold of 10^{-7} a.u. The use of a very small field strength requires a tight convergence.

An excellent agreement with the analytical values was obtained with a quartic polynomial fit and a tight convergence. In this case, the choice of the field strengths was of minor importance: both sets of the field strengths 0.001 and 0.0002 a.u. or 0.001 and 0.002 a.u lead to values of the polarizability components nearly identical to the analytical ones.

5.1.2. CCSD(T) results

Extrapolating the analysis of the MP2 results, we choose as reference values those given by the finite-field method and a quartic polynomial fit. The best convergence that could be reached (10^{-9} a.u.) was of sufficient accuracy. Using the field strengths of 0.002 and 0.001 a.u. and the basis S+ gives 21.59 and 40.22 a.u. for the α_{zz} and α_{xx} components, respectively [7].

With the tight threshold (10^{-9} a.u.), the error given by the parabolic fit was 0.17, 0.04 and 0.002 a.u. for the field strengths of 0.002, 0.001 and 0.0002 a.u., respectively. The error became 0.17, 0.1 and 1.3 a.u., respectively, with the convergence threshold of 10^{-7} a.u. Of course, similar errors also occur for the non-diagonal components that should strictly be zero. This clearly confirms that a weak convergence threshold cannot be used with the small field strength of 0.0002 a.u. and the parabolic fit. A quadratic polynomial fit, however, can give rather accurate values with this weak convergence threshold: the error was only 0.05 a.u. for the sets of the field strengths 0.002 and 0.001 a.u.

This analysis shows that the general qualitative behaviour is correctly described, though not accurately, with the default convergence threshold (10^{-7} a.u.), a parabolic fit and a field strength of 0.001 a.u. The largest error found (0.1 a.u.) gives an estimate of the order of magnitude of the error involved, which may be somewhat dependent on the basis set. To get a better accuracy, a tighter convergence and a quartic polynomial fit have to be employed. In very delicate cases, it can be wise to explore if a fifth-order polynomial fit should be used, as done in Ref. [22].

5.2. Electron correlation contribution

5.2.1. Electron correlation effect

As written in the previous section the best value of the perpendicular polarizability component of OH^- with the basis S+ was 36.15 and 40.22 a.u. at the MP2 and CCSD(T) levels, respectively. The SCF

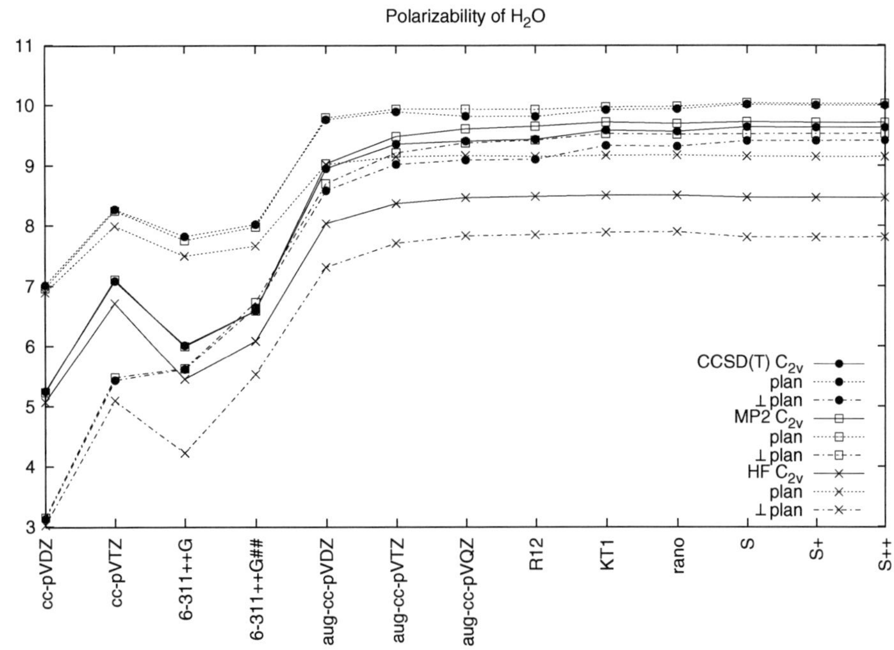

Fig. 1. Static dipole polarizability components of H_2O (in a.u., basis sets given in abscissa).

value was 20.92 a.u. This shows the great importance of the correlation contribution to the polarizability of this system, and also the effect of the method used for the correlation treatment. We will thus analyze the correlation contribution at both the MP2 and CCSD(T) levels for the hydroxide anion and the water molecule, and will further comment on the MPn series convergence.

The electron correlation contribution to the static dipole polarizability components is not of similar importance for water and for the hydroxide anion. We gathered in Fig. 1 the HF, MP2 and CCSD(T) values obtained for the water molecule with some basis sets used in Refs. [4,8,22]. Figure 2 gathers similar results for OH^-, taken from Refs. [8,22]. Figures 1 and 2 allow to analyse both the correlation and the basis set effects, that may be closely interrelated.

Figure 1 shows that for the water molecule the electron correlation effect roughly varies from 7 to 18 % of the total correlated value according to the component and the basis set, except for the cc-pVDZ basis that is not adequate to describe the polarizability, in particular its correlation part (see Section 5.3). The MP2 and CCSD(T) values are quite close, for any component and any basis set. Though not quantitatively large, the electron correlation has a qualitative effect on the anisotropy of the polarizability components: they are significantly anisotropic at the HF level, and become nearly isotropic at the correlated level.

Figure 2 shows that (again except for the cc-pVXZ bases) the electron correlation effect is very basis set dependent, and much larger for $\alpha_\perp$ than for the $\alpha_\|$ component. It can exceed the HF part for the perpendicular component. The MP2 and CCSD(T) values are again close, but the CCSD(T) values becomes significantly larger than the MP2 ones for the $\alpha_\perp$ component when the best basis sets are employed (see Section 5.3). The electron correlation contribution enhances the anisotropy of the polarizability tensor: this effect is the reverse of that noted for the water molecule.

It is clear from Figs 1 and 2 that the electron correlation has a quite different effect in the hydroxide anion and the water molecule. This concerns the quantitative importance as well as the qualitative change in the anisotropy of the polarizability tensor. The conclusions obtained for the water molecule cannot

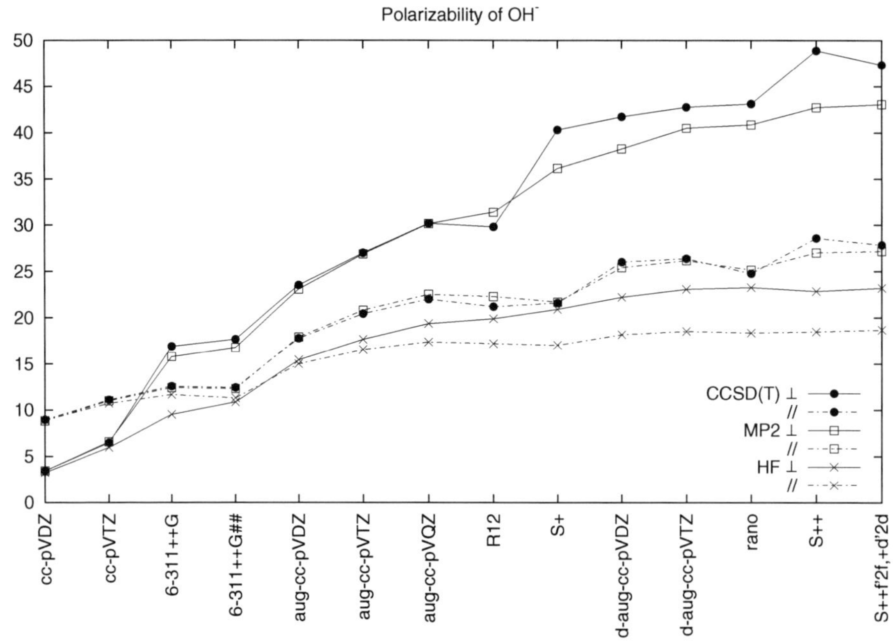

Fig. 2. Static dipole polarizability components of OH^- (in a.u., basis sets given in abscissa).

be extrapolated to the anion. In particular, a proper description of the $\alpha_\perp$ component of the anion is especially delicate.

5.2.2. Convergence of the MPn series

The convergence of the Møller-Plesset expansion deserves a special attention. Before the development of the CCSD(T) approximation, which may serve as a "reference" in this field, the MP4 level was considered as the most accurate. However, the convergence of the MPn series was questioned in the calculations of the polarizability of OH^- [8,9] and in the related work on two-body [72] and three-body [73] intermolecular interactions energies in the $OH^-(H_2O)_2$ system. An oscillating pattern was found in all these cases, and a probable divergence of the series was suspected for some structures of the water-ion system. It is thus relevant to recall the main outlines of this important question. We collected in Table 2 some of the results previously published for the static dipole polarizability components of the hydroxide anion [8] and of water [4]. Several basis sets were selected, showing that this problem is not basis set specific.

The results reported in Table 2 confirm again that the theoretical treatment of the water molecule polarizability components cannot be extrapolated to the hydroxide anion, though we may found similarities in the convergence of the MPn series for the two systems. In both cases, the MPn series exhibit an oscillatory convergence pattern, with MP2 and MP4(SDQ) values close to the CCSD(T) results. The MP3 values are systematically smaller and the MP4 ones larger than the MP2 and the CCSD(T) results. However, though the largest discrepancy between the MP3 and MP4 values is smaller than 8 % for the water molecule, it reaches 50 % for the hydroxide anion. Neither the MP3 nor the MP4 results can be retained as acceptable values for the anion. All basis sets lead to the same behaviour, as illustrated by the three ones selected here. Furthermore, it is not possible to guess if the MPn series will converge, or will diverge. Pluta et al. [9] suggested that, given the opposite sign of the MP3 and MP4 contributions,

Table 2
Convergence of the MPn series for the OH^- and H_2O static dipole polarizability components. Values in a.u.

		OH⁻ (Ref. [8])				H_2O (Ref. [4])[a]	
		S++	R-ANO	d-aug-cc-pVTZ		KT1	
$\alpha_{\parallel}$	HF	18.50	18.40	18.54	α_{C2v}	HF	9.18
	MP2	27.03	25.17	26.17		MP2	9.98
	MP3	21.06	20.28	20.55		MP3	9.64
	MP4(DQ)	23.33	21.82	22.46		MP4(DQ)	9.68
	MP4(SDQ)	28.64	26.61	26.20		MP4(SDQ)	9.82
	MP4	33.26	28.45	30.73		MP4	10.05
	CCSD(T)	28.59	24.85	26.38		CCSD(T)	9.93
$\alpha_{\perp}$	HF	22.86	23.30	23.09	α_{plan}	HF	7.90
	MP2	42.72	40.85	40.49		MP2	9.54
	MP3	29.11	28.05	27.97		MP3	8.80
	MP4(DQ)	34.26	32.22	32.14		MP4(DQ)	8.90
	MP4(SDQ)	46.78	40.64	40.69		MP4(SDQ)	9.13
	MP4	56.91	50.47	50.42		MP4	9.53
	CCSD(T)	48.79	43.16	42.72		CCSD(T)	9.34
	HF				$\alpha_{\perp\text{plan}}$	HF	8.52
	MP2					MP2	9.73
	MP3					MP3	9.18
	MP4(DQ)					MP4(DQ)	9.26
	MP4(SDQ)					MP4(SDQ)	9.45
	MP4					MP4	9.77
	CCSD(T)					CCSD(T)	9.59

[a]The CCSD values are 9.77, 9.07 and 9.37, respectively, for the 3 components of the water molecule.

the fifth-order correction may again contribute with an altered sign and cancel some of the fourth-order terms. This would be compatible with the convergence towards the CCSD(T) results. Thus, calculations beyond the MP2 level in the MPn scheme are not recommended.

5.3. *Basis sets artefacts*

The difference in the theoretical treatments of the static dipole polarizability components of water and of the hydroxide anion is particularly sensitive to the choice of the basis set. This is illustrated in Figs 1 and 2.

Let us first examine the water case (Fig. 1). As already stated [8], the diagram can be divided into two parts, with a limit given by the aug-cc-pVDZ basis. The left part corresponds to bases that miss dedicated diffuse functions (valence and polarization) to properly describe the polarizability components. They are reported on this Figure to illustrate the role of such diffuse and polarization functions. Such bases are sometimes found in the literature to treat the $OH^-(H_2O)$ system. However, we must be aware that only cancellations of errors may hide their deficiencies [8]. The bases reported on the right part of the diagram do include such dedicated functions. They lead to similar results, though the aug-cc-pVDZ basis somewhat underestimates two of the components.

The situation is not so favorable for the hydroxide anion (Fig. 2). The aug-cc-pVXZ series strongly underestimates the values of the perpendicular component. The R12 basis results lie in the same range, slightly smaller, in particular for the CCSD(T) values obtained from the CCSD(T)-R12/B approach [22, 74]. The S+ basis is a transition between these bases and those including more diffuse polarization functions able to describe the perpendicular component. A detailed analysis of the exponent of the most diffuse function in each symmetry was recently given for most of the bases reported in Fig. 2 [8]. It was shown that the dominant role was played by the most diffuse function of the valence and first symmetry polarization shell (that is p(O) and d(O)). Higher-symmetry functions have only a minor role. The basis set effect is more pronounced for the perpendicular than for the parallel component. "Reference values" were proposed [8] using the basis S++f'2f,+d'2d and a quartic polynomials fit: 27.14 ad 27.81 a.u. at the MP2 and CCSD(T) levels, respectively, for the parallel component, and 42.73 and 46.58 a.u., respectively, for the perpendicular component. It is interesting to note that an extrapolation of the results obtained with the d-aug-cc-pVDZ and d-aug-cc-pVTZ values would fall in this range.

This brief analysis gives additional information on the great differences in the treatment of the water molecule and the hydroxide anion. They were encountered at all steps. The choice of the basis set is not the least of the difficulties.

5.4. Do we progress in the knowledge of the OH^- and water polarizabilities?

Comparison with other works on the static dipole polarizability components was previously reported for the water molecule [4] and the hydroxide anion [8].

The experimental values for water [75] are 10.31, 9.55 and 9.91 a.u., respectively, for the three components. The theoretical values cited in Ref. [4] are taken from Ref. [36,37,43]. They vary from 9.77 to 10.29 a.u. for the xx (in-plane) component, from 9.28 to 9.64 a.u. for the out-of-plane component, and from 9.44 to 9.73 a.u. for the zz (C_{2v}) component. They are within the range of the plateau of Fig. 1. Though a still higher accuracy may be searched for, our knowledge of the water polarizability component can be considered as satisfying. It is noteworthy that the cancellation of errors may hide the difficulty to obtain a very high accuracy. As an example, we compared the CCSD(T) values obtained for the zero-point and vibrationally corrected geometries [43] (10.22, 9.40 and 9.76 a.u., respectively) with our uncorrected CCSD(T) values (10.01, 9.43, 9.64 a.u., basis S++). It is extremely difficult to guess a priori which part of the cancellation of errors leads to this excellent agreement.

No experimental values are available for the hydroxide anion. Our knowledge comes solely from the theory. Early values were reported in Ref. [8], taken from Refs. [9,51–54]. We already discussed the wide range spanned. They dramatically depend on both the level of the correlation treatment and of the basis set used. The values recommended in Ref. [8] for the $\alpha_{\parallel}$ and $\alpha_{\perp}$ components were 27.8 and 46.6 a.u. at the CCSD(T) level, using a quartic polynomial fit. They were obtained with an extended diffuse and polarized basis set of the "Sadlej type". In that paper, calculations with the d-aug-cc-pVXZ set of the "Dunning type" were not performed beyond the d-aug-cc-pVTZ quality because the convergence seemed very slow, requiring huge basis sets. In the present work, we tried to get more information about the performances of this series. We did calculations with the d-aug-cc-pV5Z basis at the MP2 level, and with and q-aug-cc-pVQZ basis at the MP2, MP4 and CCSD(T) level. The results are reported in Table 3, together with some others using "Sadlej type" bases. An inspection of this Table shows that the results obtained with the d-aug-cc-pV5Z basis are not converged, even at the MP2 level. The q-aug-cc-pVQZ and S++f'2f,+d'2d bases lead to very close results at the MP2 level for both components, and for the parallel component at the CCSD(T) level. The agreement is not so good for the MP4 results of the two components and for the CCSD(T) perpendicular component. They differ by 3 or 4 per cent. However, they can be considered as qualitatively converged.

We reported in Table 3 the number of contracted basis functions, N_{cont}, for each basis. This deserves some comments. The best basis set in the two series requires 249 and 87 functions, respectively. The time for CCSD(T) calculations varying as N^7_{cont}, the ratio will be around 1500. The use of the q-aug-cc-pVQZ basis for the study of the hydroxide anion monohydrate is presently not favorable, not to say quasi impossible.

Table 3
Convergence of the static dipole polarizability components of OH^- with the Sadlej and Dunning basis sets series. Values in a.u.

Base	$N_{\rm cont}$	$\alpha_\parallel$			$\alpha_\perp$		
		MP2	MP4	CCSD(T)	MP2	MP4	CCSD(T)
aug-cc-pVDZ	32	17.92	19.09	17.77	23.09	25.88	23.53
d-aug-cc-pVDZ	45	25.42		26.02	38.25		41.69
d-aug-cc-pVTZ	94	26.17	30.73	26.38	40.49	50.42	42.72
d-aug-cc-pV5Z	268	26.84			41.59		
q-aug-cc-pVQZ	249	27.20	32.33	27.52	42.87	54.22	45.08
S++	47	27.03	33.26	28.59	42.72	56.91	48.79
S++2f,+2d	75	27.27	33.35	28.07	43.04	56.71	47.42
S++f'2f,+d'2d	87	27.15	33.34	27.82	42.73	56.70	46.67

We wish to end this section by briefly considering the vibrational effects on the polarizability components of the hydroxide anion. We estimated the magnitude of the vibrational corrections by taking the average of the components as functions of the OH distance with the ground-state vibrational wave function in the harmonic approximation. The parameters of the harmonic oscillator were taken from the experimental work of Saykally and collaborators [69]. The atomic masses of the oxygen and hydrogen atoms, 15.99491 and 1.007825 amu, were taken from Ref. [76]. The polarizability components as functions of the OH distance were computed at the MP2 level with the S++2f,+2d basis. The OH distances ranged from 0.93 Å to 0.99 Å with a step of 0.01 Å, fully covering the well region of the OH^- potential energy curve (the equilibrium distance in OH^- amounts to ≈ 0.96 Å). The polarizability components as functions of the OH distance were fitted to a quadratic function, and well-known formulas for the average values of powers of the distance with the harmonic oscillator wave functions were used [77].

The vibrational corrections to the perpendicular and parallel components of the OH^- polarizability are small. They amount to –0.398 and –0.184 a.u., respectively. They are of the same order of magnitude as the vibrational corrections to the components of the polarizability of water, except that the sign is reversed. However, one should note that the polarizability components of the hydroxide ion are about three or four times larger than those of water, so in the case of the OH^- ion the effect of the vibrational averaging is of modest importance, as it represents only 0.85% and 0.66% of the equilibrium values for the perpendicular and parallel components, respectively

6. Concluding remarks

The present analysis emphasized the differences in *ab initio* calculations of the static dipole polarizability tensor of the water molecule and of the hydroxide anion. The need for a better knowledge of such differences is of particular importance in the study of the $OH^-(H_2O)_n$ clusters. It may be tempting to extrapolate to such systems the techniques that are efficient in the study of neutral dimers. Indeed, this is rather often found in the literature. Great care, however, must be taken when the ion is present. The difficulty may be hidden in the artificial cancellation of errors. We recently stressed [8] that an overestimation of the (attractive) basis-set superposition error may compensate the underestimation of the dispersion energy related to the polarizability.

A correct treatment of the polarizability must especially take into account two crucial points:

1. The need for much more diffuse valence and polarization functions in the basis set of the hydroxide anion than for the water molecule. This was illustrated in Figs 1 and 2.

2. The oscillating behaviour of the MPn series, which is much more pronounced for the hydroxide anion than for the water molecule. In the former case, the convergence of the MPn series is doubtful, and we may question the validity of the early MP4 values reported in the literature. Of course, cancellations of errors may also occur in this case, an overestimation due to the MP4 level being compensated by the use of a basis set that is not polarized enough.

Less crucial problems may, nevertheless, be sources of inaccuracies. Our analysis leads to recommend the CCSD(T) approach. In many cases, the MP2 values of the static dipole polarizability components are quite close to the CCSD(T) ones. They can be a good approximation for a neutral molecule. This is not the case for the hydroxide anion: significant differences between the MP2 and the CCSD(T) values are found for the perpendicular component computed with the best basis sets. This may lead to inaccuracies in the anisotropy of the polarizability tensor. The CCSD(T) reference values proposed in Ref. [8] for OH^- (27.8 and 46.6 a.u. at the CCSD(T) level for the $\alpha_{\|}$ and $\alpha_{\perp}$ components) are probably closed to the convergence. The vibrational corrections in this case represent less than one per cent of the values.

Acknowledgments

The calculations were performed in Strasbourg (on our laboratory work stations and at the Centre Universitaire Régional de Ressources Informatiques CURRI), at the University of Warsaw, in Grenoble at the CECIC (centre d'experimentations pour le calcul intensif en chimie de Grenoble) and at the Centre IDRIS (Projects 980306 and 990306) of the CNRS (Centre National de la Recherche Scientifique, Orsay, France). The CNRS, the University Louis Pasteur of Strasbourg, the Universite Joseph Fourier of Grenoble and the University of Warsaw are acknowledged for providing computer facilities. Two of us (RM and EK) were supported by the French Ministère des Affaires Etrangères and by the Polish Scientific Research Council KBN (Projet concerté de coopération scientifique et technique entre la France et la Pologne, project n° 7137). R.M. is grateful to the French government and the University Louis Pasteur for a PAST position in Strasbourg.

References

[1] F. London, *Trans. Faraday Soc.* **33** (1937), 8.
[2] G. Maroulis, D.Xenides, U. Hohm and A. Loose, *J. Chem. Phys.* **115** (2001), 7957.
[3] A.J. Thakkar and A.K. Das, *J. Mol. Struct. (Theochem)* **547** (2001), 233.
[4] G. Maroulis, *Chem. Phys. Lett.* **289** (1998), 403.
[5] A. Kumar, W.J. Meath, P. Bündgen and A.J. Thakkar, *J. Chem. Phys.* **105** (1996), 4927.
[6] S.A.C. McDowell and W.J. Meath, *Can. J. Chem.* **76** (1998), 483.
[7] G. Weck, A. Milet, R. Moszynski and E. Kochanski, *J. Mol. Struct. (Theochem)* **591** (2002), 141.
[8] G. Weck, A. Milet, R. Moszynski and E. Kochanski, *J. Phys. Chem. A* **106** (2002), 12084; **107** (2003), E4008.
[9] T. Pluta, A.J. Sadlej and R.J. Bartlett, *Chem. Phys. Lett.* **143** (1988), 91.
[10] J.E. Del Bene, *Int. J. Quantum Chem.: Quantum Biology Symp.* **14** (1987), 27.
[11] J.E. Del Bene, *J. Phys. Chem.* **92** (1988), 2874.
[12] J.E. Del Bene and I. Shavitt, *Int. J. Quantum Chem.: Quantum Chem. Symp.* **23** (1989), 445.
[13] R.V. Stanton and K.M. Mertz, Jr., *J. Chem. Phys.* **101** (1994), 6658.
[14] J.E. Del Bene and I. Shavitt, *J. Mol. Struct. (Theochem)* **307** (1994), 27.
[15] A.R. Grimm, G.B. Bacskay and A.D.J. Haymet, *Mol. Phys.* **86** (1995), 369.
[16] S.S. Xantheas, *J. Am. Chem. Soc.* **117** (1995), 10373.
[17] A.T. Pudzianowski, *J. Chem. Phys.* **102** (1995), 8029.
[18] J.R. Pliego, Jr. and J.M. Riveros, *J. Chem. Phys.* **112** (2000), 4045.
[19] M. Masamura, *J. Mol. Struct. (Theochem)* **498** (2000), 87.

[20] M. Masamura, *J. Comp. Chem.* **22** (2000), 31.
[21] M. Masamura, *Theor. Chem. Acc.* **106** (2001), 301.
[22] C.C.M. Samson and W. Klopper, *J. Mol. Struct.* (*Theochem*) **586** (2002), 201.
[23] G.P. Arrighini, C. Guidotti and O. Salvetti, *J. Chem. Phys.* **52** (1970), 1037.
[24] S.P. Liebmann and J. Moskowitz, *J. Chem. Phys.* **54** (1971), 3622.
[25] M. Jaszunski, A. Okninski and A.J. Sadlej, *Acta Phys. Polonica A* **41** (1972), 595.
[26] K. Thomsen and P. Swanstrøm, *Mol. Phys.* **26** (1973), 751.
[27] H.J. Werner and W. Meyer, *Molec. Phys.* **31** (1976), 855.
[28] P. Swanstrøm, W.P. Kraemer and G.H.F. Diercksen, *Theor. Chim. Acta* **44** (1977), 109.
[29] A.J. Sadlej, *Chem. Phys. Lett.* **47** (1977), 50.
[30] A.J. Sadlej, *Mol. Phys.* **34** (1977), 731.
[31] G. Segher and E. Kochanski, *Int. J. Quantum Chem.* **17** (1980), 955.
[32] G.H.F. Diercksen, V. Kellö and A.J. Sadlej, *J. Chem. Phys.* **79** (1983), 2918.
[33] A.J. Sadlej, *Coll. Czech. Chem. Commun.* **53** (1988), 1995.
[34] M. Urban and A.J. Sadlej, *Theor. Chim. Acta* **78** (1990), 189.
[35] G. Maroulis, *J. Chem. Phys.* **94** (1991), 1182.
[36] H. Sekino and R.J. Bartlett, *J. Chem. Phys.* **98** (1993), 3022.
[37] Y. Luo, H. Agren, O. Vahtras, P. Jørgensen, V. Spirko and H. Hettema, *J. Chem. Phys.* **98** (1993), 7159.
[38] A.J. Russell and M.A. Spackman, *Mol. Phys.* **84** (1995), 1239.
[39] S.A.C. McDowell, R.D. Amos and N.C. Handy, *Chem. Phys. Lett.* **235** (1995), 1.
[40] S. Klein, E. Kochanski, A. Strich and A.J. Sadlej, *Theor. Chim. Acta* **94** (1996), 75.
[41] S. Klein, E. Kochanski, A. Strich and A.J. Sadlej, *J. Phys. Chem. A* **101** (1997), 4799.
[42] C.V. Caillie and R.D. Amos, *Chem. Phys. Lett.* **291** (1998), 71.
[43] D. Spelsberg and W. Meyer, *J. Chem. Phys.* **108** (1998), 1532.
[44] J.M. Stout and C.E. Dykstra, *J. Phys. Chem. A* **102** (1998), 1576.
[45] E.K. Dalskov and S.P.A. Sauer, *J. Phys. Chem. A* **102** (1998), 5269.
[46] O. Christiansen, J. Gauss and J.F. Stanton, *Chem. Phys. Lett.* **305** (1999), 147.
[47] C.V. Caillie and R.D. Amos, *Chem. Phys. Lett.* **328** (2000), 446.
[48] K. Ruud, P.-O. Åstrand and P.R. Taylor, *J. Chem. Phys.* **112** (2000), 2668.
[49] O. Quinet, B. Champagne and B. Kirtman, *J. Comp. Chem.* **22** (2001), 1920.
[50] J.R. Maple and C.S. Ewig, *J. Chem. Phys.* **115** (2001), 4981.
[51] G. Maroulis and D.M. Bishop, *Mol. Phys.* **57** (1986), 359.
[52] L. Adamowicz, *J. Chem. Phys.* **89** (1988), 6305.
[53] D.P. Chong and S.R. Langhoff, *J. Chem. Phys.* **93** (1990), 570.
[54] G.H.F. Diercksen and A.J. Hernandez, *J. Mol. Struct.* (*Theochem*) **254** (1992), 191.
[55] H.D. Cohen and C.C.J. Roothaan, *J. Chem. Phys.* **43** (1965), S34.
[56] J.A. Pople, J.W. McIver, Jr., and N.S. Ostlund, *J. Chem. Phys.* **49** (1968), 2960.
[57] J.E. Gready, G.B. Bacskay and N.S. Hush, *Chem. Phys.* **23** (1977), 9.
[58] G.H.F. Diercksen and A.J. Sadlej, *J. Chem. Phys.* **75** (1981), 1253.
[59] F. London, *Z. Phys.* **63** (1930), 245.
[60] P.-O. Widmark, P.-A. Malmqvist and B.O. Roos, *Theor. Chim. Acta* **77** (1990), 291.
[61] T.H. Dunning, Jr., *J. Chem. Phys.* **90** (1989), 1007.
[62] R.A. Kendall, T.H. Dunning, Jr., and R.J. Harrison, *J. Chem. Phys.* **96** (1992), 6769.
[63] D.E. Woon and T.H. Dunning, Jr., *J. Chem. Phys.* **94** (1994), 2975.
[64] M. Jaszunski, A. Rizzo and P. Jørgensen, *Theor. Chem. Acc.* **106** (2001), 251.
[65] M. Medved, M. Urban, V. Kellö and G.H.F. Diercksen, *J. Mol. Struct.* (*Theochem*) **547** (2001), 219.
[66] R. Krishnan, J.S. Binkley, R. Seeger and J.A. Pople, *J. Chem. Phys.* **72** (1980), 650.
[67] A.K. Rappé and E.R. Bernstein, *J. Phys. Chem. A* **104** (2000), 6117.
[68] W. Klopper and H.P. Lüthi, *Mol. Phys.* **96** (1999), 559.
[69] J.C. Owrutsky, N.H. Rosenbaum, L.M. Tack and R.J. Saykally, *J. Chem. Phys.* **83** (1986), 5338.
[70] Interatomic Distances, The Chemical Society of London, Special Publication No. 11, 1958; Supplement, Special Publication No. 18, 1965.
[71] M.J. Frisch, G.W. Trucks, H.B. Schlegel, G.E. Scuseria, M.A. Robb, J.R. Cheeseman, V.G. Zakrzewski, J.A. Montgomery, Jr., R.E. Stratmann, J.C. Burant, S. Dapprich, J.M. Millam, A.D. Daniels, K.N. Kudin, M.C. Strain, O. Farkas, J. Tomasi, V. Barone, M. Cossi, R. Cammi, B. Mennucci, C. Pomelli, C. Adamo, S. Clifford, J. Ochterski, G.A. Petersson, P.Y. Ayala, Q. Cui, K. Morokuma, D.K. Malick, A.D. Rabuck, K. Raghavachari, J.B. Foresman, J. Cioslowski, J.V. Ortiz, A.G. Baboul, B.B. Stefanov, G. Liu, A. Liashenko, P. Piskorz, I. Komaromi, R. Gomperts, R.L. Martin, D.J. Fox, T. Keith, M.A. Al-Laham, C.Y. Peng, A. Nanayakkara, M. Challacombe, P.M.W. Gill, B. Johnson, W. Chen, M.W.

Wong, J.L. Andres, C. Gonzalez, M. Head-Gordon, E.S. Replogle and J.A. Pople, *Gaussian* **98** (1998), Revision A.7, Gaussian, Inc., Pittsburgh PA.

[72] N. Turki, A. Milet, O. Ouamerali, R. Moszynski and E. Kochanski, *J. Mol. Struct.* (*Theochem*) **577** (2002), 239.

[73] N. Turki, A. Milet, A. Rahmouni, O. Ouamerali, R. Moszynski, E. Kochanski and P.E.S. Wormer, *J. Chem. Phys.* **109** (1998), 7157.

[74] J. Noga, W. Klopper and W. Bartlett, eds, Recent Advances in Coupled-Cluster Methods, *Recent Advances in Computational Chemistry*, (vol. 3), World Scientific, Singapore, 1997, pp. 1–48.

[75] W.F. Murphy, *J. Chem. Phys.* **67** (1977), 5877.

[76] *Handbook of Chemistry and Physics*, R.C. Weast and M.J. Asth, eds, (CRC Press, Boca Raton 1981).

[77] A.D. Buckingham, *J. Chem. Phys.* **36** (1962), 3096.